Institute of Cancer Research Library

Sutton

Please return this book by
the last date stamped below

13/11	24. NOV. 1995
29 Nov	-9. APR. 1997
30. JAN. 1991	30. JUN. 1998
11. JUN. 1991	24. SEP. 1999
-2. DEC. 1991	23. MAR. 2000
30. APR. 1992	
23. JUN. 1992	
27. JUL. 1992	
-3. SEP. 1992	
-9. FEB. 1995	
20. OCT. 1995	

MOLECULAR BIOLOGY OF CANCER GENES

ELLIS HORWOOD SERIES IN MOLECULAR BIOLOGY
Series Editor: Dr A. J. TURNER, Department of Biochemistry, University of Leeds

Goodfellow and Moss	COMPUTER MODELLING OF BIOMOLECULAR PROCESSES
Tata, J.	HORMONAL SIGNALS:
	Cellular and Molecular Basis of Hormone Action and Communication
Turner, A. J.	MOLECULAR AND CELL BIOLOGY OF MEMBRANE PROTEINS
Eyzaguirre, J.	HUMAN BIOCHEMISTRY AND MOLECULAR BIOLOGY, Vols. 1-3
Parker & Katan	MOLECULAR BIOLOGY OF ONCOGENES AND
	CELL CONTROL MECHANISMS
Sluyser, M.	MOLECULAR BIOLOGY OF CANCER GENES

MOLECULAR BIOLOGY OF CANCER GENES

Editor
M. SLUYSER
Division of Tumour Biology
Netherlands Cancer Institute, Amsterdam

ELLIS HORWOOD
NEW YORK LONDON TORONTO SYDNEY TOKYO SINGAPORE

First published in 1990 by
ELLIS HORWOOD LIMITED
Market Cross House, Cooper Street,
Chichester, West Sussex, PO19 1EB, England
A division of
Simon & Schuster International Group

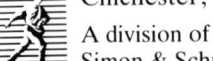

© Ellis Horwood Limited, 1990

All rights reserved. No part of this publication may be reproduced, stored in a retrieval system, or transmitted, in any form, or by any means, electronic, mechanical, photocopying, recording or otherwise, without the prior permission in writing, from the publisher

Typeset in Times by Ellis Horwood Limited
Printed and bound in Great Britain
by The Camelot Press, Southampton

British Library Cataloguing in Publication Data

Molecular biology of cancer genes.
1. Man. Cancer. Molecular biology
I. Sluyser, M. (Mels), *1930–*
616.99′407
ISBN 0–13–599614–7

Library of Congress Cataloging-in-Publication Data available

Table of contents

Preface 7

1 Genetic stability and instability in tumours 9
J. Volpe

2 Chromosome fragile sites for cancer research 24
S. Watanabe, T. Furuya and H. Ochi

3 Cytogenetic abnormalities in human small cell lung cancer 45
P. H. Rabbitts and J. J. Waters

4 Onco-suppressor genes 62
M. L. M. Anderson and D. A. Spandidos

5 *Ras* oncogenes in human cancers 82
N. Lemoine

6 Anti-cancer elements in oncogenes 119
H. Honkawa, R. Kikuno, M. Takahashi, K.-I. Tezuka and T. Hashimoto-Gotoh

7 The oncogenic potential of fibroblast growth factor genes 134
S. Ortega and K. A. Thomas

8 Transcription factors in normal and malignant cells 150
P. Herrlich, H. Ponta, B. Stein, C. Jonat, S. Gebel, H. König, R. O. Williams,
V. Ivanow and H. J. Rahmsdorf

9 Aberrant steroid-thyroid receptors as oncogenes 169
M. Sluyser

Table of contents

10 Analysis of c-*myc* amplification and expression in xenotransplanted human prostatic carcinoma cells 193
M. Fukumoto and O. Midorikawa

11 Oncogenes in human lung cancer 207
F. J. Kaye, S. K. Barksdale, J. W. Harbour and J. D. Minna

12 Neural cell adhesion molecule (NCAM) and human small cell lung cancer . 223
R. J. A. M. Michalides, K. C. E. C. Moolenaar, R. E. Kibbelaar and W. J. Mooi

13 Regulation of cytokine gene expression 237
M. Turner and M. Feldmann

14 The molecular biology of multidrug resistance 263
L. Veinot-Drebot and V. Ling

Index 290

Preface

The title of this book is 'Molecular Biology of Cancer Genes' rather than 'Molecular Biology of Oncogenes' because it deals with not only genes that cause cancer, but also those that are involved in other ways in the neoplastic process. A more detailed study of these gene systems can perhaps help us to unravel the intricate mechanisms that play a role in malignant transformation, onco-suppression, tumour progression, multidrug resistance, cytogenetic abnormalities and the interaction that cancer genes have with external factors such as hormones and growth factors.

The general introduction on genetic instability in tumours, is followed by several chapters on chromosomal abnormalities. Onco-suppression is reviewed, of which the classic example is the retinoblastoma gene. The next chapters concern oncogenes and their interaction with growth factors, steroid-receptor-like proteins with oncogenic potential, and transcription factors in normal and malignant cells. Then special examples of transforming genes are described, and the final chapter reviews the molecular biology of drug resistance.

We hope that this book will be a useful guide to workers in the field, and also serve as an overview to those interested in the mechanisms that are at the basis of cancer.

Amsterdam 1990   Mels Sluyser

1

Genetic stability and instability in tumours

John Volpe
Department of Experimental Radiotherapy, The University of Texas, M. D. Anderson Cancer Center, 1515 Holcombe Blvd, Houston, TX 77030, USA

1. CANCER AND THE HISTORY OF GENETIC INSTABILITY

Ever since Boveri [1] published his study in 1914, it has been known that an aberration of the genome was associated with cancer. Boveri presented the first genetic hypothesis for the origin of cancer by postulating that this abberation may be responsible. His hypothesis was first supported by studies on the inheritance of cancer (or an increased risk) in man [2,3] and by the fact that physical and chemical mutagenic agents could cause tumours in man and in animals [4–6]. However, the further understanding of cancer and the further testing of Boveri's hypothesis has been slow and difficult.

Not until Makino [7,8] studied ascites tumour cells in 1952 was it known that tumour cells could be distinctly non-diploid and could vary considerably in chromosome number, even among the cells of the same tumour. The extent of this karyotypic instability of tumour cells became readily apparent with the advent of chromosome-banding techniques in the late 1960s. In relating tumour karyotypic instability to the molecular level, German and associates [9] may have been instrumental by discovering that the instability of DNA, in patients with Bloom syndrome, could be determined in the cellular genome. Knudson [10] furthered our understanding of cancer by hypothesizing that two mutations were responsible for retinoblastoma and that one mutation could be inherited in the hereditary cases of retinoblastoma. A short time later, Loeb and associates [11] discovered that one of the proteins responsible for maintaining the DNA of cells, DNA polymerase, was working improperly in leukaemic lymphocytes. Seen in retrospect and from a historical viewpoint, it appeared as a logical step that Nowell [12] in 1976 developed the modern theory that genetic instability was the driving force of clonal evolution and tumour progression, although he appears to have derived his hypothesis more from karyotypic studies. Nonetheless, after six decades of cancer research, Boveri's hypothesis that a genetic aberration causes cancer still has not been proved or disproved.

All of the above facts and theories have suggested a role for genes and genetic instability in cancer, although none has approached the molecular level except that of Loeb and associates [11]. Recently, specific genes, oncogenes and anti-oncogenes (or tumour-suppressor genes) have been implicated in tumour formation [13–16]. Many other genes or their protein products have also been added to a rapidly growing list of genes that appear to play some role in cancer, including proteases, other degradative enzymes, angiogenesis factors, immune blocking factors, vascular permeability factors, growth factors, and the inherited cancer factors in man [3,4,17–24].

Interestingly, two characteristics of tumours have no widely accepted genetic explanation. Investigators have not determined (a) why malignant tumour cells are genetically unstable or (b) why some tumours remain benign while others become malignant. Clearly, genetic instability is a trait of the genome and, although genetics cannot totally explain cancer, it would be an irony if genetics could not explain something of its own discipline. It should also be noted that a non-genetic explanation for both of the above topics has been proposed recently by Rubin [25,26]. My goal here is to advance a genetic hypothesis on the molecular level, detailing the instability of tumour cellular DNA as the reason some tumours become malignant and capable of attaining a highly metastatic state. It is hoped that the differences between Rubin's [25,26] theory and the genetic theory outlined here will stimulate research into these fundamental topics of malignant neoplasia.

2. GENETIC INSTABILITY AND CELLULAR EVOLUTION

Without genetic instability, a metastatic tumour cell probably cannot be produced. This is because all malignant cells appear to be derived from normal cells and, without genetic instability, the genome of a cell can only change through spontaneous (background) mutation (including spontaneous point mutations, spontaneous chromosome rearrangements, and spontaneous deletions and additions of DNA). Spontaneous mutation in normal cells can produce slow changes in the genome over extended periods of time. However, although spontaneous mutation may be able to change a normal cell into a tumour cell (especially a benign tumour cell) if given the time, spontaneous mutation probably is too slow to produce the changes necessary for a metastatic tumour cell. This is because a metastatic cell appears to be the product of many changes in the gene expression of a normal cell [27,28], too many to be accounted for by spontaneous mutation alone during the lifetime of a host. Whereas background mutation can only create slow changes with time, a genetically unstable cell line can produce multiple genomic errors in a relatively short period. These multiple errors of the genome may result in diverse cellular variants that could lead to tumour formation and subsequent metastasis, if the correct set of errors were to accumulate in a line.

The errors produced by genetic instability (and probably also by mutation) are probably more or less random with respect to tumour formation. That is, they are not goal oriented, with malignancy being the end product. Rather, multiple errors occur at random in a line, some of which may accumulate and lead by accident to tumour formation (especially when subjected to repeated carcinogenesis). Eventually, one of the diversified variants created by genetic instability (or repeated carcinogenesis and, at a lower frequency, by spontaneous mutation) may develop a growth advantage. When this happens, selection and cellular evolution may occur within the line. As selection and evolution are goal oriented, the accumulation of further genetic aberrations in the line probably will cease being random and begin to accumulate towards those variants that have a growth advantage, which may lead to tumour formation and malignant progression.

The accumulated genomic changes in a cell line caused by genetic instability may be augmented by several other factors. For an initiated stem cell line to undergo further genome changes, the line cannot become differentiated or extinct; therefore, cell division enhancers or differentiation blockers may increase the probability of malignant tumour formation. Increased cell numbers increase the probability of a genetic change occurring in the line. Increased time probably increases the percentage of changes that are expressed or observed (i.e. more tumours tend to be found 2 years after initiation than 1 year after). Mutagenesis further augments genomic alterations. Finally, studies of tumour promotion [29,30] and the subcutaneous embryo transplantation [31] suggest that the environment plays some poorly understood role in cancer development.

3. ONCOGENES AND STABILITY GENES

From studies of cell hybridization, DNA transfection, and viral transformation of cells, several genes have been found that appear to be important in oncogenesis [14]. These genes have been termed anti-oncogenes (or tumour-suppressor genes) and oncogenes. The molecular role and function of the anti-oncogenes are not well understood, but the oncogenes appear to be involved in the regulation, development, and growth of normal cells [13,14,32]. No comprehensive review of the oncogenes and anti-oncogenes will be given here, as they have been reviewed elsewhere [13,14,32,33]. For simplicity, these genes will here be assumed to convey in some manner tumorous properties and a growth advantage on cells misexpressing them. For carcinogenesis, the misexpression of one or more of these oncogenes or anti-oncogenes may be necessary. However, a necessary step may not be a sufficient step. Carcinogenesis gives rise to both benign and malignant tumours, and there is little if any difference in the expression of at least the oncogenes between these two very different classes of tumours [34,35]. For this reason, many researchers, including many of the proponents of the oncogene hypothesis, believe that the oncogenes do not totally explain cancer [13,14,36,37].

As there are genes important in the molecular regulation of cellular growth, so there are also genes involved in the molecular control of the genome's stability. (This concept of stability genes is a modest extension of earlier discussions [12,38,39].) The genes important in regulating the stability of the genome are those genes that repair, protect, recombine, and replicate DNA and ensure that each daughter cell gets a complete set of chromosomes in mitosis. There are even certain developmentally controlled genes that amplify and rearrange the DNA of specific cells during the development of at least some species. Many of these genome stability genes are known, whereas others have not been identified except, perhaps, for their function (see Table 1).

Table 1 — Genes involved in the stability of the genome[a]

Genes that repair or protect DNA [77–79]
    Helicase (uvrD)
    Proteins that bind to base pair mismatches (mutS) [78]
    Single-strand binding proteins
    Proteins that split UV-induced pyrimidine dimers (photolysase) [80]
    Proteins that remove inappropriate methyl groups (alkyl transferase)
    Proteins that cleave damaged bases (glycosylase)
    Endonucleases, exonucleases, and ligase
    Proteins that help prevent damage to DNA (catalase)

Genes involved in nuclear division [81]
    Genes coding for the mitotic spindle or its equivalent
    Centromeric DNA sequences (cen)
    Proteins that bind to centromeric DNA
    Other genes involved in karyokinesis

General recombination genes [82]
    DNA crossover sites and their flanking inverted repeats [83]
    Recombinases or invertases (RecA)
    Genes that enhance recombination (recombinase enhancer)
    Recombinase enhancer binding site on DNA
    DNA denaturing proteins (RecBC) [84]
    Recombination cut site on DNA (chi) [85]
    Endonucleases and ligase

Site-specific recombination genes
    Phage integration and excision proteins [82]
    DNA sequences of phage integration [82]
    Cross-complementing and chyming DNA invertases (Hin) [82,83]
    Transposon resolvase (transposase) [86,87]
    DNA binding site of transposase [86,87]
    Regulatory proteins of transposition [86,87]

DNA replication genes
    DNA replication site on a chromosome (replicons) [81,88]
    DNA sequences of linear DNA replication (telomere sequences) [81]
    Proteins that bind to telomere DNA sequences [81]
    Polymerases and accessory proteins (polI) [89]
    Helicases and helix destabilizing proteins [89]
    Enzymes catalysing RNA priming of DNA (primase) [90]
    Proteins that increase polymerase processivity [82]
    Nucleases, ligase, and decatenase [82]
    Termination site of replication (terC) [91]

[a] The table presents the function or class of genes involved in genome stability. Specific genes or gene classes may be named in parentheses. A reference is also given.

Although some have theorized that the misexpression of one or more oncogenes may result in neoplastic transformation [13,32,40], oncogene activation by itself (i.e. without the aberration of at least one stability gene) may result in little more than a benign tumour (for exceptions, see below in sections 5 and 8). This is because a normal (genomically stable) cell line with an activated oncogene can only evolve by spontaneous mutation. Such a slowly evolving line may be able to form a benign tumour but would be unlikely to acquire the necessary and multiple changes of metastatic malignancy. Therefore, I propose that it is unlikely for a normal cell to attain the metastatic state unless at least two (or more) genes, an oncogene and a stability gene, are misexpressed. Indeed, stability genes may not be necessary for the less malignant neoplastic growths, but they may greatly increase the likelihood of such growths' occurring.

This oncogene–stability gene hybrid hypothesis as outlined above may require two clarifications. First, the production of malignant tumours from oncogene transfection studies does not disprove the role of stability genes in carcinogenesis. Oncogenes are often inserted into *in vitro* cell lines that appear to be somewhat unstable genomically [41,42]. Also, gene transfection itself may cause a cell line to become unstable [32,37], perhaps by insertional mutagenesis of a stability gene. Second, genome instability genes by themselves do not cause a cell line to develop malignant or even tumorous properties; rather, they allow for the continuous creation of new clones, some of which may have malignant properties. The created clonal variation may be able to tap numerous properties encoded in the cellular genome that would be useful to a growing and evolving population of cells [43]. These diverse properties would be much less likely to be acquired by a genetically stable cell line and would allow for cellular evolution within a tumour.

4. THE INITIATION OF A TUMOUR

Generally, a cell must undergo an initiation process in order for a tumour to form. As stated above, a benign tumour cell may be initiated from an error affecting one or more oncogenes [13]. However, a metastatic tumour cell may be initiated by one of three methods. The first is similar to benign tumour initiation: an oncogene becomes misexpressed. Initially, a benign tumour line may form. In a few of the benign tumours at some time after initiation, a cell may spontaneously start to misexpress one or more stability genes. This second aberration in the stability gene or genes may occur through spontaneous mutation or repeated carcinogenesis. The aberrant expression of a stability gene or genes may result in genetic instability within a focus of the benign tumour, allowing for malignant tumour progression. This ontogeny of a tumour may explain the rapid tumour progression that occasionally occurs in benign tumours [30,44–46].

The second method by which a metastatic tumour cell may be initiated occurs when a stability gene is affected. On cellular expansion, such an initiated cell may form a genetically unstable but non-tumorous cell line. Such a line probably would not have a reproductive advantage over another stem cell line until genetic instability had created a variant that had a growth advantage. Given time and cell division, genetic instability may create a variant with a selective advantage and tumorous properties, such as by activating one or more oncogenes. If this variant were to form a tumour, such a tumour would be unstable and would probably progress rapidly. This method of metastatic tumour initiation may be illustrated by malignant tumours that arise from non-tumerous tissue [46] or from genetically unstable but benign-appearing tumours [47].

Tumour initiation by the second method is probably not as common as initiation by the first method, even though the number of stability genes may be approximately the same as the number of oncogenes. An initiation event probably affects stability genes about as often as it changes oncogenes, but a tumour may be less likely to form from a misexpressing stability gene than a misexpressing oncogene. This is because the initiation event appears to have a low probability of forming a tumour relative to the number of altered cells per number of tumours that form [48,49], such that a tumour is unlikely to form unless an altered cell undergoes much cell division or has a reproductive advantage. As an activated oncogene may give a cell a reproductive advantage and an aberrant stability gene may not (at least initially), then a cell line misexpressing an oncogene may be less likely to differentiate or become extinct than a line misexpressing a stability gene. A line composed of one or a few cells and having no growth advantage may be unlikely to activate an oncogene (or other genes necessary to obtain a growth advantage and tumorous properties), even if it has genetic instability. A line composed of many cells is more likely to attain the changes necessary for malignancy.

If one or more stability genes were misexpressed, one or more oncogenes were activated, and a tumour were to form, then one would predict that such a tumour would have a high probability of progressing to malignancy. This is because genetic instability allows the tumour cells to evolve early and rapidly. Some support for this prediction can be found in the Bloom syndrome disease of man (see below).

The third way a metastatic tumour cell may originate would occur if an initiation event were to change the expression of at least one oncogene and one stability gene within the same cell. If such a double aberration–mutation were to occur, subsequent tumour formation and progression would be much more likely than if a single aberration had occurred. This is because such a double initiating event would bestow both a growth advantage and an evolving capacity on the initiated cell. However, according to the laws of chance, such a double initiating event should occur much less frequently than a single initiation event.

5. GENETIC INSTABILITY AND ITS COMPLEXITIES

As presented above, genetic instability has been assumed and described as distinct from genomic stability. This simple model has been adequate thus far, since metastatic tumours tend to be more genomically unstable than benign tumours [47,50]. However, there is really a spectrum of instability in neoplasia, and there are many exceptions. For example, some malignant teratocarcinomas appear to be more stable than some clinically benign tumours [51]. Even a genetically stable (normal) cell may spontaneously mutate and rearrange its chromosomes [52].

Another simplification in the mechanisms described above is that genetic instability was assumed to have a constant rate of change with time. It is well known that genetic instability may increase [12,38], but it may also decrease [53]. If a tumour line is not well adapted to its environment, or if the line is not expressing many of the genes necessary for maximum growth, then there probably is natural selection for increased genetic instability. This is because a new variant with a reproductive advantage is more likely to be created by a more unstable line than a less stable line. In contrast, as a tumour line becomes better adapted to its environment, the benefit from increased genetic instability may decrease. The level of genetic instability may even decrease if the selection pressure for genomic fidelity ever becomes stronger than the selection pressure for increased instability. We know that some selection pressure for genome fidelity exists, because an unstable genome produces more cells that die or are less fit than does a stable genome (see below).

Another complication to the hypothesis posed is that different cells appear to express different gene programs [54]. Some cells of some tissues (i.e. melanoblastic, embryonic, and especially embryonic gonadal cells) already appear to express many of the genes necessary for metastatic malignancy, so that relatively few additional changes may be needed for malignancy. Other cells (i.e. fat and skin cells) appear to express few of the genes necessary for metastatic cancer, such that relatively many changes may be needed for malignancy. This may explain some of the tissue-specific ease of malignant progression.

One further complication, excellently discussed elsewhere, is that the environment may affect the expression of genes in a cell [55,56]. Thus far, only the metastatic potential of the average cell of a clone has been discussed, although the environment may cause the metastatic potential of all other cells of a clone to deviate from the mean [57,58]. Unfortunately, a tumour cell can be in different microenvironmental settings in its host and in its tumour, and these environmental settings may change with time [59].

6. GENETIC INSTABILITY AND ITS CONSEQUENCES

A tumour cell population consisting of genetically diverse cells should evolve with time, unless all the diversity is neutral with respect to selection [52,60]. Those clones that reproduce faster, survive host immune responses better, and have larger proportions of stem cells should, over time, come to predominate in the tumour. Such a population of tumour cells may tend to evolve gradually. This is because the newly created variants that will in turn come into predominance will probably be derived from the original predominating parental line(s). Therefore, the properties and the chromosomes of a tumour should tend to change in a stepwise manner [44,61]. However, a new varaint with a reproductive advantage may arise from one of the non-predominating clones in a tumour. If this happens, and if the minority cell line in question is distantly related to the predominating lines(s), then the evolution of the tumour may be discrete [41,44].

The advantage of genetic instability is that it allows the tumour cells to evolve and to become autonomous of their host sooner than would be possible with spontaneous mutation. In short, genetic instability allows a population of tumour cells to diversify and to change radically within the lifetime of the host. The advantage is greater when the tumour cells are not well adapted for growth in their environment or when the environment has changed drastically. Genetic instability may bestow on the population a variant capable of rapid growth in the given environment even though most other clones may not be capable of such growth. This may be the case when an *in vivo* tumour is confronted with chemotherapy or with an *in vitro* environment [62].

Throughout this discussion, the advantages of a genetically unstable tumour have been discussed; however, genetic instability has its disadvantages. Of the multitude of new tumour variants created by genetic instability, only a few are better fit than the parental line; most are either neutral or less fit [52,61]. Probably some tumour cells have such devastating changes in their genome that they are doomed to die [63]. Sarraf and Bowen [64] have in fact observed some tumour cells that appear to be well supplied by the blood vasculature but nevertheless die. These disadvantageous variants of a genetically unstable tumour cell population are akin to the genetic load of a sexually reproducing population. The genetic load of a population is the less fit variants created by sexual reproduction, recombination and, to a much lesser extent, mutation. The term 'genetic load' derives from the fact that these disadvantageous variants are a load on the population, representing the extent of departure from a perfect genetic constitution [52]. In the case of tumours, the greater the genetic instability, the greater is the genetic load (the greater the proportion of disadvantageous variants the population has). It is interesting that many of the tumour therapies (radiotherapy and chemotherapy) increase the number of genetic changes in tumour cells as well as kill these cells through cytotoxic mechanisms. Thereby, the tumour cells are confronted with two difficulties: a decrease in cell number from cell killing and an increase in the genetic load from DNA damage.

7. SIGNIFICANCE WITH REGARD TO TUMOUR THERAPY

Radiotherapy and chemotherapy are potent antitumour modalities, yet they are known carcinogens–mutagens. One may wonder whether this apparent contradiction is a coincidence or whether some mechanism underlies it. The hypothesis of stability genes suggests that there may be an underlying molecular mechanism. If tumour cells have malfunctioning stability genes so that the fidelity of their genome is poorly maintained, one would expect such cells to be more sensitive to cell killing by a carcinogen–mutagen than their normal counterparts, if all other things are equal. This is because a normal cell may be more able to tolerate the alterations in its gene expression and to repair the treatment-induced DNA damage than a cell that is poorly maintaining its DNA. This is especially true if a tumour cell had been misexpressing some of the DNA repair genes.

However, probably different tumours and maybe even separate cells within the same tumour may incorrectly express different stability genes. The chromosome instability syndromes in man teach us that different syndromes (and hence different stability genes) have different sensitivities to different mutagens: xeroderma pigmentosum cells are hypersensitive to ultraviolet light [65,66] but are not especially susceptible to ionizing radiation [67], whereas cells from patients with ataxia telangiectasia are hypersensitive to ionizing radiation but are not especially sensitive to ultraviolet light [68,69]. If this were true with tumour cells, it would mean that a therapy that is highly effective against one tumour might be of little benefit to another, depending on which stability genes had been altered.

Further research into the stability genes and their particular sensitivities to certain therapeutic agents may benefit the existing protocols of chemotherapy and radiotherapy. In fact, the use of these carcinogenic–mutagenic agents against neoplastic disease may mean that the proposed defect of stability genes in cancer has already been exploited to some degree. If true, perhaps some day a pathologist will be able to identify the affected stability gene or genes from a tumour biopsy and use such knowledge in prescribing a therapeutic agent to exploit the particular defect of each malignant tumour.

8. SUPPORT FOR THE STABILITY GENE HYPOTHESIS

The stability gene hypothesis may be considered separate from but complementary to the oncogene hypothesis [13,14,16,32] of cancer. Both hypotheses have a large body of support but, unlike the oncogene hypothesis, the stability gene hypothesis has mostly indirect support. The stability genes are proposed to be those genes involved in karyokinesis and the repair, recombination, protection, and replication of DNA. It has been proposed here that these genes may be misexpressed in tumours and that they may play a major role in tumour formation, malignant progression, and the evolution of drug resistance in neoplasia.

Some patients with certain inherited diseases (such as the chromosome instability syndromes) appear to have cells containing aberrant stability genes [40,70]. These cells and other cells with unstable genomes (i.e. established cell lines) have an increased risk of becoming tumorigenic. At least in the case of Bloom syndrome tumours, these genetically unstable tumour cells also appear to have a greatly increased risk of malignant progression [40,70]. Therefore, stability gene misexpression appears to increase the likelihood of tumour formation and subsequent malignant progression of the affected cells.

Some support for the stability gene hypothesis has even approached the molecular level. Mutagenic treatment has been found to reduce the fidelity of DNA polymerase of the treated cells [71]. On transformation, murine embryo fibroblasts have been found to become genetically unstable and less efficient in repairing DNA [72]. In fact, polymerase from leukaemic lymphocytes has been found to have a tenfold greater error rate than that from normal lymphocytes [11].

Rubin [25,26] has offered an alternative hypothesis, that cancer is an aberration of development and that the associated genome irregularities are derived from the environment of the tumour milieu. Although Rubin's [25] hypothesis has not been as complementary as the stability gene hypothesis to the chromosome instability syndrome and the oncogene data, his hypothesis can more easily explain why a 6-day-old murine embryo cell can form teratocarcinomas [73]. To explain these teratocarcinomas, the stability gene hypothesis would have to assume that the oncogenes–anti-oncogenes are also involved (as stability genes alone do not confer tumorous properties), that the embryonic cells are already expressing many of the genes necessary for malignancy, and that the environment affects the expression of the cell's genes (since embryonic cells may form an embryo or a teratocarcinoma, depending on whether the cells are inside or outside the uterus). That the environment can affect the expression of the genes of a cell has been reviewed elsewhere [55]. However, unlike Rubin's [25] non-genetic hypothesis, the stability gene hypothesis can easily explain the following: (a) why genetically stable teratocarcinoma cells can participate in normal embryogenesis and unstable teratocarcinoma cells cannot [51], and (b) why a malignant tumour clone is genetically unstable under both *in vivo* and *in vitro* conditions and a more benign tumour cell or a normal host cell of a tumour is genetically stable by comparison [41,47,74].

Although the stability gene hypothesis has little concrete support, it derives its power (especially when it is wedded to the concept of oncogenes–anti-oncogenes) from its ability to explain on a detailed and molecular level some of the diverse and otherwise perplexing tumour data contained in the literature. Among other situations, it can explain the following: why some cell lines require long periods of culture *in vitro* before becoming tumorigenic (those misexpressing only a stability gene or genes) whereas other lines (those misexpressing one or more of both the oncogenes and stability genes) do not [75,76]; why benign tumours are more common than malignant tumours when they are initiated with a low carcinogen dose and treated with a promoting agent (see above); why some tumours have relatively unstable genomes (those misexpressing one or more stability genes) and other tumours (those with normal or more normal stability gene expression) have relatively stable genomes [47]; why some tumours diversify early, evolve, undergo malignant progression, and develop drug resistance (those misexpressing one or more stability genes) and other tumours (those with normal stability genes) do not or do so at much slower rates [46,62].

9. SUMMARY

It is proposed here that the genes involved in karyokinesis (nuclear division) and DNA repair, replication, protection, and recombination may be responsible for the genetic instability of neoplastic cells. When these genes, collectively called stability genes, are aberrantly expressed, they may increase the probability that a cell will acquire tumorous properties. However, the incorrect expression of these stability genes may be more important in malignant progression, since the aberrations in these genes may be responsible for the creation of diverse clonal variants that allow for extensive evolution and progression within the tumour cell population.

ACKNOWLEDGEMENTS

I am very grateful to Luka Milas for invaluable discussion and criticism. This work was supported in part by funding from the American Legion Auxiliary.

NOTE ADDED IN PROOF

Day *et al*. (*Nature (London)* **288**, 724, 1980) have found that a number of tumors are more sensitive to anti-neoplastic agents than their corresponding normal tissues. Some of this increased sensitivity appears to be due to the reduced ability of neoplastic cells to repair O-6-methylguanine (Yagi and Day, *Mutation Res.* **184**, 223, 1987), and some may be due to the defective suppression of cellular growth and DNA synthesis when the genome is damaged (Lambert and Borek, *J. Natl. Cancer Inst.* **80**, 1492, 1988). DNA ligase has also been found to be deficient in certain leukemias (Rusquet, *Cander Res.* **48**, 4038, 1988).

REFERENCES

[1] Boveri, T. (1929). *The Origin of Malignant Tumors*, Williams & Wilkins, Baltimore, MD. (Translated by M. Boveri from the German edition of 1914)

[2] Marks, P. A. (1985). Introduction: genetics, cell differentiation, and cancer. In P. A. Marks (ed.) *Genetics, Cell Differentiation and Cancer*, Academic Press, San Diego, CA, pp. 1–11

[3] Harnden, D. G. (1987). Genetics and cancer. Mechanisms of susceptibility. In: A. P. Maskens, P. Ebbesen & A. Burny (eds) *Concepts and Theories in Carcinogenesis, Proceedings of the 4th Symposium of the European Organiza-*

tion for Cooperation in Cancer Prevention Studies, Excerpta Medica, Amsterdam, pp. 21–36
[4] Russell, W. L. (1951). X-ray induced mutations in mice. *Cold Spring Harbor Symp. Quant. Biol.* **16** 327–330
[5] McCann, J. & Ames, B. N. (1976). Detection of carcinogens as mutagens in the Salmonella/microsome test: assay of 300 chemicals: discussion. *Proc. Natl. Acad. Sci. USA* **73** 950–954
[6] Barrett, J. C., Hesterberg, T. W., Oshimura, M. & Tsutsui, T. (1985). Role of chemically induced mutagenic events in neoplastic transformation of Syrian hamster embryo cells. In: J. C. Barrett & R. W. Tennant (eds) *Carcinogenesis—a Comprehensive Survey, vol. 9, Mammalian Cell Transformation Mechanisms of Carcinogenesis and Assays for Carcinogens*, Raven Press, New York, pp. 123–138
[7] Makino, S. (1952). A cytological study of the Yoshida sarcoma, an ascites tumor of white rats. *Chromosoma* **4** 649–674
[8] Makino, S. (1952). Cytological studies on cancer. III. The characteristics and individuality of chromosomes in tumor cells of the Yoshida sarcoma which contribute to the growth of the tumor. *Gann* **43** 17–34
[9] German, J., Archibald, R. & Bloom, D. (1965). Chromosomal breakage in a rare and probably genetically determined syndrome of man. *Science* **148** 506–507
[10] Knudson, A. G. (1971). Mutation and cancer: statistical study of retinoblastoma. *Proc. Natl. Acad. Sci. USA* **68** 820–823
[11] Loeb, L. A., Springgate, C. F. & Battula, N. (1974). Errors in DNA replication as a basis of malignant changes. *Cancer Res.* **34** 2311–2321
[12] Nowell, P. C. (1976). The clonal evolution of tumor cell populations. *Science* **194** 23–28
[13] Bishop, J. M. (1983). Cellular oncogenes and retroviruses. *Annu. Rev. Biochem.* **52** 301–354
[14] Bishop, J. M. (1985). Trends in oncogenes. *Trends Genet.* **1** 245–249
[15] Land, H., Parada, L. F. & Weinberg, R. A. (1983). Tumorigenic conversion of primary embryo fibroblasts requires at least two cooperating oncogenes. *Nature (London)* **304** 596–602
[16] Hall, A. (1987). Oncogenes. In: A. P. Maskens, P. Ebbesen & A. Burny (eds) *Concepts and Theories in Carcinogenesis, Proceedings of the 4th Symposium of the European Organization for Cooperation in Cancer Prevention Studies*, Excerpta Medica, Amsterdam, pp. 87–92
[17] Nicolson, G. L. & Poste, G. (1982). Tumor cell diversity and host responses in cancer, part 1. Properties of metastatic cells. *Curr. Probl. Cancer* **7**(6) 1–83
[18] Nicolson, G. L. & Poste, G. (1983). Tumor cell diversity and host responses in cancer, part 2. Host immune responses and therapy of metastases. *Curr. Probl. Cancer* **7**(7) 1–42
[19] Garbisa, S., Pozzatti, R., Muschel, R. J., Saffiotti, U., Ballin, M., Goldfarb, R. J., Khoury, G. & Liotta, L. A. (1987). Secretion of type IV collagenolytic protease and metastatic phenotype: induction by transfection with c-Ha-*ras* but not c-Ha-*ras* plus Ad2-E1a. *Cancer Res.* **47** 1523–1528
[20] Liotta, L. A., Tryggvason, K., Garbisa, S., Hart, I., Foltz, C. M. & Shafie, S.

(1980). Metastatic potential correlates with enzymatic degradation of basement membrane collagen. *Nature* (London) **284** 67–68
[21] Folkman, J. (1974). Tumor angiogenesis factor. *Cancer Res.* **34** 2109–2113
[22] Cianciolo, G. J. & Snyderman, R. (1986). Effects of tumor growth on host defenses. *Cancer Metastasis Rev.* **5** 15–27
[23] Senger, D. R., Perruzzi, C. A., Feder, J. & Dvorak, H. F. (1986). A highly conserved vascular permeability factor secreted by a variety of human and rodent tumor cell lines. *Cancer Res.* **46** 5629–5632
[24] Salomon, D. S. & Perroteau, I. (1986). Growth factors in cancer and their relationship to oncogenes. *Cancer Invest.* **4** 43–60
[25] Rubin, H. (1985). Cancer as a dynamic developmental disorder. *Cancer Res.* **45** 2935–2942
[26] Rubin, H. (1987). The sources of heritable variation in cellular growth capacities. *Cancer Metastasis Rev.* **6** 85–89
[27] Nicolson, G. L., Labiche, R. A., Frazier, M. L., Blick, M., Tressler, R. J., Reading, C. L., Irimura, T. & Rotter, V. (1986). Differential expression of metastasis-associated cell surface glycoproteins and mRNA in a murine large cell lymphoma. *J. Cell. Biochem.* **31** 305–312
[28] Hall, L., Henney, A., Ralphs, D. N. L., Herries, D. G. & Craig, R. K. (1986). Changes in gene expression in established human mammary tumor cell lines when compared with normal breast and breast tumor tissue. *Cancer Res.* **46** 5786–5794
[29] Kennedy, A. R. (1984). Promotion and other interactions between agents in the induction of transformation *in vitro* in fibroblasts. In: T. J. Slaga (ed.) *Mechanisms of Tumor Promotion*, vol. III, *Tumor Promotion and Carcinogenesis In Vitro*, CRC Press, Boca Raton, FL, pp. 13–56
[30] Slaga, T. J. (1984). Mechanisms involved in two-stage carcinogenesis in mouse skin. In: T. J. Slaga (ed.) *Mechanisms of Tumor Promotion*, vol. II, *Tumor Promotion and Skin Carcinogenesis*, CRC Press, Boca Raton, FL, pp. 1–16
[31] Pierce, G. B., Shikes, R. & Fink, L. M. (1978). *Cancer: a Problem of Developmental Biology*, Prentice-Hall, Englewood Cliffs, NJ
[32] Nicolson, G. L. (1987). Tumour cell instability, diversification and progression to the metastatic phenotype: from oncogene to oncofetal expression. *Cancer Res.* **47** 1473–1487
[33] Klein, G., Bregula, U., Wiener, F. & Harris, H. (1971). The analysis of malignancy by cell fusion. I. Hybrids between tumour cells and L cell derivatives. *J. Cell Sci.* **8** 659–672
[34] Balmain, A., Ramsden, M., Bowden, G. T. & Smith, J. (1984). Activation of the mouse cellular Harvey-*ras* gene in chemically induced benign skin papillomas. *Nature (London)* **307** 658–660
[35] Spandidos, D. A. & Kerr, I. B. (1984). Elevated expression of the human *ras* oncogene family in premalignant and malignant tumours of the colorectum. *Br. J. Cancer* **49** 681–688
[36] Weinberg, R. A. (1982). Oncogenes of spontaneous and chemically induced tumors. *Adv. Cancer Res.* **36** 149–163
[37] Liotta, L. A. (1986). Molecular biology of metastases: a review of recent approaches. *Eur. J. Cancer Clin. Oncol.* **22** 345–348

[38] Nowell, P. C. (1986). Mechanisms of tumor progression. *Cancer Res.* **46** 2203–2207

[39] Hsu, T. C. (1983). Genetic instability in the human population: a working hypothesis. *Hereditas* **98** 1–9

[40] German, J. (1983). Bloom's syndrome. X. The cancer proneness points to chromosome mutation as a crucial event in human neoplasia. In: J. German (ed.) *Chromosome Muttion and Neoplasia*, Alan R. Liss, New York, pp. 347–357

[41] Hsu, T. C. (1961). Chromosomal evolution in cell populations. *Int. Rev. Cytol.* **12** 69–161

[42] Greig, R. G., Koestler, T. P., Trainer, D. L., Corwin, S. P., Miles, L., Kline, T., Sweet, R., Yokoyama, S. & Poste, G. (1985). Tumorigenic and metastatic properties of normal and *ras*-transfected NIH/3T3 cells. *Proc. Natl. Acad. Sci. USA* **82** 3698–3701

[43] Burns, F. J., Albert, R. E. & Altshuler, B. (1984). Cancer progression in mouse skin. In: T. J. Slaga (ed.) *Mechanisms of Tumor Promotion*, vol. II *Tumor Promotion and Skin Carcinogenesis*, CRC Press, Boca Raton, FL, pp. 17–40

[44] Noble, R. L. & Hoover, L. (1975). A classification of transplantable tumors in Nb rats controlled by estrogen from dormancy to autonomy. *Cancer Res.* **35** 2935–2941

[45] Isaacs, J. (1982). Mechanisms for and implications of the development of heterogeneity of androgen sensitivity in prostatic cancer. In: A. H. Owens, D. S. Coffey & S. B. Baylin (eds) *Tumor Cell Heterogeneity: Origins and Implications*, Academic Press, Orlando, FL, pp. 99–114

[46] Foulds, L. (1965). Multiple etiologic factors in neoplastic development. *Cancer Res.* **25** 1339–1347

[47] Sandberg, A. A. (1980). *The Chromosomes in Human Cancer and Leukemia*, North-Holland, New York

[48] Thomassen, D. G., Nettesheim, P., Gray, T. E. & Barrett, J. C. (1985). Quantitation of the rate of spontaneous generation and carcinogen-induced frequency of anchorage-independent variants of rat tracheal epithelial cells in culture. *Cancer Res.* **45** 1516–1524

[49] Miller, D. R., Viaje, A., Aldaz, C. M., Conti, C. J. & Slaga, T. J. (1987). Terminal differentiation-resistant epidermal cells in mice undergoing two-stage carcinogenesis. *Cancer Res.* **47** 1935–1940

[50] Cin, P. D., Rao, U., Turc-Carel, C. & Sandberg, A. A. (1988). Translocation [7;21] in a lipoma. *Cancer Genet. Cytogenet.* **30** 17–22

[51] Mintz, B. & Cronmiller, C. (1981). Mett-1: a karyotypically normal *in vitro* line of developmentally totipotent mouse teratocarcinoma cells. *Somatic Cell Molec. Genet.* **7** 489–505

[52] Strickberger, M. W. (1985). *Genetics*, 3rd edn, Macmillan, New York

[53] Wang, R., Hsu, T. C. & Kendal, W. S. (1987). Repeated tandem translocations in a clone and subclones of B16-F10 murine melanoma. *Cancer Genet. Cytogenet.* **29** 81–89

[54] Browder, L. W. (1980). *Developmental Biology*, Saunders, Philadelphia, PA, pp. 34–56, 534–583

[55] Truman, D. E. S. & Clayton, R. M. (1982). Introductory review: strategies of

regulation: signals, receptors and effectors. In: R. M. Clayton & D. E. S. Truman (eds) *Stability and Switching in Cellular Differentiation*, Plenum, New York, pp. 245–251

[56] Schirrmacher, V. (1980). Shifts in tumor cell phenotypes induced by signals from the microenvironment: relevance for the immunobiology of cancer metastasis. *Immunobiology* **157** 89–98

[57] Volpe, J. P. G. (1988). Genetic instability of cancer. Why a metastatic tumor is unstable and a benign tumor is stable. *Cancer Genet. Cytogenet.* **34** 125–134

[58] Falconer, D. S. (1981). *Introduction to Quantitative Genetics*, Longman, New York

[59] Weiss, L. (1985). *Principles of Metastasis*, Academic Press, Orlando, FL

[60] Dracopoli, N. C., Alhadeff, B., Houghton, A. N. & Old, L. J. (1987). Loss of heterozygosity at autosomal and x-linked loci during tumor progression in a patient with melanoma. *Cancer Res.* **47** 3995–4000

[61] Sandberg, A. A. (1982). Chromosomal changes in human cancers. Specificity and heterogeneity. In: A. H. Owens, D. S. Coffey & S. B. Baylin (eds) *Tumor Cell Heterogeneity: Origins and Implications*, Academic Press, Orlando, FL, pp. 367–397

[62] Poste, G. (1986). Pathogenesis of metastatic disease: implications for current therapy and for the development of new therapeutic strategies. *Cancer Treat. Rep.* **70** 183–199

[63] Hu, F., Wang, R. & Hsu, T. C. (1987). Clonal origin of metastasis in B16 murine melanoma: a cytogenetic study. *J. Natl. Cancer Inst.* **78** 155–163

[64] Sarraf, C. E. & Bowen, I. D. (1986). Kinetic studies on a murine sarcoma and an analysis of apoptosis. *Br. J. Cancer* **54** 989–998

[65] Marshall, R. R. & Scott, D. (1976). The relationship between chromosome damage and cell killing in UV-irradiated normal and xeroderma pigmentosum cells. *Mutat. Res.* **36** 397–400

[66] Maher, V. M., Ouellette, L. M., Curren, R. D. & McCormick, J. J. (1976). Frequency of ultraviolet light induced mutation is higher in xeroderma pigmentosum variant cells than in normal human cells. *Nature (London)* **261** 593–595

[67] Hsu, T. C. (1987). A historical outline of the development of cancer cytogenetics. *Cancer Genet. Cytogenet.* **28** 5–26

[68] Taylor, A. M. R., Metcalfe, J. A., Oxford, J. M. & Harnden, D. G. (1976). Is chromatid-type damage in ataxia telangiectasia after irradiation at G_0 a consequence of defective repair? *Nature (London)* **260** 441–443

[69] Taylor, A. M. R., Rosney, C. M. & Campbell, J. B. (1979). Unusual sensitivity of ataxia telangiectasia cells to bleomycin. *Cancer Res.* **39** 1046–1050

[70] German, J., Bloom, D. & Passarge, E. (1984). Bloom's syndrome. XI. Progress report for 1983. *Clin. Genet.* **25** 166–174

[71] Chan, J. Y. H. & Becker, F. F. (1979). Decreased fidelity of DNA polymerase activity during N-2-fluorenylactamide hepatocarcinogenesis. *Proc. Natl. Acad. Sci. USA* **76** 814–818

[72] Elliot, G. C. & Johnson, R. T. (1983). DNA repair in mouse embryo fibroblasts. I. Decline in ultraviolet-induced incision rate accompanies cell transformation. *J. Cell Sci.* **60** 267–288

[73] Stevens, L. C. (1983). The origin and development of testicular, ovarian, and

embryo-derived teratomas. In: B. Clarkson & R. Baserga (eds) *Cold Spring Harbor Conferences on Cell Proliferation. 10. Control of Proliferation in Animal Cells*, Cold Spring Harbor, New York, pp. 23–36

[74] Levan, G. & Mitelman, F. (1976). G-banding in Rous rat sarcomas during serial transfer: significant chromosome aberrations and incidence of stromal mitoses. *Hereditas* **84** 1–14

[75] Tsutsui, T. & Ts'o, P. O. P. (1986). Requirement of long progression time for the expression of neoplastic phenotypes following direct perturbation to specific region(s) of DNA of syrian hamster embryo cells. *Cancer Res.* **46** 3533–3537

[76] Contreras, G., Bather, R., Furesz, J. & Becker, B. C. (1985). Activation of metastatic potential in African green monkey kidney cell lines by prolonged *in vitro* culture. *In Vitro Cell. Dev. Biol.* **21** 649–652

[77] Lahue, R. S., Su, S., Welsh, K. & Modrich, P. (1986). Analysis of methyl-directed mismatch repair *in vitro*. In: R. McMacken & T. J. Kelly (eds) *DNA Replication and Recombination: Proceedings of a UCLA Symposium*, Alan R. Liss, New York, pp 125–134

[78] Su, S.-S. & Modrich, P. (1986). *Escherichia coli* mutS-encoded protein binds to mismatched DNA base pairs. *Proc. Natl. Acad. Sci. USA* **83** 5057–5061

[79] Hanawalt, P. C. & Sarasin, A. (1986). Cancer-prone hereditary diseases with DNA processing abnormalities. *Trends Genet.* **2** 124–129

[80] Sutherland, B. M., Chamberlin, M. J. & Sutherland, J. C. (1973). Deoxyribonucleic acid photoreactivating enzyme from *Escherichia coli*. *J. Biol. Chem.* **248** 4200–4205

[81] Blackburn, E. H. (1985). Artificial chromosomes in yeast. *Trends Genet.* **1** 8–11

[82] Cozzarelli, N. R. (1986). Summary of the conference. In: R. McMacken & T. J. Kelly (eds) *DNA Replication and Recombination: Proceedings of a UCLA Symposium*, Alan R. Liss, New York, pp. 1–16

[83] Johnson, R. C. & Simon, M. I. (1985). *Hin*-mediated site-specific recombination requires two 26 bp recombination sites and a 60 bp recombinational enhancer. *Cell* **41** 781–791

[84] Taylor, A. & Smith, G. R. (1980). Unwinding and rewinding of DNA by the RecBC enzyme. *Cell* **22** 447–457

[85] Smith, G. R., Kunes, S. M., Schultz, D. W., Taylor, A. & Triman, K. L. (1981). Structure of Chi hotspots of generalized recombination. *Cell* **24** 429–436

[86] O'Hare, K. (1985). The mechanism and control of P element transposition in *Drosophila melanogaster*. *Trends Genet.* **1** 250–254

[87] Fink, G. R., Bocke, J. D. & Garfinkel, D. J. (1986). The mechanism and consequences of retrotransposition. *Trends Genet.* **2** 118–123

[88] Buhk, H.-J. & Messer, W. (1983). The replication origin of *E. coli*: nucleotide sequence and functional units. *Gene* **24** 265–279

[89] Alberts, B. M. (1985). Protein machines mediate the basic genetic processes. *Trends Genet.* **1** 26–30

[90] Nagley, P. (1985). RNA transcription and DNA synthesis at the origins of DNA replication. *Trends Genet.* **1** 290–291

[91] Nagley, P. (1986). Termination of replication of bacterial chromosomes. *Trends Genet.* **2** 221–222

2

Chromosome fragile sites for cancer research

Shaw Watanabe, Takashi Furuya and **Hisako Ochi**
Epidemiology Division, National Cancer Centre Research Institute, 5-1-1 Tsukiji, Chuo-ku, Tokyo 104, Japan

1. INTRODUCTION

Several types of genetic alterations have been indentified in human cancer. Enhanced oncogene expression and loss of alleles at loci on certain chromosomes seem to be common genetic alterations in most cancers [1,2,4,5,9–13]. Oncogene alterations have been considered to be important in carcinogenesis, but recent studies revealed that most oncogene activation is more related to tumour progression [3]. On the contrary, loss of alleles may be a critical event to unmask the recessive genetic changes for carcinogenesis.

Loss of allelic heterozygosity is shown in chromosome 13q in retinoblastoma, 11p in Wilms' tumour, and others [6–8]. Recessive genes may play a more general role in familial risk situations, since the loss of activity of one allele could be transmitted vertically through the germ line [11]. Examples of this include childhood cancers or cancers with heredity, such as colon cancer from familial polyposis, renal cell carcinoma, and others [12,13]. The problem is whether or not a similar mechanism is in operation in common cancers in adults and why the loss of alleles occurs in the particular tissue giving rise to cancer.

Loss of heterozygosity at a specific chromosomal location in a common human malignancy was first reported by Fearon et al. [12] who found that 40% of bladder carcinomas lose heterozygosity at 11p, suggesting that recessive tumour genes might play a more general role in human carcinogenesis. Recently, Yokota et al. [13] reported the loss of heterozygosity on chromosomes 3p, 13q and 17p in small cell lung carcinoma, 11p, 13q and 17p in breast cancer, 3p in uterine cancer, 5q, 17p and 18q in colon cancer, etc.

The mechanism of the loss of heterozygosity in any particular tissue is unclear. Fragile sites on chromosomes are consistent with loci of oncogenes or of translocations in tumour cells [14,15]. Therefore, we consider that some fragile sites are

Ch. 2]   **Chromosome fragile sites for cancer research**   25

important in oncogenesis, causing breaks in chromosomes leading to deletion and rearrangements leading to cancer. If so, screening individuals for the expression of fragile sites would become one of the means for identifying persons at risk for developing particular types of cancer.

2. WHAT ARE FRAGILE SITES?

Fragile sites are expressed as gaps, breaks or triradials on the chromosomes by proper induction [16]. Small gaps which appeared in unstained regions of banded mitotic figures tended to be overlooked, so that two-step staining (G banding after destain of the initial Giemsa staining) was necessary [17,54]. Frequency and mode of induction are the practical criteria available for use now in classifying fragile sites [18].

Fragile sites can be divided into two major groups; those which are rare and those which are common. The cytological appearance of fragile sites is variable, but there are several essential properties of rare fragile sites [19]: (a) there is a non-staining gap of varying width that usually involves both chromatids; (b) the fragile site is always at exactly the same point on the chromosome in an individual or a kindred; (c) the fragile site is inherited in a Mendelian codominant fashion; (d) the fragile site exhibits fragility under appropriate conditions of induction as evidenced by acentric fragments, deleted chromosomes, triradial figures, etc. (Fig. 1). 24 rare fragile sites

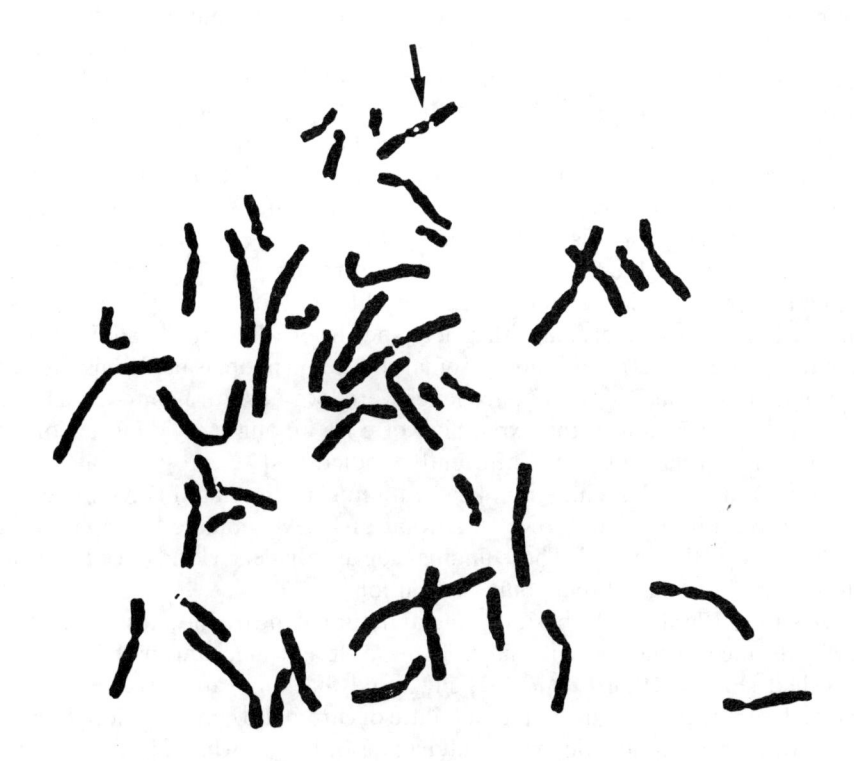

Fig. 1 — Gap on chromosomes (arrow).

are reported in Sutherland and Mattei [20] (Human Gene Mapping, 9th version); 18 are folate sensitive (group 1), 3 are distamycin A sensitive (group 2), 2 are bromodeoxyuridine (BrdU) requiring fragile sites (group 3) and 1 is unclassified (Fig. 2). The folate-sensitive fragile sites are those at 2q11.2, 2q13, 2q22.3, 6p23, 7p11.2, 8q22.3, 9p21.1, 9q32, 10q23.3, or 10q24.2, 11q13.3, 11q23, 12q13.1, 12q24.13, 16p12.3, 19p13, 20p11.23, 22q13 and Xq27.3. Group 2 fragile sites are those at 8q24.1, 16q22.1 and 17p12, and group 3 fragile sites are at 10q25.2 and 12q24.2. The unclassified rare fragile site on chromosome 8 at 8q13 was tentatively assigned to group 4. We recently found a new rare fragile site by induction of distamycin A at 3p25 (Fig. 3). More rare fragile sites may be found in the future, and differences between rare and common classifications may become arbitrary.

The common fragile sites are taken to be any point on a chromosome which breaks non-randomly when cells have been exposed to a clastogen. Yunis et al. [21] reported 110 different bands in which non-random breakage was induced by one of 16 different clastogens. 58 out of 110 bands were listed in Human Gene Mapping 8 [22]. The most consistently studied clastogen is aphidicolin, which is an inhibitor of DNA polymerase α [23]. It induces fragile sites most effectively on chromosomes, and addition of caffeine enhances its action. 55 common fragile sites were recognized at the Eighth International Workshop in Human Gene Mapping, but this number reached 80 in Human Gene Mapping 9. In addition to aphidicolin, fluorodeoxyuridine (FUdR), 5-azacytidine (Aza) and BrdU can cause common fragile sites [23,24]. Common fragile sites are weakly induced by folate- and thymidine-depleted medium. The most common of the fragile sites is 3p14; in addition, 2q31, 6q26, 7p32, 16q23 and Xp22 are rather common. Smeets et al. [25] demonstrated that 70 of 70 persons carried the fragile site at 3p14. Some other fragile sites are at least as frequent as the site at 3p14. Further details will be dealt with in a later section.

Fragile sites on chromosomes have been studied in relation to specific structural changes, such as translocation, deletion and inversion, especially in haematolymphoid malignancies. Yunis and Soreng [26] pointed out that coincidence was observed in 20 of 51 c-fra and 6 of 15 rare-fra among 31 specific breakpoints frequently appearing in neoplastic cells. Hecht and Sutherland [27] also reported that 8 rare-fra were concordant with cancer-specific breakpoints. In addition to these coincidences between fra and chromosomal changes in neoplasms, Yunis and Soreng [26] pointed out that 7 of the 17 oncogenes were located at or near c-fra. However, the degree of variation in the expression of c-fra among individual persons or in different organs has not been sufficiently studied [28]. Expression rates of rare-fra(X)(p27) in fibroblasts and in pokeweed-mitogen-stimulated B lymphocytes are known to be different, and lower than that of PHA-stimulated peripheral blood lymphocytes (PBL) [29–31]. These findings suggest that each kind of cell may have a different frequency and breakpoint distribution.

We know little about the basic molecular nature of the fragile sites and little about the significance of most fragile sites. Three different mechanisms have been distinguished (Fig. 4). Laird et al. [32] suggested that c-fra appeared as a result of delayed DNA synthesis with resultant failure of chromatid condensation. Some kind of injury in DNA replication may result in gaps or breaks which, if they were fixed, could become fragile sites. It is not known whether all c-fra occur through delayed DNA synthesis. It is generally accepted that genes for housekeeping enzymes, which

Chromosome fragile sites for cancer research

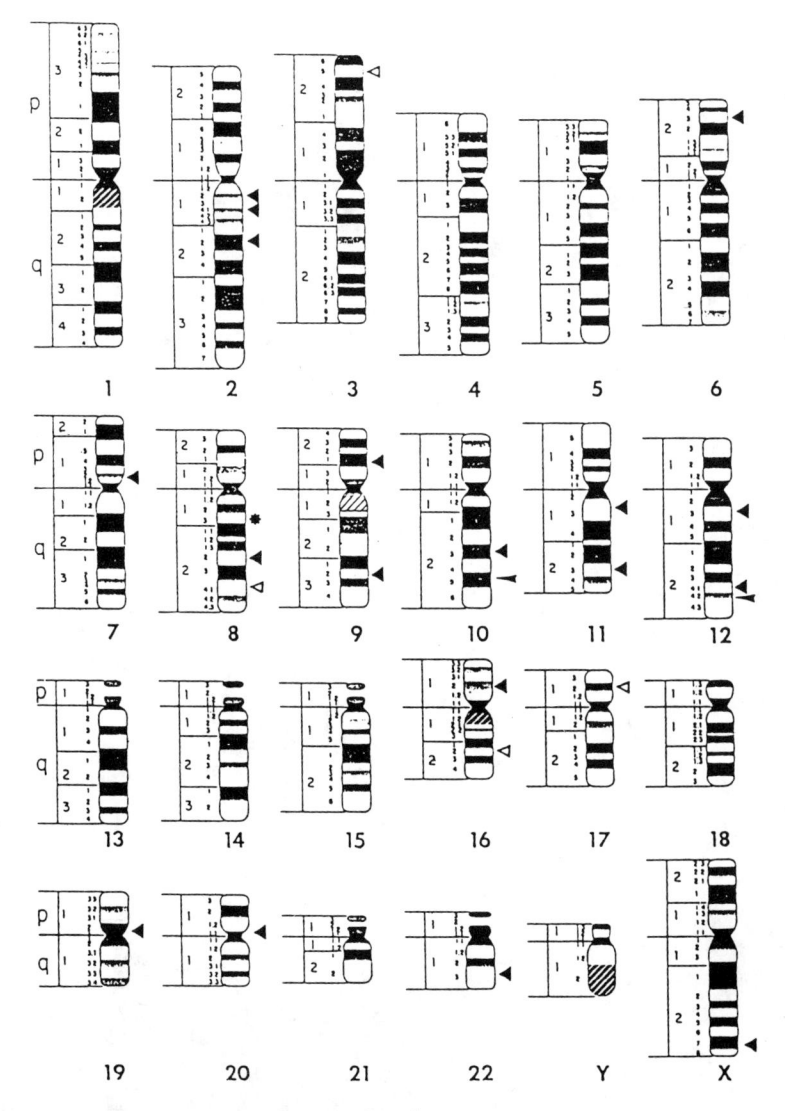

◄ : Folic acid sensitive fragile sites (18 sites)

◁ : Distamycin A inducible fragile sites (4 sites)

⃖ : BrdU dependent fragile sites (2 sites)

• : Unclassified (1 site)

Fig. 2 — Sites of rare fragile sites on chromosomes. Modified from Human Gene Mapping 9.

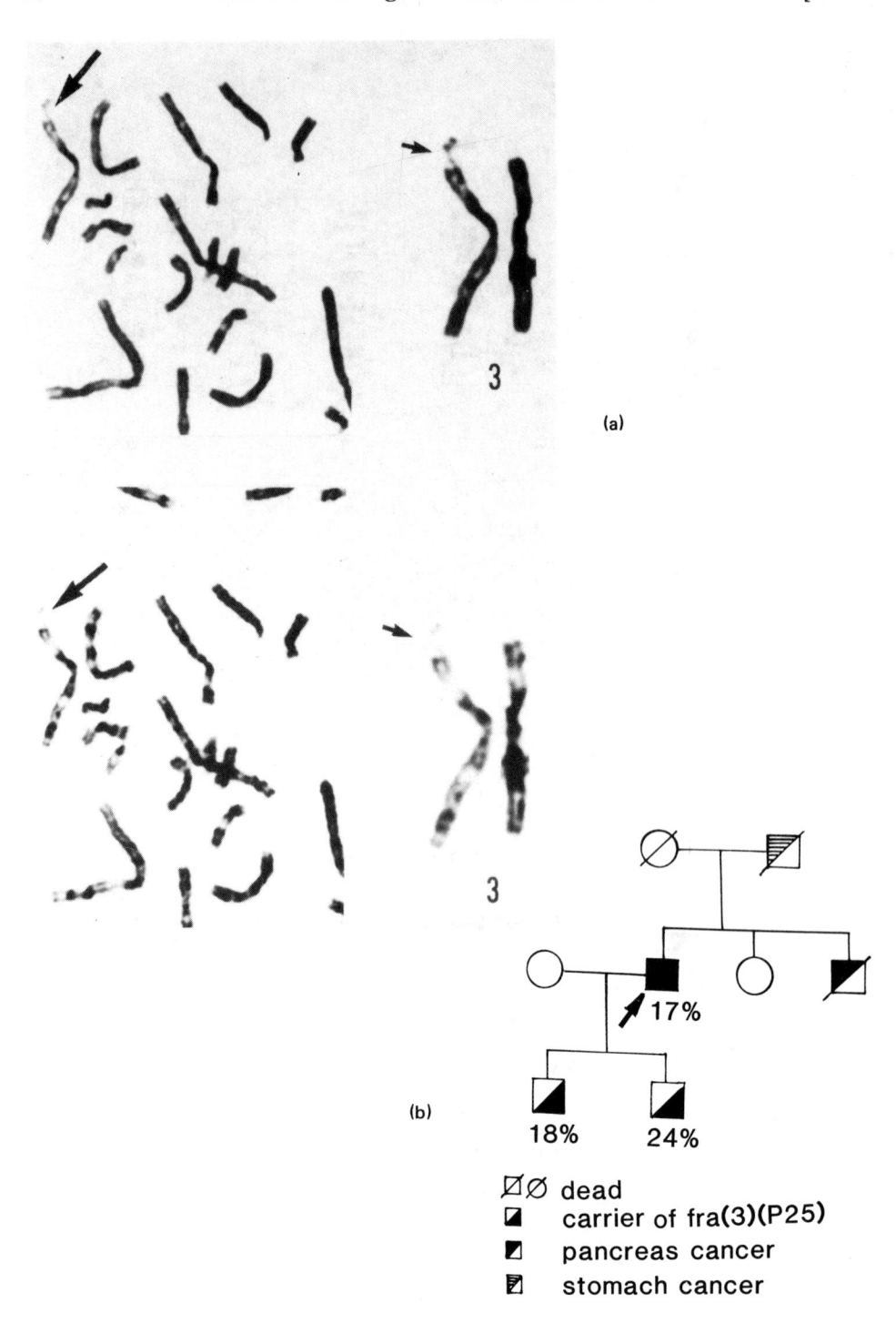

Fig. 3 — New rare fragile site: (a) fra(3)(p25) from a patient with pancreatic cancer (upper, Giemsa stain; lower, G banding); (b) family pedigree — both sons are carriers of fra(3)(p25).

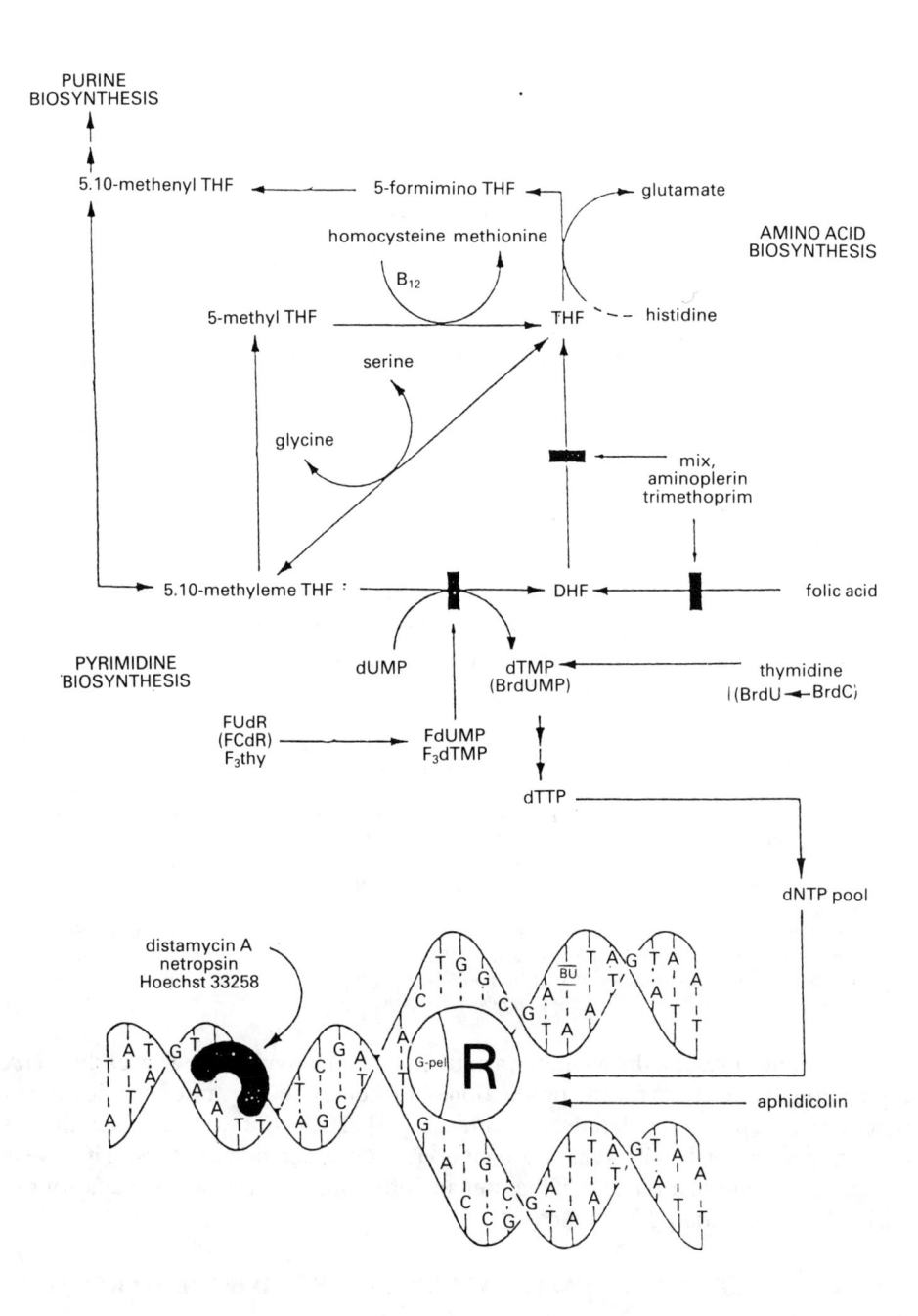

Ch. 2] Chromosome fragile sites for cancer research 29

Fig. 4 — (*Continued on next page*)

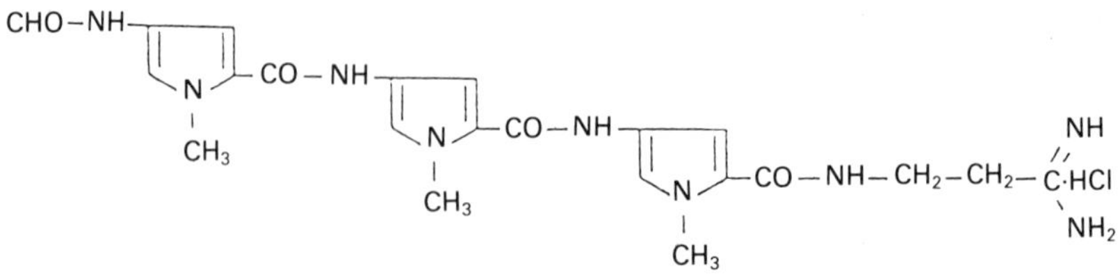

Fig. 4 — Scheme for the action of known chemical factors that affect fragile site expression. Folate pathway enzymes are (1) dihydrofolate reductase, (2) serine hydroxymethyltransferase, (3) methylene-THF reductase, (4) methionine synthetase, (5) glutamate formininotransferase, (6) formimino-THF cyclodeaminase, (7) methylene-THF dehydrogenase, (8) thymidylate synthetase and (9) thymidine kinase. Solid blocks represent sites of enzyme inhibition. R in hypothetical DNA molecule represents the replication enzyme complex containing DNA polymerase α Distamycin A, netropsin and Hoeschst 33258 are shown binding to consecutive A–T base pairs. (From *Fragile Sites on Human Chromosomes*, Oxford Monograph on Medical Genetics, 1985, p. 81.)

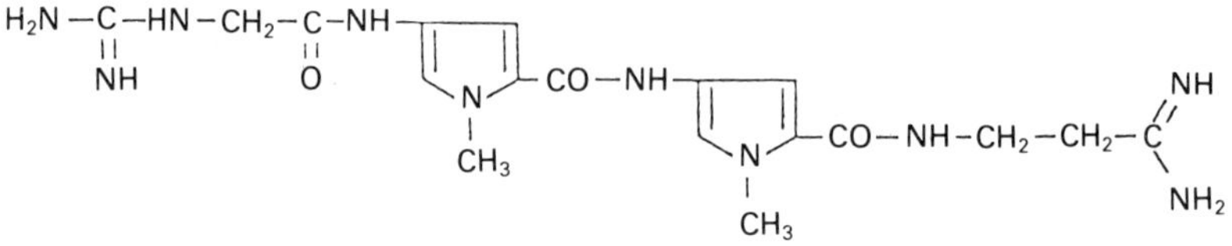

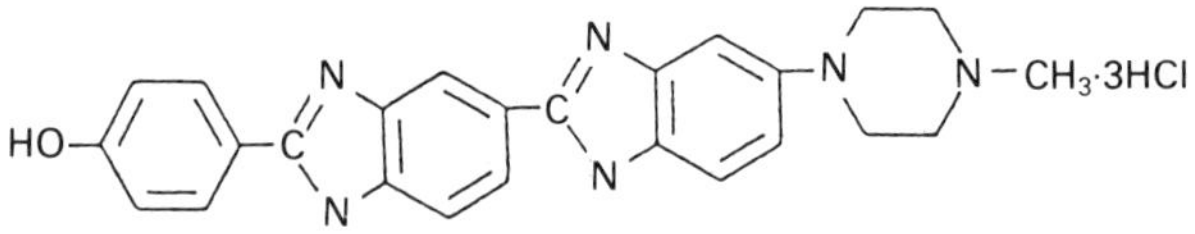

are commonly present throughout various tissues, are synthesized in early DNA replication, while genes for the production of proteins that are more specific for cell function are replicated in late DNA synthesis [33]. Therefore we can expect that, if the expressions of fragile sites are common in different tissues, these sites bear common cellular function, and, if the sites are different, they are more specific for the particular differentiated cell function.

3. FRAGILE SITES IN NORMAL LYMPHOCYTES AND BONE MARROW CELLS

Fragile sites are mainly studied by means of PBL stimulated with phytohaemagglutinin. c-fra of cells in different tissues have not been compared at all; only special sites, such as the X chromosome in fragile X syndrome, have been studied using skin fibroblasts and B lymphocytes [29–31]. We examined c-fra of bone marrow cells and peripheral blood lymphocytes from 6 healthy male students (Table 1). In PBL, 3p14 was the most frequent break, recognized in 36.9% of all breaks, followed by 16q23 and 6q26, with frequencies of 10.6% and 9.0% respectively of all breaks. Breaks at 1p32 and 2q31 were each observed in 2.8% of the total. All other breakpoints appeared at frequencies of less than 2.0% of total breaks. No break was found in chromosomes 15, 18, 19 and Y. Expressions of breaks in these frequent sites showed no difference among the six individuals examined and were stable by re-examination, although the frequency was variable (Fig. 5). For example, subject 5 had 17.6% of total breaks at site 6q26, while the frequency was only 5.3% in subject 6; 4 of 6 subjects showed no break at band 3q27, while subject 1 had breaks at this site with a frequency of 4.6%. Other less frequent breakpoints also showed considerable intersubject variation.

Common fragile sites in bone marrow cell were most frequent at 4q21–q25, accounting for 13.4% of all breaks. Further determination of 4q21–q25 was very different, because of unclear banding in this particular region by the G banding method. Recently, Morgan *et al.* [34] defined this site to be 4q23. They also described 7q11.32 as BM-specific c-fra. c-fra 16q23 followed with 9% of all breaks; 7q11.2, 3p14 and 7q22 formed the next group, with 7.2% 6.8% and 5.9% respectively of all breaks. Breaks at band 3q27 were present in 3.6% of all breaks, followed by 2q31 and 6q26 in 2.3%. The other breakpoints accounted for less than 2.0% of all breaks. A breakpoint having more than four breaks indicated a significantly high frequency ($p<0.05$). Chromosomes 15, 17, 18, 19, 21 and Y had no breaks. Several points, such as 4q31, showed considerable variation, but no significant individual differences among the 6 persons were observed in BM.

Three different categories were distinguished between PBL and BM; in one the breakpoints were present in both PBL and BM at high frequency, in the second the breakpoints were present in both PBL and BM but with a different frequency, and in the third the breakpoints were only present in PBL or in BM. Bands 2q31, 3p14, 6q26 and 16q23 commonly appeared in both PBL and BM with a frequency of more than 2%. Breaks at 3p14 and 6q26, however, were significantly more frequent in PBL, while no significant difference between PBL and BM was present at bands 2q31 and 16q23. Breaks in both 2p and 5q were significantly frequent in PBL, 2.8% ($p<0.05$), while breaks at 4q21–q25, 7q11.2, 7q22 and 13q21 were predominant in BM. Although there was no statistically significant difference ($0.05<p<0.1$), breaks in chromosome 11 tended to be more frequent in BM (4%). Similar trends in c-fra of BM cells are reported by induction with FUdR [34] and by methotrexate and thymidine induction [35].

The frequencies of c-fra recognized by us were higher than those reported by Glover *et al.* [23] and Tedeschi *et al.* [36], although the frequent sites, such as 3p14, 16q23, 6q23 and 2q31, were common. The frequency of breaks in BM was lower than that in PBL. As BM were not stimulated by PHA, spontaneous mitosis could be more resistant to aphidicolin. According to the hypothesis of Goldman *et al.* [33] breakpoints common to both PBL and BM may be related to more fundamental functions, while the specific breakpoints expressed only in PBL or BM may be related to specific functional genes in each cell lineage. Breaks at 2q31, 3p14, 6q26

Table 1 — Sites and numbers of breaks in metaphases of peripheral blood lymphocytes and bone marrow cells

Site[a]	Lymphocytes Subject 1	2	3	4	5	6	Total	%[b]	Bone marrow cells Subject 1	2	3	4	5	6	Total	%[b]
1p32	1	3	—	1	3	3	11	2.8	—	—	1	1	—	1	3	1.4
2q31	3	2	3	1	1	1	11	2.8	—	1	1	2	1	—	5	2.3
3p14***	35	21	27	25	12	23	143	36.9	6	2	3	3	—	1	15	6.8
3q27	4	—	—	—	1	—	5	1.3	1	2	1	2	—	2	8	3.6
4q21-25[c]***	—	—	1	—	—	2	3	0.8	5	2	3	8	5	7	30	13.4
5q13-35*	—	1	1	4	4	1	11	2.8	—	—	—	—	—	—	—	0.0
6q26**	8	4	6	5	9	3	35	9.0	—	1	1	1	—	2	5	2.3
7q11.2***	—	2	—	—	—	—	2	0.5	5	1	2	3	4	1	16	7.2
7q22**	1	—	—	1	2	—	4	1.0	5	3	1	1	2	1	13	5.9
13q21*	—	—	—	—	—	—	—	0.0	—	2	1	1	—	—	4	1.8
16q23	8	7	7	10	5	4	41	10.6	7	2	2	3	4	2	20	9.0
Total[d]***	87	60	72	61	51	57	388	100.0	47	45	22[e]	48	29	33	224	100.0

[a] Sites of breaks with a frequency of more than 2% of total breaks in peripheral blood lymphocytes and/or bone marrow cells or with a statistically significant difference in frequencies between peripheral blood lymphocytes and bone marrow cells.
[b] Percentage of the number of breaks at each site per number of total breaks.
[c] Total number of breaks at each band because of insufficient banding of this region in metaphases.
[d] Totals include number of unidentified breaks.
[e] Including one chromatid exchange.
*, $p<0.05$; **, $p<0.01$; ***, $p<0.001$; by χ^2 test.

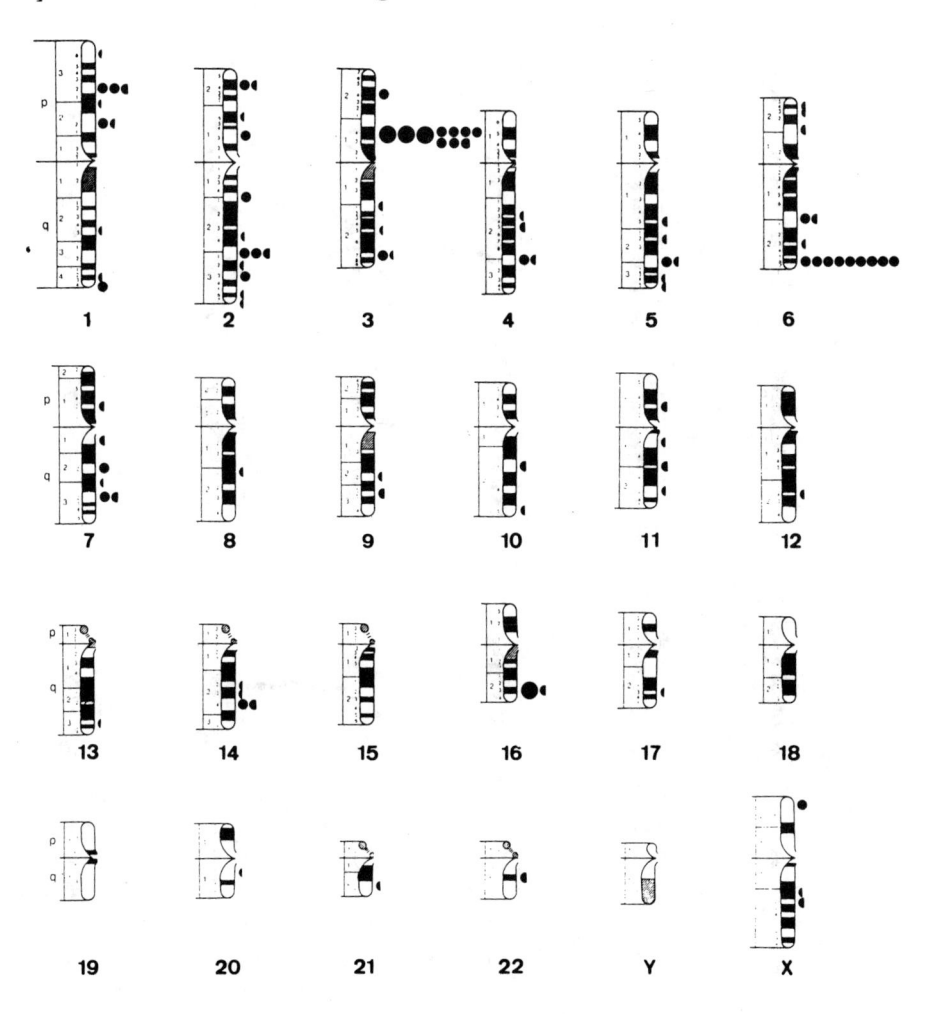

Fig. 5 — Distribution and frequency of c-fra induced by aphidicolin. (a) Peripheral blood lymphocytes stimulated by PHA. ●, 10%, and •, 1%, of total breaks on the chromosomes.

and 16q23 belonged to the former, and those at 4q21–q25, 5q15–q35, 7q11.2, 7q22 and 13q21 to the latter.

4. FRAGILE SITES IN LEUKAEMIA PATIENTS

Are those different breakpoint categories related to the site of translocation or deletion in leukaemia and lymphoma? Many breakpoints in PBL and BM from healthy volunteers did not coincide with breakpoints observed in leukaemias and lymphoma. Only a few of of these breakpoints are seen in malignancies, such as 4q21 occurring in acute leukaemia (most likely monocytoid and/or myeloid) [37,38], and 11q23 occurring in myelomonocytic leukaemia and monocytic leukaemia [14,

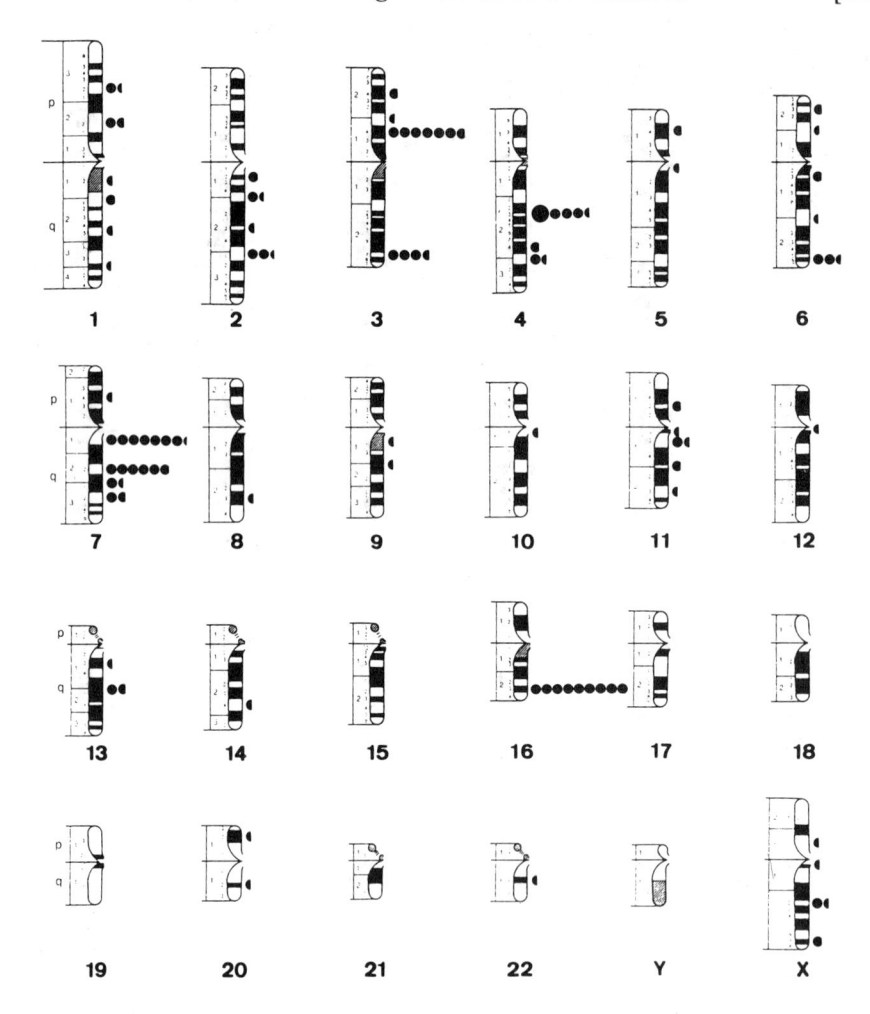

Fig. 5 — Distribution and frequency of c-fra induced by aphidicolin. (b) Bone marrow cells without stimulation. ●, 10%, and •, 1%, of total breaks on the chromosomes.

39–41]. Frequent sites of breaks, such as 3p14, 6q26 and 16q23, have not been reported to be frequent breakpoints in haematolymphoid malignancies, except for 3p14 in acute lymphoblastic leukaemia on very rare occasions [42]. Furthermore, site 5q12–q35, which was not present in BM, is often involved in myelodysplastic syndrome [43–45], acute non-lymphoblastic leukaemia [39–41] and secondary acute leukaemia [45]. Sites 7q11 and 7q22, which were frequent in BM but not in PBL, are the breakpoints occurring in adult T cell leukaemia [46,47]. These findings suggest that neoplastic change could take place when breaks (or preceding changes in DNA) occur at sites which do not express breaks in normal cells.

In leukaemia and lymphoma, coincidence of sites between fra and breaks in neoplastic cells has been observed at several locations, such as 8q22, 11q23, and

16q22 [14,48–50]. Among these, only rare-fra(16)(q22) has been observed in the same patients with leukaemia (M4Eo) in both PHA-stimulated PBL and leukaemic cells [49,50]. Although Simmers et al. [51] reported that rare-fra at 16q22 was not coincident with the breakpoint in the chromosomal rearrangement at 16q22 in AMMoL, the conclusion must be confirmed by further study. Such a direct relation of c-fra has not yet been observed in other regions [14,49,52].

We examined BM and PBL samples obtained at the same time from 15 patients before any treatment for leukaemia [17]. The types of leukaemias were four M2, four M3, one M4, one M6, four ALL and one refractory anaemia with excess blasts according to FAB classification. The list of patients and leukaemias is summarized in Table 2. In PBL from leukaemic patients, the most frequent breakpoint was the 3p14

Table 2 — Leukaemia patients and karyotypes of leukaemic cells

Case	Age	Diagnosis	Bone marrow NCC ($\times 10^4/\mu l$)	Blast (%)	Karyotypes
1	31	ALL	30.0	94	46,XY
2	71	ALL	19.0	95	46,XY
3	54	ALL	20.0	97	46,XY
4	80	ALL	29.1	93	45,XY,4q+,+7,−8,−15,18q+, +20,−22,−22,+Mar
5	73	M2	37.4	38	46,XY
6	14	M2	15.0	67	46,XX,t(8;21)
7	42	M2	27.0	17	46,XX,t(8;21)
8	5	M2	22.8	37	45,X,−Y,t(8;21)
9	43	M3	46.2	93	46,XX
10	46	M3	3.9	57	46,XY,t(15;17)
11	71	M3	32.0	90	46,XY,t(15;17)
12	24	M3	43.2	82	46,XY,15q+,i(17q−)
13	48	M4	169.3	96	46,XX
14	44	M6	107.0	52	46,XX
15	74	RAEB	11.0	11	40,XY,−5,−7,+8,+9,−11,−13, −16,−18,−19,−20,−21,+2Mar

ALL: Acute lymphocytic leukaemia.
M2–M6: The FAB classification.
RAEB: Refractory anaemia with excess of blasts.

band which appeared in 30.2% of total breaks; 16q23 and 6q26 followed with frequencies of 11.8% and 9.7% respectively of total breaks (Table 3). Breaks at bands 1p22, Xp22, 7q32 and 14q24 were observed in 3.4%, 3.3%, 2.2% and 2.2% respectively of total breaks. Frequencies at all other breakpoints were less than 2% of total breaks. Compared with frequencies from control PBL, 1p22 had a higher frequency and 3p14 had a lower frequency in PBL from ALL patients ($p<0.05$), although Rao et al. [52] reported no significant difference between c-fra of ALL patients and c-fra of normal controls. All other breakpoints did not show significant differences in sites from control PBL. Generally, c-fra from ALL patients resembled normal PBL, and c-fra from AML patients seemed to be different from both normal PBL and normal BM.

Table 3 — Distribution of breaks in peripheral blood lymphocytes from patients with acute leukaemia

Bands	PBL from HP	BM cells from HP	PBL from ALL patients	PBL from AML patients
1p32	2.4	1.4	0.9	0.1***
1p22	1.2	1.8	3.4*	1.2
2q31	2.5	2.3	0.9	1.6
3p14	37.2	6.8***	30.2*	22.0***
3q27	1.0	3.6**	0.9	1.2
4q21–25	0.4	13.4***	0.3	2.1***
6q26	8.0	2.3**	9.7	2.9***
7q11.2	0.2	7.2***	0.0	0.6
7q22	1.0	5.9***	1.2	2.2*
7q32	1.1	1.8	2.2	4.3***
14q24	1.5	0.5	2.2	2.3
16q23	11.6	9.0	11.8	7.3**
Xp22	1.1	3.7***	0.0	3.3
Total number of breaks	1305	224	321	681

Percentage of total breaks; PBL: peripheral blood lymphocytes; HP: healthy persons; BM: bone marrow ; ALL: acute lymphoblastic leukaemia; AML: acute myeloblastic leukaemia.
*, $p<0.05$; **, $p<0.01$; ***, $p<0.001$; by χ^2 test in comparison with PBL from HP.

In the seven AML patients, band 3p13 was observed in 22.0% of total breaks, and 16q23 (7.3%) and 7q32 (4.3%) followed. The frequencies of breaks at bands Xp22, 6q26, 14q24, 7q22 and 4q21–q25 were 3.7%, 2.9%, 2.3%, 2.2% and 2.1% respectively. All other breakpoints had frequencies of less than 2.0% of total breaks. Bands 4q21–q25, 7q22, 7q32 and Xp22 had higher frequencies and 1p32, 3p14, 6p26 and 16q23 had lower frequencies than those of control PBL (7q22, $p<0.05$; 16q23, $p<0.01$; the rest, $p<0.001$). Cases with complex chromosome abnormality tended to have increased numbers of c-fra (Fig. 6).

The occurrence of c-fra on chromosomes 8 and 21 from patients with AML with t(8;21) was noteworthy. There was no break on 8p, but 11 breaks out of 421 were observed of three breakpoints on 8q (Table 4). This ratio was significantly higher than those from either the other leukaemias without t(8;21) or the controls ($p<0.001$). Breaks at 8q22 were especially high ($p<0.001$). The other translocation site of chromosome 21 caused no break by aphidicolin induction. The c-mos oncogene was mapped on chromosomes 8q22 by Neel et al. [54]. As cells with fra(8)(q22) belonged to T lymphocytes, which constituted a different cell lineage from leukaemia cells with t(8;21), this result suggested that fragility at 8q22 at the stem cell level may be related to the occurrence of chromosomal rearrangement in M2 with t(8;22) leukaemia, because no increase in the frequency of breaks at 21q22 was observed [54].

Chromosome fragile sites for cancer research

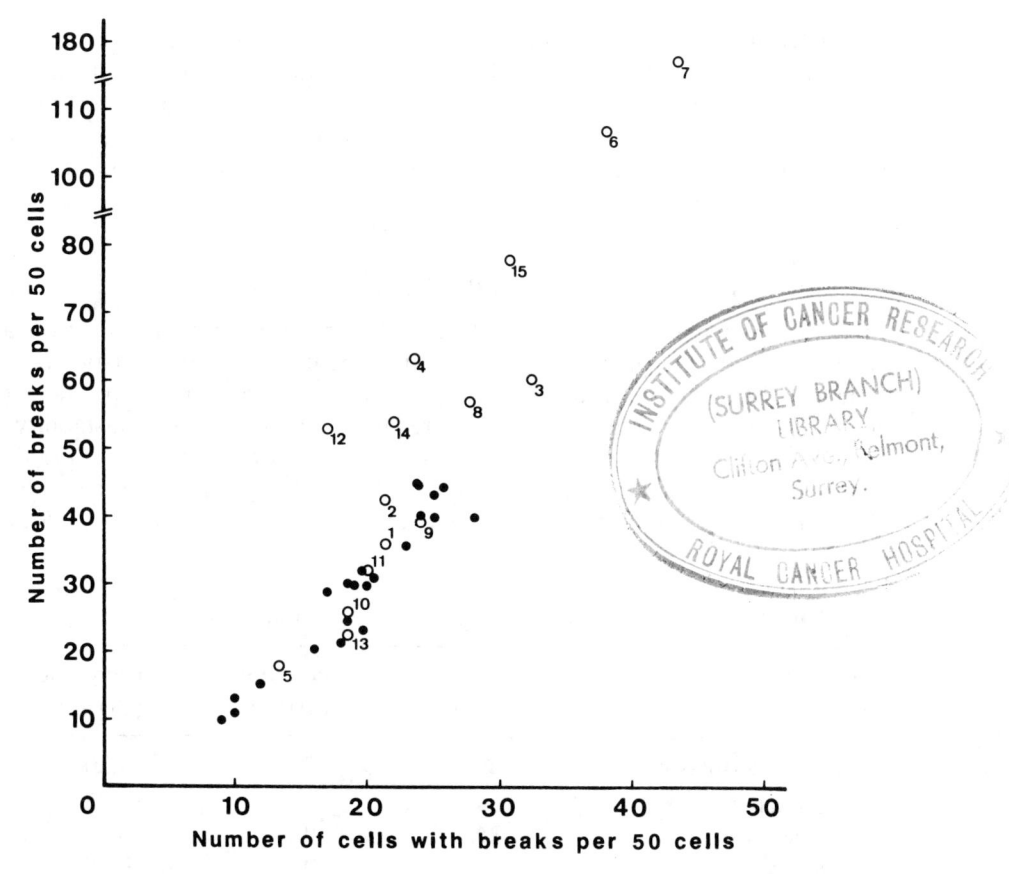

Fig. 6 — Numbers of c-fra induced by aphidicolin in lymphocytes from leukaemia patients and healthy controls. Leukaemia patients with abnormal karyotype show a greater number of c-fra than those without karyotypic change or controls. Numbers next to circles are the case numbers in Table 2.

Table 4 — Frequency of breaks at 8q13–24 per 100 metaphases in peripheral blood lymphocytes

Bands	M2 with t(8;21)	Leukaemias without t(8;21)	Healthy persons
8q13	0.0	0.1	0.0
8q21	0.5	0.0	0.0
8q22	3.2*	0.1	0.3
8q23	0.0	0.0	0.1
8q24	1.4*	0.1	0.1
Total	5.1*	0.4	0.5

*, $p<0.001$ by χ^2 test in comparison with healthy persons.

Although c-fra at 15q22 and 17q21 in PBL did not increase in acute promyelocytic leukaemia with t(15;17) (q22;q21) by aphidicolin, fragility could be detected at the same bands as translocation sites of leukaemic cells in PBL by other induction methods. All these findings suggested that instability of chromosomes, even though the cellular lineages were different, seemed to be related to the occurrence of chromosomal rearrangement.

5. FRAGILE SITES IN CANCER PATIENTS

Increased frequency of breaks in PBL has been reported in patients with ovarian cancer [37] and lung cancer [55], and these findings also suggested a relation between cancer and fragility of chromosomes. Stable induction of c-fra at specific sites in specific cells suggested that the c-fra is a reliable marker for genetic instability.

First, we consider that the rare-fra may be a marker for genetic susceptibility to cancer. We examined the frequency of rare-fra in 327 various cancer patients, and the frequency was not different from that of the general population (Table 5). Lung

Table 5 — Frequency of rare fragile site expression among cancer patients

Diseases	Number of patients	Number of carriers	Frequency (%)
Malignant neoplasms	327	17	5.20
Pancreas	3	1	33.33
Stomach	24	0	0.00
Colon–rectum	22	0	0.00
Lung	85	8	9.41
Breast	93	3	3.23
Uterus	33	1	3.03
Ovary	10	1	10.00
Bladder	39	2	5.13
Borderline lesion	10	1	10.00
Benign tumour	29	3	10.34
Healthy control[a]			6.0

[a] According to Takahashi et al. [57].

cancer patients and ovarian cancer patients (Table 6), however, showed significantly high rates of rare-fra. 5 of 40 adenocarcinoma cases of the lung showed fragility at 16p12, 16q22 and 17p12 (Table 6). 3 of 28 squamous cell carcinoma patients showed fragility at 11q13 and 16q22 [56]. The frequency compared with that for the general population, reported by Takahashi et al. [57], was more than double in both adenocarcinoma and squamous cell carcinoma. It is interesting that 17p12 aggregated in adenocarcinoma cases. On the contrary, heritable fragile sites were

Table 6 — Frequency of heritable fragile sites among lung cancer patients

Histology	Number	Folate sensitive 11q13	16p12	Distamycin A 16q22	17p12	Total	%
Adeno	40	—	1	1	3	5	12.5
Squamous	23	1	—	2	—	3	10.7
Large	9	—	—	—	—	0	0
Small	3	—	—	—	—	0	0
Other	5	—	—	—	—	0	0
Carcinoid	2	—	—	—	1	1	50
Total (%)	87	1 (1.14)	1 (1.14)	3 (3.45)	4 (4.59)	9 (10.20)	
General population (%)		0.20	0.00	1.42	3.08		4.70

Table 7 — Frequency of heritable fragile sites among breast cancer patients

Histology	Number	Distamycin A 8q24	16q22	17p12	BrdU 10q15	Total	%
Adenocarcinoma	90	0	1	1	1	3	3.3
Benign lesions	10	1	1	0	0	2	20
Total (%)	100	1 (1.0)	2 (2.0)	1 (1.0)	1 (1.0)	5	(5.0)
General population (%)		0.71	1.42	3.08	0.29		5.5

rare among 90 breast cancer patients [56]. Although 5 carriers were detected, 2 had benign lesions (Table 7). The overall frequency is the same as that in the general population; the sites were 10q22 in one, 16q22 in two, 17p12 in one, and 8q24 in two. We found a new rare-fra at 8q24 in a breast tumour case and 3p25 in a pancreatic cancer patient [58]. The elder brother of the latter patient had died of pancreatic cancer, so that r-fra(3)(p25) may be related to the occurrence of pancreas cancer. *Raf*1 is present at 3p25.

The involvement of c-fra in cancer patients is still under study. Preliminary data showed that the frequency and the site were influenced by certain host factors, such as the age of menarche [59]. The relationship between c-fra in lymphocytes and neoplastic cells *per se* is not known but, in addition to the lung and ovarian cancer patients, about two-thirds of breast cancer patients showed a higher frequency of c-fra (Fig. 7). Moreno *et al.* [60] found that c-fra in 22q12, 11p13 and 11p14 were only observed in lymphocytes from meningioma cases, although 3p14, 16q23, Xp22,

Common Fragile Sites in Normal and Breast Cancer Patients

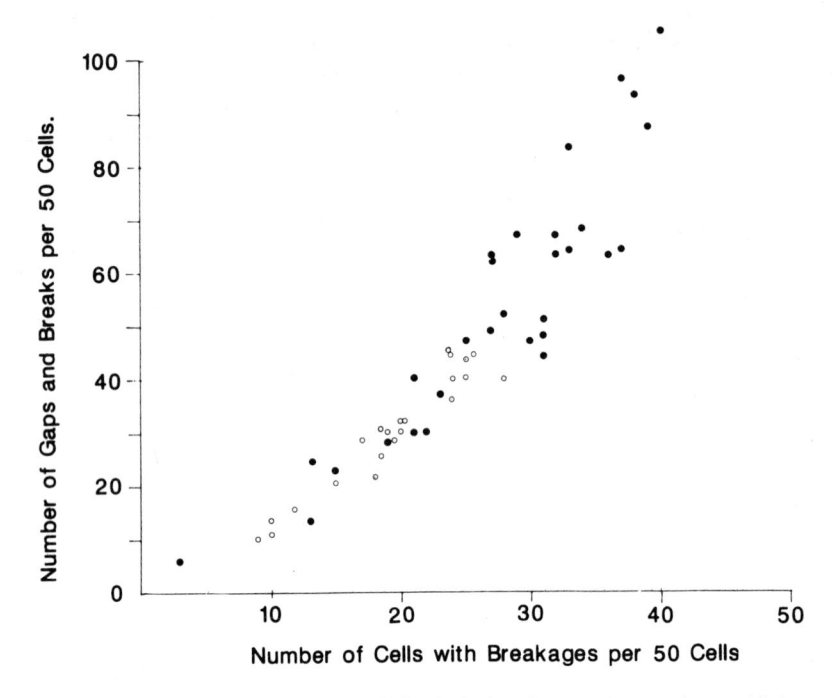

Fig. 7 — Numbers of c-fra induced by aphidicolin in lymphocytes from patients with breast cancer. About two thirds of patients show a higher number of c-fra than controls.

2q33, 1p22, 6q26, 7q32, 10q22 and 14q24 were found in the control group. Oncogene PDGFB is present at 22q12, and deletion and/or monosomy of 22q is frequent in meningioma cell.

Marty et al. [61] reported that the frequency of sister chromatid exchanges (SCE) are significantly higher in lymphocytes from patients with uterine cervical cancer, followed by lymphocytes in cases of percancerous lesions, than in those of age-matched healthy women. SCE is known to be influenced by habits such as cigarette smoking, so the association might not be direct. c-fra seems to be less influenced, so that it may be a more stable biomarker for predisposition to cancer. Analyses of the expression of c-fra, in particular of tissue cells from which malignant cells appear, may clarify the mechanism of loss of heterozygosity or chromosomal deletion in the early stage of carcinogenesis.

6. CONCLUSION

Recent development in the study of carcinogenesis and genetic epidemiology has clarified that both carcinogenic stimuli and genetic predisposition to cancer act for tumour growth. Loss of heterozygosity is commonly found in many cancers, and it has become to be considered as an initial trigger for carcinogenesis. The mechanism of the DNA deletion, however, has not been clarified. Chromosome instability of fragility of the DNA chain during replication is an attractive hypothesis for the initial event of DNA deletion. In this regard, fragile sites on chromosomes would be a useful marker for the predisposition to cancer or other genetically determined diseases. More analyses for fragile sites in different tissue cells in comparison with neoplastic cells should open a new field of cancer research.

REFERENCES

[1] Harris, C. C. (1987). Human tissues and cells in carcinogenesis research. *Cancer Res.* **47** 1–10
[2] Law, D. J., Olshwang, S., Monpezat, J. P., Lifrancois, D., Jagelman, D., Petrelli, N., Thomas, G. & Feinberg, A. P. (1988). Concerted nonsystemic allelic loss in human colorectal carcinoma. *Science* **241** 961–965
[3] Tsuda, H., Shimosato, Y., Upton, M. P., Yokota, J., Terada, M., Ohira, M., Sugimura, T. & Hirohashi, S. (1988). Retrospective study on amplification of N-myc and c-myc genes in pediatric solid tumors and its association with prognosis and tumor differentiation. *Lab. Invest.* **59** 321–324
[4] Hansen, M. F., Koufos, A., Gallie, B. L., Phillips, B. A., Fodstad, O., Bragger, A., Gedde-Dahl, T. & Cavenee, W. K. (1985). Osteosarcoma and retinoblastoma: a shared chromosomal mechanism revealing recessive predispostion. *Proc. Natl. Acad. Sci. USA* **82** 6216–6220
[5] Cannon-Albright, L. A., Skolnick, M. H., Bishop, D. T., Lee, R. G. & Burt, R. W. (1988). Common inheritance of susceptibility to colonic adenomatous polyps and associated colorectal cancers. *N. Engl. J. Med.* **319** 533–537
[6] Xu,. H.-J., Hu, S.-X., Hashimoto, T., Takahashi, R. & Bennedict, W. F. (1989). The retinoblastoma susceptibility gene product: a characteristic pattern in normal cells and abnormal expression in malignant cells. *Oncogene* (in press)
[7] Reeve, A. E., Sih, S., Raizis, A. & Feinberg, A. P. (1989). Loss of allelic heterozygosity at a second locus on chromosome 11 in sporadic Wilms' tumor. *Mol. Cell. Biol.* (in press)
[8] Seizinger, B. R., Rouleau, G. A., Ozelius, L. J., Lance, A. H., Farmer, G. E., Lamiell, J. M., Haines, J., Yuen, J. W. M., Collins, D., Majoor-Krakauer, D., Bonner, T., Mathew, C., Rubenstein, A., Halperin, J., McConkie-Rosell, A., Green, J. S., Schimke, R., Oostra, B., Aronin, N., Smith, D. I., Drabkin, H., Waziri, M. H., Hobbs, W. J., Martuza, R. L., Conneally, P. M., Hsia, Y. E. and Gusella, J. F. (1988). Von Hippel-Lindau disease maps to the region of chromosome 3 associated with renal cell carcinoma. *Nature (London)* **332** 268–269
[9] Hansen, M. F. & Cavenee, W. K. (1987). Genetics of cancer predisposition. *Cancer Res.* **47** 5518–5527
[10] Sandberg, A. A., Turc-Carel, C. & Gemmill, R. M. (1988). Chromosomes in solid tumours and beyond. *Cancer Res.* **48** 1049–1059
[11] Kundson, A. G., Jr. (1986). Genetics of human cancer. *An. Rev. Genet.* **20** 231–251
[12] Fearon, E. R., Feinberg, A. P., Hamilton, S. H., & Vogelstein, B. (1985). Loss

of genes on the short arm of chromosome 11 in bladder cancer. *Nature (London)* **318** 377–380
[13] Yokota, J., Wada, M., Shimosato, Y., Terada, M. & Sugimura, T. (1987). Loss of heterozygosity on chromosomes 3, 13, and 17 in small cell carcinoma and on chromosome 3 in adenocarcinoma of the lung. *Proc. Natl. Acad. Sci. USA* **84** 9252–9256
[14] Le Beau, M. M. (1986). Chromosomal fragile sites and cancer-specific rearrangements. *Blood* **67** 849–858
[15] Le Beau, M. M. & Rowley, J. D. (1984). Heritable fragile sites in cancer. *Nature (London)* **308** 607–608
[16] Sutherland, G. R. & Hecht, F. (1985). *Fragile Sites on Human Chromosomes*, Oxford University Press, New York
[17] Furuya, T., Ochi, H. & Watanabe, S. (1989). Common fragile sites on chromosomes in bone marrow cells and peripheral blood lymphocytes from healthy persons and leukemia patients. *Cancer Genet. Cytogenet.* (in press)
[18] Hecht, F. (1986). Rare, polymorphic, and common fragile sites: a classification. *Hum. Genet.* **74** 207–208
[19] Sutherland, G. R. (1979). Heritable fragile sites in human chromosome 2. Distribution, phenotype and cytogenetics. *Am. J. Hum. Genet.* **31** 121–135
[20] Sutherland, G. R. & Mattei, J. F. (1987). Report of the committee on cytogenetic markers. *Cytogenet. Cell Genet.* **46** 316–324
[21] Yunis, J. J., Soreng, A. L. & Bowe, A. E. (1987). Fragile sites are targets of diverse mutagens and carcinogens. *Oncogene* **1** 59–69
[22] Human Gene Mapping 8 (1985). Eighth International Workshop on Human Gene Mapping. *Cytogenet. Cell Genet.* **40**
[23] Glover, T. W., Berger, C., Coyle, J. & Echo, B. (1984). DNA polymerase alpha inhibition by aphidicolin induces gaps and breaks at common fragile sites in human chromosomes. *Hum. Genet.* **67** 136–142
[24] Hecht, F., Tajara, E. H., Lockwood, D., Sandberg, A. A. & Hecht, B. K. (1988). New common fragile sites. *Cancer Genet. Cytogenet.* **33** 1–9
[25] Smeets, D. F. C. M,. Scheres, J. M. J. C. & Hustinx, T. W. J. (1986). The most common fragile site in man is 3p14. *Hum. Genet.* **72** 215–220
[26] Yunis, J. J. & Soreng, A. L. (1984). Constitutive fragile sites and cancer. *Science* **226** 1199–1204
[27] Hecht, F. & Sutherland, G. R. (1984). Fragile sites and cancer breakpoints. *Cancer Genet. Cytogenet.* **12** 179–181
[28] Craig-Holmes, A. P., Strong, L. C., Goodacre, A. & Pathak, S. (1987). Variation in the expression of aphidicolin-induced fragile sites in human lymphocyte cultures. *Hum. Genet.* **76** 134–137
[29] Broobwell, R., Daniel, A., Turner, G. & Fishbum, J. (1984). The fragile X(q27) form of X-linked mental retardation: FUdR as an inducing agent for fra(X)(q27) expression in lymphocytes, fibroblasts, and amniocytes. *Am. J. Med. Genet.* **13** 139–148
[30] Marchese, C. A., Lin, M. S. & Wilson, M. G. (1984). Comparison of expression of the fragile site at Xq27 in T and B lymphocytes. *Hum. Genet.* **67** 213

[31] Sutherland, G. R. & Baker, E. (1986). Induction of fragile sites in fibroblasts. *Am. J. Hum. Genet.* **38** 573–575

[32] Laird, C., Jaffe, E., Karpen, G., Lamb, M. & Nelson, R. (1987). Fragile sites in human chromosomes as regions of late-replicating DNA. *Trends. Genet.* **3** 274–281

[33] Goldman, M. A., Holmquist, G. P., Gray, M. C., Caston, L. A. & Nag, A. (1984). Replication timing of genes and middle repetitive sequences. *Science* **224** 686–692

[34] Morgan, R., Morgan, S. S., Hecht, B. K. & Hecht, F. (1988). Fragile sites at 4q23 and 7q11.23 unique to bone marrow cells. *Cancer Genet. Cytogenet.* **31** 47–53

[35] Przylepa, K. A. & Wenger, S. L. (1988). Chromosome breaks and fragile sites in leukemic bone marrow cells. *Cancer Gent. Cytogenet.* **33** 35–38

[36] Tedeschi, B., Porfirio, B., Vernole, P., Caporossi, D., Dallapiccola, B. & Ni'oletti, B. (1987). Common fragile sites: their prevalance in subjects with constitutional and acquired chromosomal instability. *Am. J. Med. Genet.* **27** 471–482

[37] Kocova, M., Kowalczyk, J. R. & Sandberg, A. A. (1985). Translocation 4;11 acute leukemia: three case reports and review of the literature. *Cancer Genet. Cytogenet.* **16** 21–32

[38] De Braekeleer, M. & Lin, C. C. (1986). 4;11 translocation-associated acute leukemia: a comprehensive analysis. *Cancer Genet. Cytogenet.* **21** 53–66

[39] Yunis, J. J., Bloomfield, C. D. & Ensrud, K. (1981). All patients with acute non-lymphocytic leukemia may have a chromosomal defect. *New Engl. J. Med.* **305** 135–139

[40] Sandberg, A. A. (1986). The chromosomes in human leukemia. *Semin. Hematol.* **23** 201–217

[41] Yunis, J. J. (1983). The chromosomal basis of human neoplasia. *Science* **221** 227–236

[42] Williams, D. L., Look, A. T., Melvin, S. L., Robertson, P. K., Dahl, G., Flake, T. & Stass, S. (1984). New chromosomal translocations correlate with specific immunophenotypes of childhood acute lymphoblastic leukemia. *Cell* **36** 101–109

[43] Yunis, J. J., Rydell, R. E., Oken, M. M., Arnesen, M. A., Mayar, M. G. & Lobell, M. (1986). Refined chromosome analysis as an independent prognostic indicator in *de novo* myelodysplastic syndrome. *Blood* **67** 1721–1730

[44] Gyger, M., Perreault, C., Pichette, R., Forest, L. & Lussier, P. (1986). Interstitial deletion of the long arm of chromosome 5(5q−) in leukemia and other hematological disorders: clinical and biological relevance of variable breakpoint pattern. *Leuk. Res.* **10** 9–15

[45] Perger, R., Bloomfield, C. D. & Sutherland, G. R. (1985). Human gene mapping 8. *Cytogenet. Cell Genet.* **40** 490

[46] Miyamoto, K., Tomita, N., Ishii, A., Nonaka, H., Kondo, T., Tanaka, T. & Kitajima, K. (1984). Chromosome abnormalities of leukemia cells in adult patients with T-cell leukemia. *J. Natl. Cancer Inst.* **73** 353–362

[47] Fifth International Workshop on Chromosomes in Leukemia–Lymphoma (1987). Correlation of chromosome abnormalities with histological and immunologic characteristic in non-Hodgkin's lymphoma and adult T cell leukemia–lymphoma. *Blood* **70** 1554–1564
[48] Arthur, D. C., Aaseng, S. M. & Bloomfield, C. D. (1985). Acute myelomonocytic leukemia with marrow eosinophilia (M4Eo) and inv(16)(p13q22) in a patient with a heritable fragile site 16q22. *Cytogenet. Cell Genet.* **40** 572
[49] de Braekeleer, M. (1987). Fragile sites and chromosomal structural rearrangements in human leukemia and cancer. *Anticancer Res.* **7** 417–422
[50] Hecht, F. & Glover, T. W. (1984). Cancer chromosome breakpoints and common fragile sites induced by aphidicolin. *Cancer Genet. Cytogenet.* **13** 185–188
[51] Simmers, R. N., Sutherland, G. R., West, A. & Richards, R. I. (1987). Fragile sites at 16q22 are not at the breakpoint of chromosomal rearrangement in AMMoL. *Science* **236** 92–94
[52] Rao, P. N., Heerema, N. A. & Palmer, C. G. (1988). Expression of fragile sites in childhood acute lymphoblastic leukemia patients and normal controls. *Hum. Genet.* **79** 329–334
[53] Neel, B. G., Jhanwar, S. C., Chaganti, R. S. K. & Hayward, W. S. (1982). Two human c-oncogenes are located on the long arm of chromosome 8. *Proc. Natl. Acad. Sci. USA* **79** 7842–7846
[54] Furuya, T., Ochi, H. & Watanabe, S. (1989). Expression of fragile site 8q22 in peripheral blood lymphocytes from patients with acute leukaemia M2 having t(8;21)(q22;q22). *Jpn. J. Clin. Oncol.* **19** 23–25
[55] Nordenson, I., Beckman, L., Linden, S. & Sternberg, N. (1984). Cromosomal aberrations and cancer risk. *Hum. Hered.* **34** 76–81
[56] Watanabe, S., Ochi, H., Kobayashi, Y., Tsugane, S., Arimoto, H. & Kitagawa, K. (1987). Frequency of multiple primary cancers and risk factors for lung and breast cancer patients. In: R. W. Miller, Watanabe, S., Fraumeni, J., Sugimura, S. & Sugano, H. (eds) *Unusual Occurrences as Clues to Cancer Etiology.* Japan Scientific Societies Press, Tokyo
[57] Takahashi, E., Hori, T. & Murata, M. (1988). Population cytogenetics of rare fragile sites in Japan. *Hum. Genet.* **78** 121–126
[58] Ochi, H., Watanabe, S. & Yamamoto, H. (1988). New heritable fragile site on chromosome 8 induced by distamycin A. *Jpn. J. Cancer Res.* **79** 145–147
[59] Ochi, H., Watanabe, S., Furuya, T. & Tsugane, S. (1988). Chromosome fragility of lymphocytes from breast cancer patients in relation to epidemiologic data. *Jpn. J. Cancer Res.* **79** 1024–1030
[60] Moreno, S., Campos, J. M., Kusak, E., Bello, M. J., Rey, J. A., Valcarcel, E., Martinez, P. & Benitez, J. (1987). Common fragile sites on chromosomes; frequency and distribution in meningioma and control populations. In: *Proc. Fourth European Conference on Clinical Oncology and Cancer Nursing*, p. 285 (Meeting abstract)
[61] Murty, V. V. S., Mitra, A. B., Luthra, U. K. & Singh, I. P. (1986). Sister chromatid exchanges in patients with precancerous and cancerous lesions of cervix uteri. *Hum. Genet.* **72** 37–42

3

Cytogenetic abnormalities in human small cell lung cancer

Pamela H. Rabbitts[*] and **Jonathan J. Waters**[†]
[*]MRC Clinical Oncology and Radiotherapeutics Unit, MRC Centre, Hills Road, Cambridge, UK; and [†]Regional Cytogenetics Laboratory, Addenbrookes Hospital, Cambridge, UK

1. INTRODUCTION

Lung tumours are the main cause of cancer deaths in men, and the second most frequent cause of cancer deaths in women in the western world. Furthermore, in developing countries the disease is reaching epidemic proportions [1]. Less than 10% of patients are alive five years after diagnosis, with about 80% of patients dying in the first year. New treatment regimes are making very little impact on these figures and it is considered unlikely that further improvements of existing treatments will be of significant benefit [2]. For this reason several research groups with clinical interests in lung cancer have pursued an understanding of the biology of these tumours with the long-term goal of designing therapies which exploit differences between lung tumour cells and the normal cells of the lung from which they arise.

Histopathological studies of lung tumours, based on light microscopy, recognize four main types, namely squamous cell carcinoma, large cell carcinoma, adenocarcinoma and small cell carcinoma. The latter must be distinguished from other lung carcinomas (N-SCLC) because clinically it requires a different approach. The biology of SCLC has largely been elucidated using cell lines established from biopsy samples and later confirmed by direct examination of tumour material. Such analyses have established that these tumours exhibit a number of neuroendocrine properties, for example dense core granules (assumed to be neurosecretory) [3], 3,4-dihydroxyphenylamine decarboxylase [4], neuron-specific enolase [5] and various regulatory peptide hormones including bombesin [6]. These features of SCLC have led to suggestions that this tumour type is derived from a different stem cell within the lung [7], or may even originate outside the lung. However, current thinking is that these four types of lung tumour derive from a pluripotent stem cell [8].

The ability to establish cell lines from lung tumour samples has allowed us to gain a better understanding of the biological properties of lung tumour cells. A selective tissue culture medium, which encourages the growth of tumour cells at the expense of fibroblasts, has been developed for SCLC biopsy samples so that cell lines can be established with a success rate of about 70% [9]. The comparatively large numbers of cell lines established from SCLC patients in various centres has made this one of the best studied solid tumours cytogenetically, although it must be remembered that the cell lines are almost always derived from metastatic rather than primary material.

2. CHROMOSOME 3 ABNORMALITIES IN SCLC

The study of chromosomes of tumours is most advanced in leukaemias and lymphomas where the tumour cells are easily obtainable and the karyotypes are often less complex. In some of these diseases the molecular structure of the observed chromosomal abnormality has been studied in great detail, and for two of them activated oncogenes have been identified at the breakpoints, namely *c abl* in chronic granulocytic leukaemia [10] and *c myc* in Burkitt's lumphoma [11]. Amongst solid tumours most detailed studies have been made of the rare tumours which show an inherited structural defect such as the deletion within the long arm of chromosome 13 seen in some retinoblastoma patients [12]. Such deletions, being constitutional, must necessarily be small if the patients are to survive, and thus are more amenable to molecular analysis than the large deletions which arise somatically in some tumours. The characteristics of retinoblastoma have been used as a model to explain how allelic loss or deletion contributes to tumour development [13]. The deletion is thought to be the site of a tumour suppressor gene, and indeed analysis of a gene found within the deleted region of 13q14 shows it to have the properties expected of a gene involved in growth control in normal retinal cells [14]. Thus cytogenetic analysis and the definition of consistent chromosomal abnormalities can often highlight areas of the genome involved in tumour development. As far as the common solid tumours are concerned there are often inherent difficulties in obtaining adequate chromosome preparations from tumour material, either because of low mitotic index, or because of necrosis or fibrotic involvement. Lung tumours of the small cell type present an additional difficulty in that these tumours are very rarely resected and therefore only very small biopsy samples of the primary tumour are available. Secondary deposits, often heavily admixed with normal cells, form the most frequent source of material available for cytogenetic analysis. For these reasons cell lines have provided a useful source of cells for chromosome studies, although, in fact, the very first series on banded chromosomes from SCLC patients was performed on samples from bone marrow aspirates [15]. 26 patients were studied with reference to abnormal chromosome number and structural aberrations. No common markers were identified, but abnormalities were found on chromosome 1 in 14 of the 18 patients who had abnormal karyotypes. However, no particular region of chromosome 1 was involved, as is often the case for chromosome 1 involvement in tumours.

In 1982 Whang-Peng and her colleagues at the National Institutes for Health published an important study which identified deletions in chromsome 3 as the hallmark of SCLC [16]. Sixteen cell lines, three short-term cultures and one direct preparation were used for karyotype analysis and all samples had at least one chromosome 3 carrying a deletion per metaphase. In the majority the abnormality occurred as a interstitial deletion with a shortest region of overlap of 3p14–23. Autologous lymphocyte cultures, where examined, had normal karyotypes, suggesting that the deletion arises as a somatic event during the development of the tumour. Abnormalities were also found in other chromosomes, namely numbers 1, 2 and 10. As in the Wurster-Hill study, although a large proportion of patients showed chromosome 1 abnormalities (14/18), no region of the chromosome was consistently involved. Interestingly, although Wurster-Hill reported no abnormalities in chromosome 3 in their study of bone marrow cells, Whang-Peng et al. consider that of the five karyotypes illustrated by Wurster-Hill two have 3p deletions [17]. This serves to emphasize the problems that can arise in the interpretation of the cytogenetics of solid tumours.

This study marked the beginning of an interest in deletions in the short arm of chromosome 3 in SCLC. Not all subsequent studies agreed with these findings as illustrated in Table 1, and indeed when Wurster-Hill and her colleagues examined

Table 1 — Cytogenetic analysis of SCLC

Reference	Cell source	del 3p
Wurster-Hill et al. [15]	Direct biopsy	0/18
Whang-Peng et al. [16]	16 cell lines; 3 short-term cultures; 1 direct biopsy	20/20
Wurster-Hill et al. [18]	Cell lines	3/14
de Leij et al. [19]	Cell lines	3/3
Yunis et al. [20]	Direct biopsy	4/4
Falor et al. [21]	Direct biopsy	3/3
Zech et al. [22]	Cell lines	2/6
Waters et al. [23]	Cell lines	6/9
Morstyn et al. [24]	Cell lines	6/9

cell lines isolated in their laboratory, although they did find a few with 3p deletions, the majority of cell lines were without deletions [18]. Figures 1 and 2 illustrate SCLC lines with and without deletions in chromosome 3 respectively. Whang-Peng et al. also studied five cell lines isolated from other lung tumours; two adenocarcinomas, two mesotheliomas and one large cell carcinoma. No chromosome 3 involvement was found and it was concluded that del 3p was a marker with potential for differential diagnosis of lung tumours. This suggestion has not been validated as some N-SC lung tumours have been shown to have deletions in choromosome 3 [22,25].

The majority of studies, whether looking at cell lines or direct preparations, have studied material from metastatic tumours, but two reports confirm that this genetic lesion is also seen in the primary tumour. Sozzi et al. were able to karyotype ten metaphases from an undifferentiated SCLC primary tumour and identified a dele-

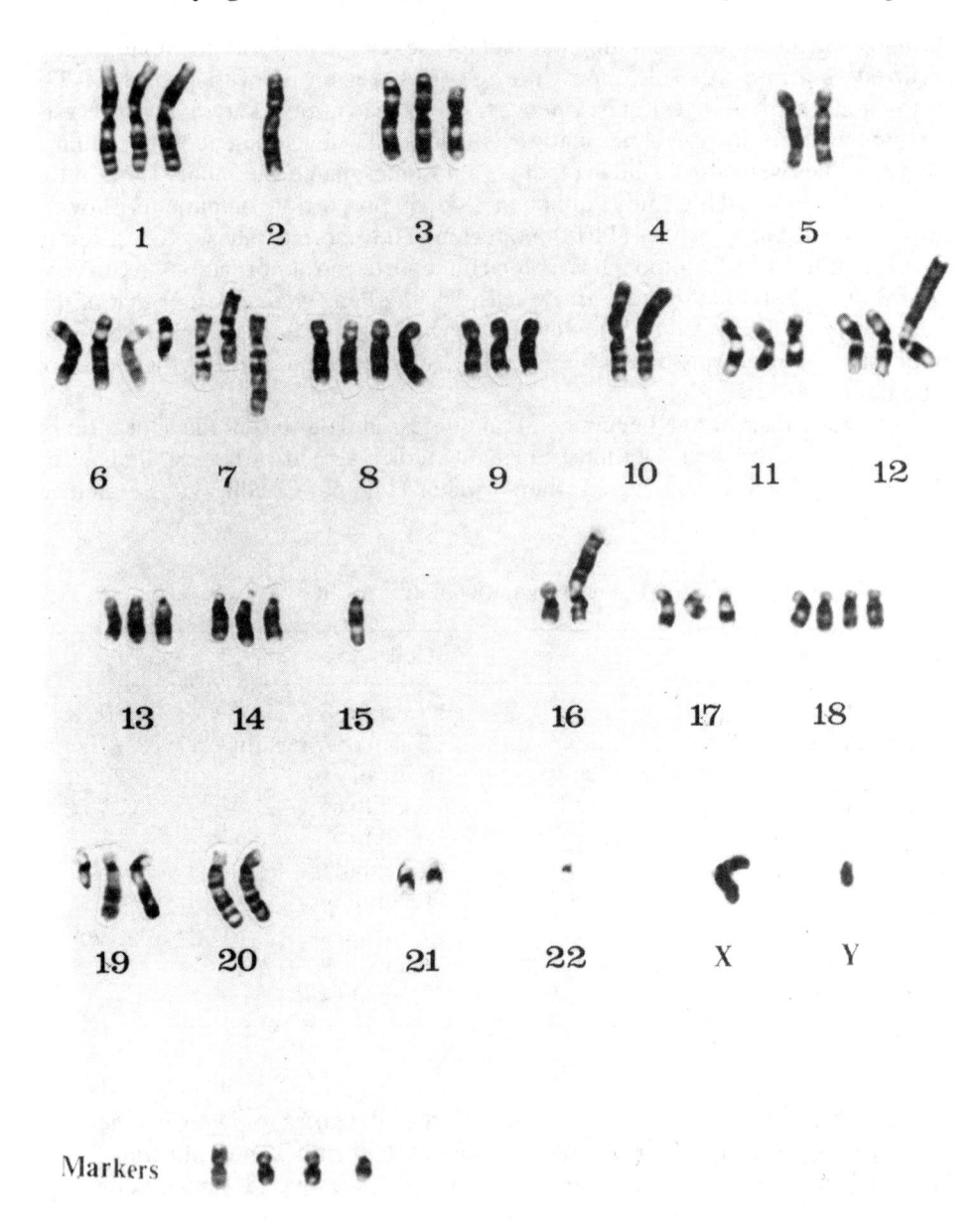

Fig. 1 — A complete karyotype of NCI H69 showing HSR on chromosome 12 and the deletion in one of the chromosome 3 homologues.

tion of 3p in seven of the metaphases [26]. Abnormalities of chromosome 1 and an iso(21q) were present in all ten metaphases. A cell line isolated from a surgically removed primary tumour from a patient who had undergone minimal chemotherapy, also showed a deletion on chromosome 3 with breakpoints at p14 to p25 [27].

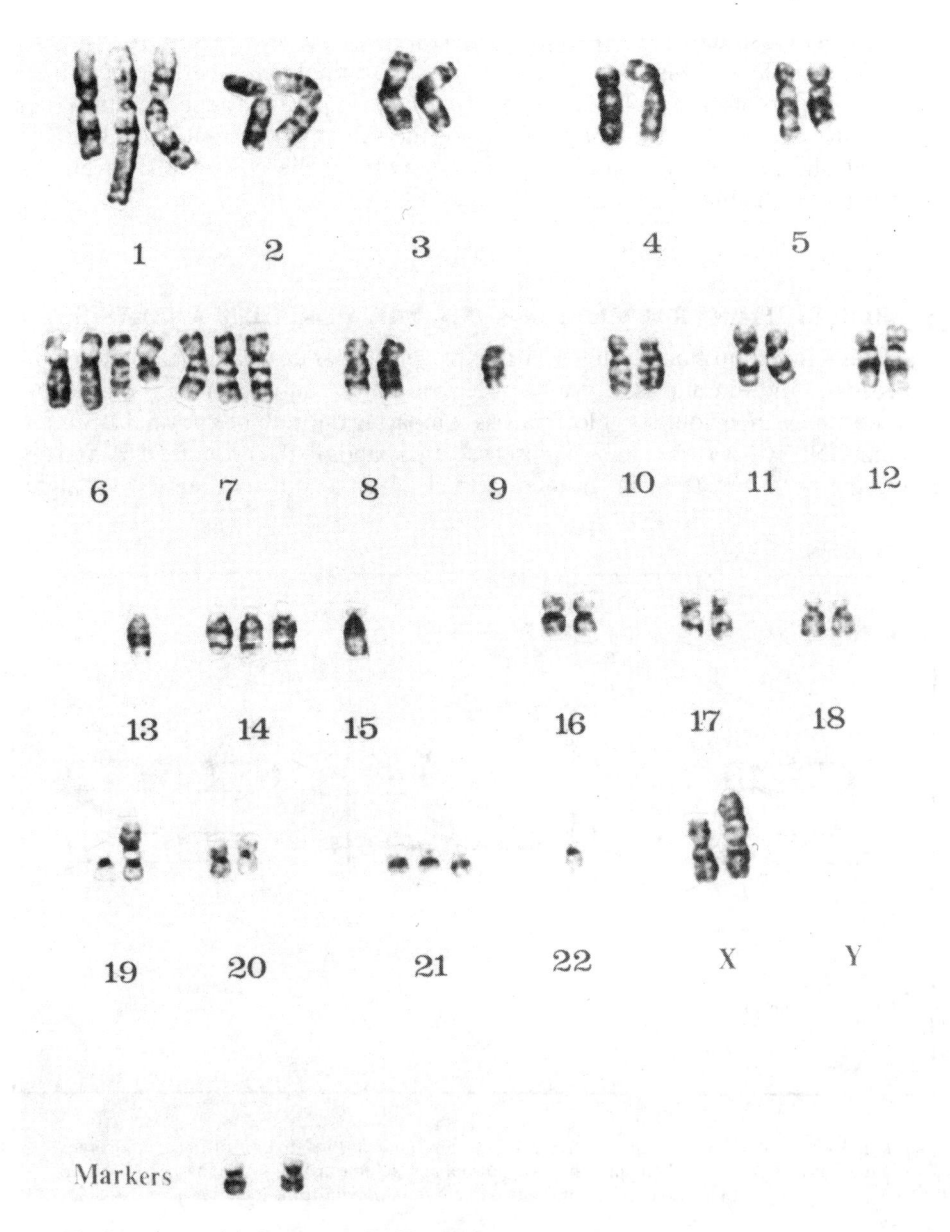

Cytogenetic abnormalities in human small cell lung cancer

Fig. 2. — A complete karyotype of NCI N417 showing an HSR on 1q and two chromosome 3s, neither having deletions in the short arm.

These various studies identified the deletion in chromosome 3 as a common feature of SCLC. From cytogenetic analyses it was noticeable that the few studies of biopsy material all showed deletions of chromosome 3 and that the discrepancies arose when the chromosomes of cell lines were karyotyped. A possible difference between the two major studies (i.e. refs [16] and [18]) was the use of different culture conditions for the establishment of the cell lines. It was not until restriction fragment length polymorphism (RFLP) analysis was used to compare the genetic information in tumours with that in the normal tissue of the same patient that loss of genetic material on 3p, sometimes detectable as deletions at the cytogenetic level, was confirmed as a highly consistent feature of SCLC.

3. RESTRICTION FRAGMENT LENGTH POLYMORPHISM ANALYSIS

If a patient is constitutionally heterozygous at a particular locus, but has lost alleles at the locus in the tumour as a result of deletion of part of one of the homologous chromosomes, then Southern blot analysis comparing the patient's normal DNA and tumour DNA will show, in the simplest case, two bands in the normal DNA and one in the tumour DNA (Fig. 3). In theory, then, given a sufficient number of highly

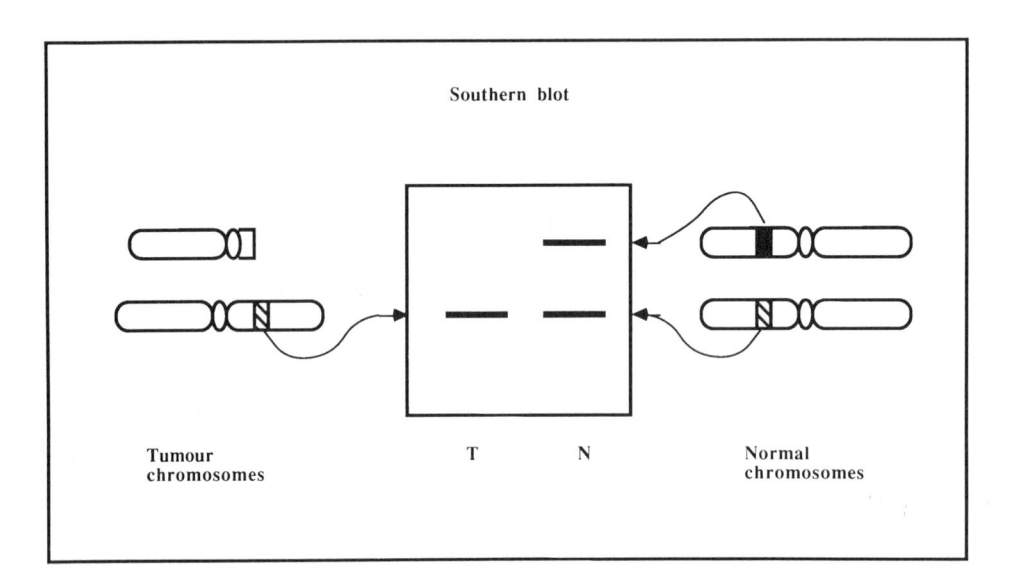

Fig. 3 — An illustration of the relationship between the deletions that occur in chromosomes and their detection by RFLP analysis. N is normal and T tumour DNA. The hatched box and black box represent the two alleles at a polymorphic locus.

polymorphic, well-spaced probes, it should be possible to 'screen' a given chromosome arm for allelic loss with far higher resolving power, and a greater degree of accuracy, than is possible with cytogenetic techniques. In practice the limited number of precisely located available probes puts some constraints on the use of this approach, but it has already provided useful information. Several independent studies of allelic loss on chromosome 3 have now been made in patients with SCLC by RFLP analysis. The results are summarized in Table 2. Of all the patients studied

Table 2 — RFLP analysis of SCLC

Reference	Patients with 3p allelic loss/total number of informative patients
Kok et al. [25]	7/7
Brauch et al. [28]	13/13
Yokota et al. [29]	7/7
Johnson et al. [30]	23/25

only two patients in the Johnson series failed to show allelic loss. One of these patients is believed to have extrapulmonary small cell carcinoma and therefore could reasonably be excluded from the series. There is, however, no such obvious explanation for lack of allelic loss in the one remaining patient. The deletion may be too small to be detected with the probes used.

4. ACCOUNTING FOR THE APPARENT DISCREPANCIES BETWEEN CYTOGENIC AND RFLP ANALYSES OF CHROMOSOME 3

The RFLP studies described above make it clear (almost) all patients with SCLC have lost alleles from chromosome 3 in their tumour. Examination of the chromosomes of SCLC patients suggests that the mechanism of this loss is probably deletion, but an apparent discrepancy arises in that a much smaller proportion of cell lines show deletion compared with the proportion of patients with allelic loss. The RFLP analyses are restricted in their ability to determine the extent of allelic loss, and hence the size of the deletion, by the paucity of highly polymorphic probes available for the short arm of chromosome 3, and by the fact that not all patients are informative (heterozygous) at all loci tested. Thus it is possible that patients could have a deletion too small to be detected cytogenetically but which could still have lost alleles over a small stretch of chromosome 3. Cell lines isolated from such patients would not have visible deletions on chromosome 3. However, no small deletions have been reported; all deletions remove at least two bands of chromosome 3 (see Fig. 4.), but this may well be because of the relatively poor quality of the published chromosome findings.

A second possible explanation for the differences between the results of RFLP and cytogenetic analyses is that the cell lines without visible deletions arise from a small subpopulation (with two normal chromosome 3s) within the tumour, which grows out during the establishment of the cell line. There is ample evidence of heterogeneity within SC tumours to support such a suggestion. RFLP analysis of polymorphic markers on chromosome 3 should detect heterozygosity in these cell lines. However, RFLP anaysis of our five cell lines without visible deletions detects no heterozygotes at the 17 polymorphic loci tested along the length of chromosome 3 (unpublished observations). Therefore the most probable explanation for our cell lines without visible deletions is that, rather than having a small degree of allelic loss, these cell lines have suffered allelic loss by some error of mitosis which results in duplication of one homologue of chromosome 3 and loss of the other. The consequences for the cell are the same; loss of vital genes so that uncontrolled growth is the result.

5. SIZE OF THE DELETION IN CHROMOSOME 3

The deletions in chromosome 3 seen in SCLC are large and variable and always encompass a part of the chromosome which could code for several hundred genes. In order to identify which of these genes might be involved in tumour suppression it is necessary to define the region of chromosome 3 which is deleted in all SC tumours so that the search is made within the smallest possible area.

As illustrated in Fig. 4 cytogenetic analyses of the minimum deleted region in

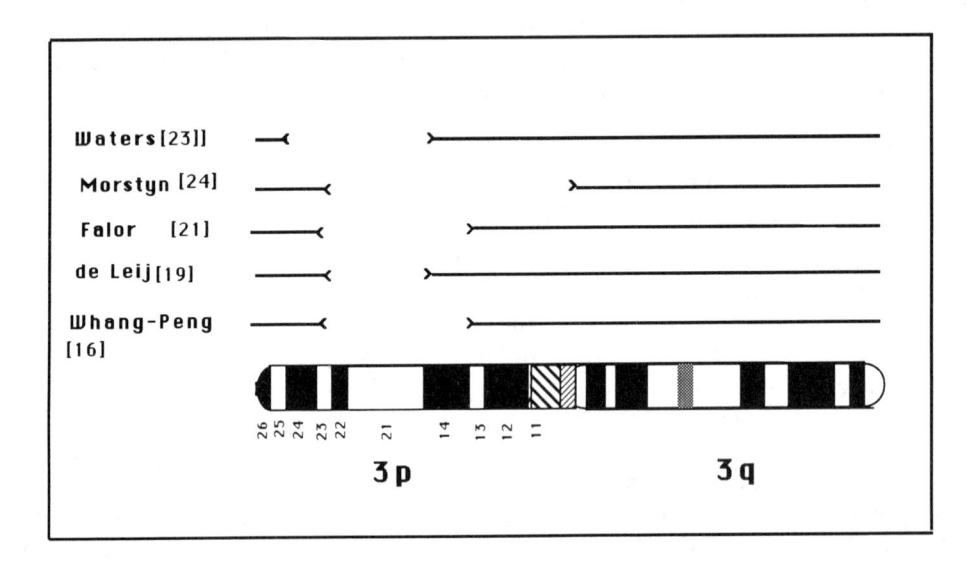

Fig. 4 — Minimum region of overlap as determined by various cytogenetic studies.

chromosome 3 in SCLC have produced similar but not identical results. All studies agree that 3p21 and 3p22 are always involved. However, the highly complex nature of the karyotypes of these tumours makes it possible that material may be translocated rather than lost. RFLP studies do not suffer from this problem although the lack of chromosome 3 probes makes it difficult to define the limits of the deletion. The polymorphic locus D3S3 (which maps to 3p13–3p14) [31] has been found to be heterozygous in several patients who show allelic loss at 3p21 and one patient in the study by Johnston et al. was heterozygous at the D3S2 locus while also showing allelic loss at 3p21 [30]. The probe which detects this polymorphic locus has not been mapped by in situ hybridization, but mapping with somatic cell hybrids suggests it is located within 3p14.2 to 3p21.1 [32,33]. The distal end of the deletion is not so well characterized by RFLP analysis. Drabkin *et al.* report one SCLC patient to be heterozygous at the Erb A2 locus which maps to 3p22–24.1 [34,35]. However, in this case the patient was not assessed for allelic loss at other loci on 3p. Taken together the RFLP data support the size of the deletion as suggested by direct chromosome analysis. Further deletion mapping may reduce the minimal deleted region but other techniques will probably be required to identify the location of the tumour suppressor gene.

6. OTHER MARKER CHROMOSOMES IN SCLC

6.1 Chromosome 1

As previously mentioned, chromosome 1 is involved in marker formation in the majority of cell lines in most of the large cytogenetic studies (Whang-Peng, 16/16; Wurster-Hill, 8/14; Morstyn, 6/9) [16, 18, 24]. No particular breakpoints were consistently noted: this behaviour of chromosome 1 has been observed in other tumours [36].

6.2 Chromosome 11

The short arm of chromosome 11 at p13 has received attention as the site of the deletion seen in patients with Wilms' tumour [37], although sequences at 11p15 also seem to be involved. Other tumours also appear to involve this region, for example ovarian tumours [38], transitional cell carcinoma of the bladder [39] and invasive ductal carcinomas of the breast [40]. Early cytogenetic studies identified clonal abnormalities in chromosome 11 in SCLC cell lines, but not in all samples studied. Wurster-Hill *et al.* had 5 of 14 cell lines with 11p involvement with breakpoints around 11p11–p12 [18] and Morstyn *et al.* four of nine cell lines with various breakpoints [24]. RFLP studies performed by Shiraishi *et al.* have shown allelic loss on chromosome 11 mainly in the p15 region in all main types of lung tumours [41]. Of the 41 patients who were informative in this study, 17 showed allelic loss (41%); 4 of the 6 SC patients showed allelic loss on 11p. Taking the tumours overall it appeared that loss might be associated with the more advanced carcinomas than those of stage 1a. Yokota *et al.* [29] also studied allelic loss in a variety of lung tumours at 24 polymorphic loci throughout the genome. Two 11p15 probes were used showing loss of 3/13 and 3/12 patients. Interestingly, one patient from whom both primary tumour and metastases were available for study had 11p allelic loss in the tumour, but not the metastases, suggesting that 11p loss occurs late and may therefore possibly be associated with tumour progression. In contrast, deletion on chromosomes 3 and 13 (see section 6.4) occurred in both the primary and secondary tumours of this patient.

6.3 Chromosome 17

In the RFLP study of Yokota *et al.* allelic loss at 17p13 was observed in all five informative SCLC patients [29]. The cytogenetic study of Morstyn *et al.* observed 6 of 9 patients to have breakpoints on 17p [24]. Band 13 on the short arm of chromosome 17 has also been implicated in the development of breast cancer [42] and colorectal cancer [43]. In both surveys around 60% of the informative patients showed loss at this locus.

6.4 Chromosome 13

Several cytogenetic analyses have identified chromosome 13 as the most frequently underrepresented chromosome in SCLC. Whang-Peng saw this in 4/4 hypodiploid lines [16]; Wurster-Hill detected underrepresentation of 13 in 17/21 lines [18] and Morstyn in 5/9 lines [24]. The cell line isolated from the primary tumour by Graziano *et al.* had only one chromosome 13 [27]. These observations have become particularly significant with the realization that the region of 13q lost in retinoblastoma is also lost in other types of tumour [44,45]. Yokota *et al.* observed allelic loss at 13q14 in all types of lung tumours with the highest proportion occurring in SC tumours (10/11) [29]. No probes for 13p were used in this study so it is not possible to determine whether whole chromosomes were lost. If the retinoblastoma gene functions in SC tumours as it does in retinoblastoma then not only should alleles be lost from one chromosome but the RB gene on the remaining homologue should be altered in some way. To test this possibility, Harbour *et al.* analysed SC DNA with respect to rearrangements and faults in expression. 1/8 SC tumour sample and 4/22 SC cell lines showed rearrangement of the RB gene. 60% of the cell lines failed to express RB mRNA in contrast to normal lung and N-SC cell lines, which showed expression in 90% of the samples [46]. Yokota *et al.* have noted a similar proportion of gene rearrangements and altered mRNA expression in the 9 SC lines they studied, and on analysis with anti RB antibody in fact none of the cell lines was making antigenically normal RB protein [47].

7. HSRs, DMs AND GENE AMPLIFICATION

Gene amplification represents an important adaptive mechanism allowing selective increased expression of certain genes. Sometimes this amplification is visible at the chromosomal level in the form of HSRs (homogeneously stained regions) (Fig. 5) or DMs (double minutes) (Fig. 6) [48]. These phenomena are associated with tumour cells but may be induced experimentally, for example in normal rodent cells exposed to high levels of methotrexate. They have never been observed constitutionally. HSRs and DMs are often seen in SC lines isolated from patients both before and after treatment with cytotoxic drugs. Members of the *myc* gene family have been found to be amplified in SCLC cell lines and in some cases the amplified gene has been localized to HSRs and DMs by *in situ* hybridization.

There has been a great deal of interest in the *myc* genes in SCLC following the observation that cell lines having the properties that might be associated with more aggressive disease were those which had *c myc* amplification [49]. The suggestion was made that *c myc* amplification might be a prognostic indicator analogous to *N myc* amplification in neuroblastoma. Subsequently two other members of the *myc* gene family, *N myc* and *L myc*, were also found to be amplified in some SCLC cell lines [50,51]. Studies of the proportion of cell lines with amplification have provided the following figures: 14/31 amplified for *c* and *N myc* (*L myc* was not tested for) [50], and in a study which did include *L myc* 13/44 cell lines were found amplified [52].

Smaller proportions were found when tumour samples were studied rather than cell lines. 5/45 *N* and *c myc* amplifications were found in samples which were from wax-embedded specimens [53]. Recently Johnson *et al.* studied 38 tumour samples and found *myc* family amplification in 6 [54]. Nine of these tumour samples were

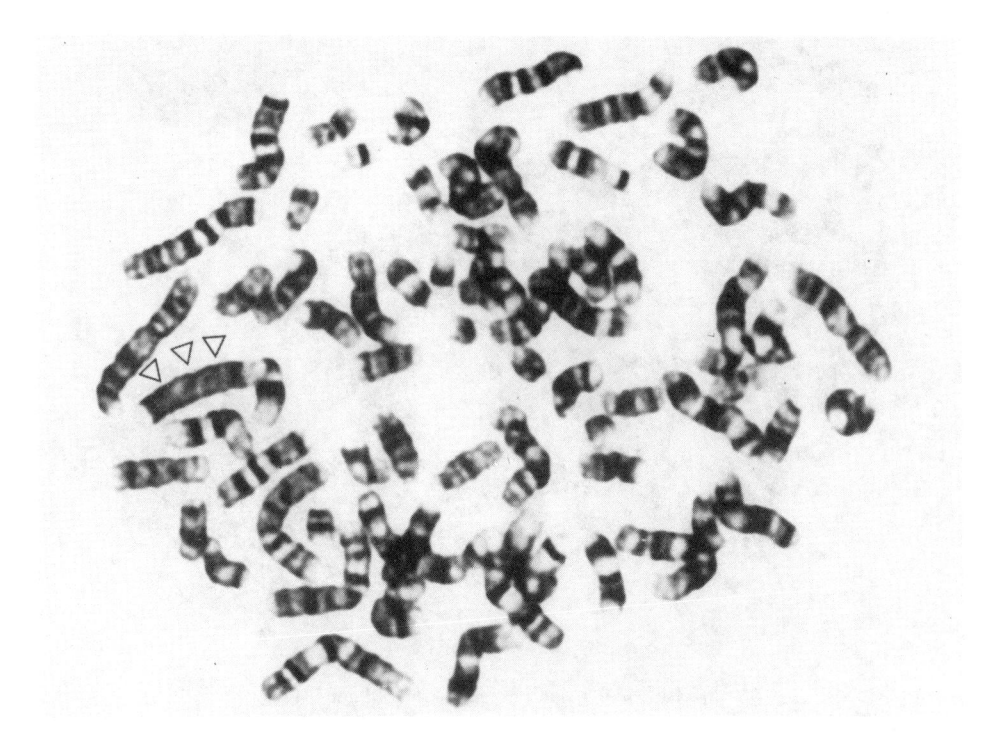

Ch. 3]   Cytogenetic abnormalities in human small cell lung cancer   55

Fig. 5 — A metaphase from cell line NCI H69. Homogeneously staining region indicated by triangles.

established as cell lines, of which two had previously carried *myc* gene amplifications. In the corresponding cell lines the same two only carried amplifications and with no obvious increase in copy number. The six patients who had amplified *myc* genes had all previously had combination chemotherapy. In a previous study Johnson *et al.* compared the *myc* family amplification status of established cell lines with the previous treatment of the patients [55]. Their finding was that 2 of 19 cell lines from untreated patients carried gene amplifications, whereas the treated group had 11 of 25 lines carrying amplifications.

There has been one report of a primary tumour which had amplified *N myc* genes, whereas some, but not all, metastases from this patient had amplified *N myc* genes. The authors take this to indicate that *N myc* amplification is associated with tumour progression rather than an initiating event. This suggestion is supported by the proportions of tumours carrying *myc* gene amplifications.

8. THE RELATIONSHIP BETWEEN SC AND N-SC LUNG TUMOURS

SC tumours of the lung, although in many ways very different from the three other main subtypes grouped together as non-small cell carcinoma (N-SCLC) of the lung, are, after years of disagreement, now considered to arise from the same cell type

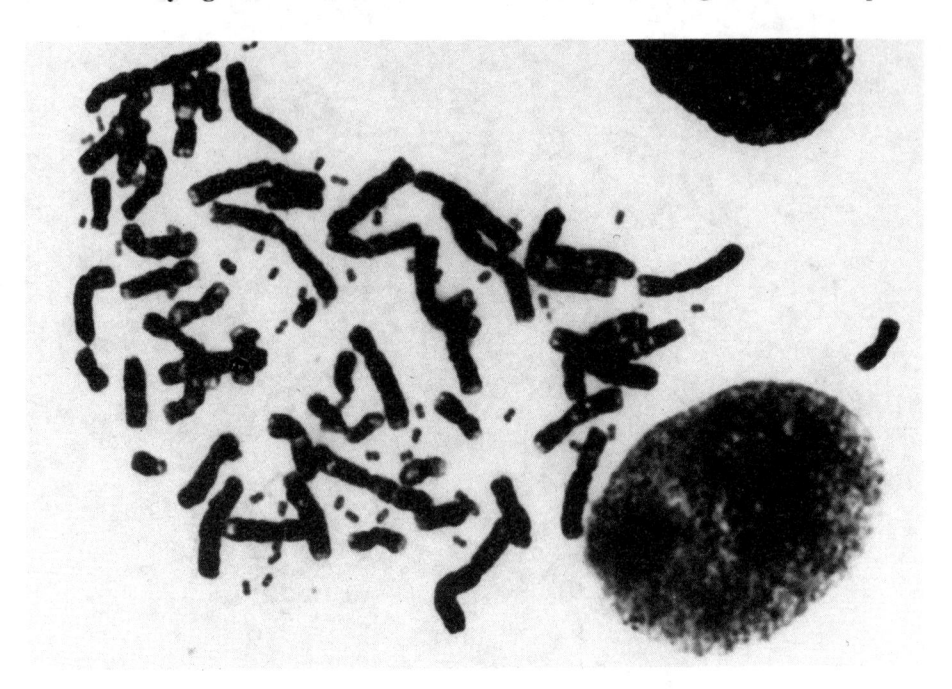

Fig. 5 — Double minutes seen in the cell line U1285.

within the bronchus [8]. It is of interest, therefore, to see what specific chromosomal aberrations they have in common with tumours of the small cell type. Cytogenetic studies of N-SC tumours have not been as extensive as SC tumours as cell lines are less readily isolated, but where they have been studied deletions in chromosome 3 have been observed in some cases [22,56]. RFLP analyses have been in complete disagreement as to whether allelic loss is a consistent feature of these tumours [25,28,29]. In our own study of 44 informative patients 75% had allelic loss on chromosome 3 [57]. Larger studies will be required to resolve this; alternatively, there may be real differences in these aberrations between, different populations of N-SCLC patients.

9. INVOLVEMENT OF CHROMOSOME 3 IN OTHER TUMOURS

Abnormalities of chromosome 3 are seen in several other neoplastic conditions, some of which are listed in Table 3. Renal cell carcinoma is of particular interest as deletions of 3 are very commonly seen in the tumours of patients with this disease [58]. As mentioned above, with reference to the retinoblastoma gene, apparently unrelated tumours can carry the same genetic defects. Such an analysis of neoplasms involving chromosome 3 awaits the isolation of a candidate tumour suppressor gene located on chromosome 3.

Table 3 — Chromosome 3 abnormalities in tumours

Tumour	Abnormality	Site
Lung (SCLC)	Deletion	3p14–3p23
Lung (N-SCLC)	Deletion	3p
Pleomorphic adenomas	Translocation	t(3:8)(p21:q12)
Ovarian cancer	Deletion	3p13–3p21
Mesothelioma	Various	3p13–3p23
Renal cell cancer	Deletion	3p13–pter
Renal cell cancer in VHL	Deletion	3p

Abbreviations: SCLC is small cell lung carcinoma; N-SCLC is non-small cell lung carcinoma; VHL is von Hippel Lindau syndrome.

ACKNOWLEDGEMENTS

We would like to thank Dr A. Gazdar of the National Cancer Institute, NIH, for the cell lines NCI H69 and NCI N417, and Dr J. Bergh of Uppsala University for the cell line U1285.

REFERENCES

[1] Silverberg, S. & Lubera, J. (1987). *Cancer statistics. CA* **37** 2–19
[2] *The Fifth World Conference on Lung Cancer, Interlaken.* In: *Cancer Top.* **7** 47–48 (1989)
[3] Bensch, K. G., Corrin, B., Parente, R. & Spencer, H. (1968). Oat cell carcinoma of the lung: its origin and relationship to bronchial carcinoid. *Cancer* **22** 1163–1172
[4] Baylin, S. B., Aberloff, J. D., Goodwin, G., Carney, D. N. & Gazdar, A. F. (1980). Activities of L-dopa decarboxylase and diamine oxidase (histaminase) in human lung cancers and decarboxylase as a marker for small (oat) cell cancer in cell culture. *Cancer Res.* **40** 1990–1994
[5] Marangos, P. J., Gazdar, A. F. & Carney, D. N. (1982). Neuron-specific enolase in human small cell carcinoma cultures. *Cancer Lett.* **15** 67–71
[6] Moody, T. W., Pert, C. B., Gazdar, A. F., Carney, D. N. & Minna, J. D. (1981). High levels of intracellular bombesin characterize human small cell lung carcinoma. *Science* **214** 1246–1248
[7] Baylin, S. B. & Gazdar, A. F. (1981). Endocrine biochemistry in the spectrum of lung cancer: implications for the cellular origin of small cell lung cancer. In: F. E. Greco, R. K. Oldham & P. A., Bunn, Jr. (eds) *Small Cell Lung Cancer*, Grune & Stratton, New York, pp. 123–143
[8] Bergh, J., Nilsson, K., Dahl, D., Andersson, L., Virtanen, I. & Lehto, V.-P. (1984). Expression of intermediate filaments in established human lung cancer cell lines. *Lab. Invest.* **51** 307–316
[9] Gazdar, A. F., Carney, D. N., Sims, H. L., Baylin, S. B., Bunn, P. A., Jr., Guccion, J. G. & Minna, J. D. (1980). Establishment of continuous clonable

cultures of small cell carcinoma of the lung which have amine precursor uptake and cell properties. *Cancer Res.* **40** 3502–3507

[10] de Klein, A., Guerts van Kessel, A., Grosveld, G., Bartram, C. R., Hagemeijer, A., Bootsman, D., Spurr, N. K., Hesterkamp, N., Groffen, J. & Stephenson, J. R. (1982). A cellular oncogene is translocated to the Philadelphia chromosome in chronic myelocytic leukaemia. *Nature (London)* **30** 765–770

[11] Dalla-Favera, R., Bregni, M., Erikson, J., Patterson, D., Gallo, R. C. & Croce, C. M. (1982). Human c-myc oncogene is located on the region of chromosome 8 that is translocated in Burkitt lymphoma cells. *Proc. Natl. Acad. Sci. USA* **79** 7824

[12] Benedict, W. F., Murphree, A. L., Banerjee, A., Spina, C. A., Sparkes, M. C. & Sparkes, R. S. (1983). Patient with 13 chromosome deletion: evidence that the retinoblastoma gene is a recessive cancer gene. *Science* **219** 973–974

[13] Knudson, A. G. (1971). Mutation and cancer: statistical study of retinoblastoma. *Proc. Natl. Acad. Sci. USA* **68** 820–823

[14] Lee, W.-H., Bookstein, R., Hong, F., Young, L.-J., Shew, J.-Y. & Lee, E. Y.-H. P. (1987). Human retinoblastoma susceptibility gene: cloning, identification and sequence. *Science* **235** 1394–1399

[15] Wurster-Hill, D. H. & Maurer, L. H. (1978). Cytogenetic diagnosis of cancer: abnormalities of chromosomes and ploidy levels in the bone marrow of patients with small cell anaplastic carcinoma of the lung. *J. Natl. Cancer Inst.* **61** 1065–1075

[16] Whang-Peng, J., Bunn, P., A., Jr., Kao-Shan, C. S., Lee, E. C., Carney, D. N., Gazdar, A. & Minna, J. D. (1982). A nonrandom chromosomal abnormality, del 3p (14–23) in human small cell lung cancer (SCLC). *Cancer Genet. Cytogenet.* **6** 119–134

[17] Whang-Peng, J., Kao-Shan, C. S. & Lee, E. C. (1982). Specific chromosome defect associated with human small-cell lung cancer: deletion 3p (14–23). *Science* **215** 181–182

[18] Wurster-Hill, D. H., Cannizzaro, L. A., Pettengill, O. S., Sorenson, G. D., Cate, C. C. & Maurer, L. H. (1984). Cytogenetics of small cell carcinoma of the lung. *Cancer Genet. Cytogenet.* **13** 303–330

[19] de Leij, L., Postmus, P. E., Buys, C. H. C. M., Elema, J. D., Ramaekers, F., Poppema, S., Brouwer, M., van der Veen, A. Y., Mesander, G. & Hauw, The T. (1985). Characterisation of three new variant type cell lines derived from small cell carcinoma of the lung. *Cancer Res.* **45** 6024–6033

[20] Yunis, J. J. (1983). The chromosomal basis of human neoplasia. *Science* **221** 227–236

[21] Falor, W. H., Ward-Skinner, R. & Wergryn, S. (1985). A 3p deletion in small cell lung carcinoma. *Cancer Genet. Cytogenet.* **16** 175–177

[22] Zech, L., Berg, J. & Nilsson, K. (1985). Karyotypic characterisation of established cell lines and short-term cultures of human lung cancers. *Cancer Genet. Cytogenet.* **15** 335–347

[23] Waters, J. J., Ibson, J. M., Twentyman, P. R., Bleehen, N. M. & Rabbitts, P. H. (1988). Cytogenetic abnormalities in human small cell lung carcinoma: cell

lines characterised for myc gene amplification. *Cancer Genet. Cytogenet.* **30** 213–223

[24] Morstyn, G., Brown, J., Noval, U., Gardner, J., Bishop, J. & Garson, M. (1987). Heterogeneous cytogenetic abnormalities in small cell lung cancer cell lines. *Cancer Res.* **47** 3322–3327

[25] Kok, K., Osinga, J., Carritt, B., Davis, M. B., van der Hout, A. H., van der Veen, A. Y., Landsvater, R. M., de Leij, L. F. M. H., Berendsen, H. H. & Buys, C. H. C. M. (1987). Deletion of a DNA sequence at the chromosomal region 3p21 in all major types of lung cancer. *Nature (London)* **330** 578–581

[26] Sozzi, G., Bertoglio, M. G., Borrello, M. G., Giani, S., Pilotti, S., Pierotti, M. & Della-Porta, G. (1987). Chromosomal abnormalities in a primary small cell cancer. *Cancer Genet. Cytogenet.* **27** 45–50

[27] Graziano, S. L., Cowan, B. Y., Carney, D. N., Bryke, C. R., Mitter, N. S., Johnson, B. E., Mark, G. E., Planas, A. T., Catino, J. J. & Comis, R. L. (1987). Small cell lung cancer cell line derived from a primary tumour with a characteristic deletion of 3p. *Cancer Res.* **47** 1412–1420

[28] Brauch, H., Johnson, B., Hovis, J., et al. (1987). Molecular analysis of the short arm of chromosome 3 in small cell and non small lung carcinoma of the lung. *N. Engl. J. Med.* **317** 1109–1113

[29] Yokota, J., Wada, M., Shimosato, Y., Terada, M. & Sugimura, T. (1987). Loss of heterozygosity on chromosomes 3, 13, and 17 in small-cell carcinoma and on chromosome 3 in adenocarcinoma of the lung. *Proc. Natl. Acad. Sci. USA* **84** 9252–9256

[30] Johnson, B. E., Sakagushi, A. Y., Gazdar, A. F., Minna, J. D., Burch, D., Marshall, A. & Naylor, S. L. (1988). Restriction fragment length polymorphism studies show consistent loss of chromosome 3p alleles in small cell lung cancer patient tumours. *J. Clin. Invest.* **82** 502–507

[31] Gerber, M. J., Miller, Y. E., Drabkin, H. A. & Scoggin, C. H. (1986). Regional assignment of the polymorphic probe D33 by molecular hybridisation. *Cytogenet. Cell Genet.* **42** 72–74

[32] Naylor, S. L., Sakagushi, A. Y., Barker, D., White, R. & Shows, T. B. (1984). DNA polymorphic loci mapped to human chromosomes 3, 5, 9, 11, 17, 18 and 22. *Proc. Natl. Acad. Sci. USA* **81** 2447–2451

[33] Harris, P., Morton, C. C., Guglielmi, P., Li, F., Kelly, K. & Latt, S. A. (1986). Mapping by chromosome sorting of several gene probes, including c-myc, to the derivative chromosomes of a 3;8 translocation associated with familial renal cancer. *Cytometry* **7** 589–594

[34] Drabkin, H., Kao, F.-T., Hartz, J., Hart, I., Gazdar, A., Weinberger, C., Evans, R. & Gerber, M. (1988). Localisation of human ERBA2 to the 3p22–3p24.1 region of chromosome 3 and variable deletion in small cell lung cancer. *Proc. Natl. Acad. Sci. USA* **85** 9258–9262

[35] Albertson, D. G., Sherrington, P. D. & Rabbitts, P. H. (1989). Location of polymorphic DNA prober frequently deleted in lung carcinoma. *Hum. Genet.* (in press)

[36] Heim, S. & Mitelman, F. (1987). In *Cancer Cytogenetics*, Alan R. Liss, New York

[37] Riccardi, V. M., Hittner, H. M., Francke, U., Unis, J. J., Ledbetter, D. & Borges, W. (1980). The aniridia–Wilms' tumour association: the critical role of chromosome band 11p13. *Cancer Genet. Cytogenet.* **2** 131–137

[38] Lee, H. J., Kavanagh, J. J., Wharton, J. T., Wildrick, D. M. & Blick, M. (1989). Allele loss at the c-Ha-ras1 locus in human ovarian cancer. *Cancer Res.* **49** 1220–1222

[39] Fearon, E. R., Feinberg, A. P., Hamilton, S. H. & Vogelstein, B. (1985). Loss of genes on the short arm of chromosome 11 in bladder cancer. *Nature (London)* **318** 377–380

[40] Theillet, C., Lidereau, R., Escot, C. et al. (1986) Loss of H-ras allele and aggressive human primary breast carcinomas. *Cancer Res.* **46** 4776–4781

[41] Shiraishi, M., Morinaga, S., Noguchi, M., Shimosato, Y. & Sekiya, T. (1987). Loss of genes in the short arm of chromosome 11 in human lung carcinomas. *Jpn. J. Cancer Res. (Gann)* **78** 1302–1308

[42] Mackay, J., Elder, P. A., Steel, C. M., Forrest, A. P. M. & Evans, H. J. (1988). Allele loss on short arm of chromosome 17 in the breast cancers. *Lancet* **II** (Part 8628) 1384–1385

[43] Law, D. L., Olschwang, S., Monpezat, J.-P., Lefrancois, D., Jagelman, D., Petrelli, N. J., Thomas, G. & Feinberg, A. P. (1988). Concerted nonsyntenic allelic loss in human colorectal carcinoma. *Science* **241** 961–964

[44] Toguchida, J., Kanji, I., Sasaki, M. S., Nakamura, Y., Ikenaga, M., Kato, M., Sugimot, M., Kotoura, Y. & Yamamuro, T. (1988). Preferential mutation of paternally derived RB gene as the initial event in sporadic osteosarcoma. *Nature (London)* **338** 156–157

[45] T'Ang, A., Varley, J. M., Chakraborty, S., Murphree, A. L. and Fung, Y.-K. T. (1988). Structural rearrangement of the retinoblastoma gene in human breast carcinoma. *Science* **242** 263–266

[46] Harbour, J. W., Lai, S.-L., Whang-Peng, J., Gazdar, A. F., Minna, J. D. & Kaye, F. J. (1988). Abnormalities in structure and expression of the human retinoblastoma gene in SCLC. *Science* **241** 353–357

[47] Yokota, J., Akiyama, T., Fung, Y.-K. T., Benedict, W. F., Namba, Y., Hanaoka, M., Wada, M., Terasaki, T., Shimosato, Y., Sugimura, T. & Terada, M. (1988). Altered expression of the retinoblastoma (RB) gene in small-cell carcinoma of the lung. *Oncogene* **3** 471–475

[48] Therman, E., (1986). In: *Human Chromosomes*, Springer, Berlin, 2nd edn, pp. 256–261

[49] Little, C. D., Nau, M. M., Carney, D. N., Gazdar, A. F. & Minna, J. D. (1983). Amplification and expression on the c-myc oncogene in human lung cancer cell lines. *Nature (London)* **306** 194–196

[50] Nau, M. M., Brooks, B. J., Carney, D. N., Gazdar, A. F., Battey, J. F., Sauville, E. A. & Minna, J. D. (1986). *Proc. Natl. Acad. Sci. USA* **83** 1092–1097

[51] Nau, M. M., Brooks, B. J., Battey, J., Sausville, E., Gazdar, A. F., Kirsch, I. R., McBride, O. W., Bertness, V., Hollis, G. G. & Minna, J. D., (1985). L-myc, a new myc-related gene amplified and expressed in human small cell lung cancer. *Nature (London)* **318** 69–73

[52] Johnson, B. E., Ihde, D. C., Makuch, R. W., Gazdar, A. F., Carney, D. N.,

Oie, H., Russell, E., Nau, M. M. & Minna, J. D. (1987). Myc family oncogene amplification in tumour cell line established from small cell lung cancer patients and its relationship to clinical status and course *J. Clin. Invest.* **79** 1629

[53] Wong, A. J., Ruppert, J. M., Eggleston, J., Hamilton, S. R., Baylin, S. B. & Vogelstein, B. (1986). Gene amplification of c-myc and N-myc in small cell carcinoma of the lung. *Science* **233** 461

[54] Johnson, B. E., Makuch, R. W., Simmons, A. D., Gazdar, A. F., Burch, D. & Cashell, A. W. (1988). *Myc* family DNA in small cell lung cancer patients' tumours and corresponding cell lines. *Cancer Res.* **48** 5163–5166

[55] Johnson, B. E., Ihde, D. C., Makuch, R. W., Gazdar, A. F., Carney, D. N., Oie, H., Russell, E., Nau, M. M. & Minna, J. D. (1982). Myc family oncogene amplification in tumour cell lines established from small cell lung cancer patients and its relationship to clinical status and course. *J. Clin. Invest.* **79** 1629–1634

[56] Ray, A., Bellow, M. J., de Campos, J. M., Kusak, E. M., Moreno, S. & Benitez, J. (1987). Deletion 3p in two lung adenocarcinomas metastatic to the brain. *Cancer Genet. Cytogenet.* **25** 355–360

[57] Rabbitts, P., Douglas, J., Daly, M., Sundaresan, V., Fox, B., Haselton, P., Wells, F., Albertson, D., Waters, J. & Bergh, J. (1989). Frequency and extent of allelic loss on the short arm of chromosome 3 in non small cell lung cancer. *Genes, Chromosome Cancer* **1** 95–105

[58] Kovacs, G., Erlandsson, R., Boldog, F., Ingvarsson, S., Muller-Brechlin, R., Klein, G. & Sumegi, J. (1988). Consistent chromosome 3p deletion and loss of heterozygosity in renal cell carcinoma. *Proc. Natl. Acad. Sc. USA* **85** 1571–1575

4

Onco-suppressor genes

Margaret L. M. Anderson* and Demetrios A. Spandidos[†,‡]
* Department of Biochemistry, University of Glasgow, Glasgow G12 8QQ, Scotland, UK
† The Beatson Institute for Cancer Research, Garscube Estate, Bearsden G61 1BD, Scotland, UK
‡ Biological Research Centre, National Hellenic Research Foundation, 48 Vas. Constantinou Ave., Athens 11635, Greece

1. INTRODUCTION

The development of a normal cell into a cancer cell is a complex, multistep process. Histopathological studies, age–incidence relationships, epidemiological studies and observations by early tumour virologists suggested the multiple step nature of tumour development long before we had any inkling of the nature of the events involved in producing a malignant cell (reviewed in ref. [1]). Estimates based on age incidence of adult cancers suggest that five to seven events are required [1].

That alterations in cellular genetic material might be involved in carcinogensis had been suggested early in the century by Boveri [2]. Many agents that cause cancer such as ionizing radiation and chemical carcinogens are mutagens [1] and there is now overwhelming evidence that accumulation of somatic mutations in cells plays an important although not exclusive role in carcinogenesis. Genetic analysis of tumours and tumour viruses has led to the identification of many genes that are involved in this process (reviewed in refs [3]–[5]).

In this chapter we will briefly review the involvement of oncogenes — genes whose aberrant expression contributes to the carcinogenic process; then we will focus on onco-suppressor genes which normally act to control cell proliferation but whose inactivation is also involved in tumorigenesis.

2. ONCOGENES

Studies on acutely transforming retroviruses showed that the viral genes now termed oncogenes were essential for tumorigenesis and that these sequences were transduced, mutated cellular genes. The normal, non-activated cellular genes are termed proto-oncogenes and are considered to be potential targets for mutation in the multi-step development of other types of cancer, a view that is supported by genetic analysis of human tumour tissue [3,4].

'Activation' can occur by various mechanisms. In general, the minimum changes necessary to activate the viral oncogene seem to be sufficient to activate the cellular proto-oncogene. Such changes include overexpression of the normal protein product (quantitative) and expression of an altered product through point mutation, truncation or translocation (qualitative) [3,6].

Amplification of proto-oncogenes has been described in a variety of human tumours. N-*myc* is amplified and overexpressed in certain tumours such as neuroblastoma and this correlates with the stage of development of the tumour and a poor prognosis [7]. Similarly, H-*ras* and *neu* (c-*erb*B-2) are overexpressed in some 70% of breast carcinomas [8,9]. Specific chromosomal translocations in several human cancers also cause activation on proto-oncogenes through unscheduled expression. In Burkitt's lymphoma, a malignancy of B lymphocytes, translocations t(8:14), t(2:8) or t(8:22) juxtapose the c-*myc* oncogene next to one of the immunoglobulin loci causing constitutive expression of the c-*myc* gene. Similar arguments can be made for the involvement of putative oncogenes such as *bcl2* and *tcl1* in B and T cell tumours respectively [6].

Qualitative changes in proto-oncogenes leading to an altered product have also been observed in human tumours. Thus some translocations cause formation of chimeric genes which encode proteins with altered properties compared with the normal proteins. In over 98% of patients with chronic myeloid leukaemia, a (9:22) translocation in the malignant cells fuses the *bcr* gene on chromosome 22 to the c-*abl* gene on chromosome 9, creating a hybrid gene [10]. The encoded protein possesses a tyrosine kinase activity lacking by both progenitor parent proteins which is thought to play a role in the development of the disease [11].

Activation of oncogenes may also occur by mutation as exemplified by members of the *ras* family of proto-oncogenes which are activated in a wide variety of human malignancies [12].

Activated oncogenes have been found in virtually every tumour type in which they have been examined. The consistent association of aberrant expression of specific oncogenes with particular tumours, e.g. c-*myc* in virtually 100% of patients with Burkitt's lymphoma, is taken as evidence of involvement of that oncogene in the development of the disease. However, for many tumours the most commonly activated oncogene is not found in nearly such as high proportion of cases, e.g. 40% of colorectal cancers and about 30% of acute leukaemias have an activated N-*ras* gene [13–15]. A causal relationship may still exist between these oncogenes and the tumours in which they are expressed and the lack of activation of these oncogenes in the other tumours of the same type may be a reflection that mutations of different genes have been involved in their formation, an explanation that is consistent with proposals that the same end point, i.e. tumour formation, can be reached by different pathways [16].

In several tumours many oncogenes are activated, but there is no consistent pattern of association [17,18]. Since genetic instability is a feature of tumours and different chromosomal changes have even been observed in different cells of the same tumour [19], activation of many of these oncogenes is likely to be a secondary effect and not involved in causing the tumour.

About 40 or so oncogenes have now been identified and they encompass a broad spectrum of functions which include growth factors, growth factor receptors, protein kinases and transcriptional activators [3–5]. In general, proto-oncogenes seem to be involved in the control of cell growth and differentiation and they may function at different stages of a pathway that transmits signals for cell division from the cell surface to the nucleus. However, their role at the molecular level has not been delineated.

3. ONCO-SUPPRESSOR GENES

Another group of genes involved in carcinogenesis shares the property that their expression suppresses the oncogenic phenotype of cancer cells. These are termed variously anti-oncogenes [20], emerogenes and platogenes [21], but we shall use the term onco-suppressor genes throughout [5]. As with oncogenes, onco-suppressor genes are a diverse group of genes. They have been identified mainly by operational criteria and our understanding of their function is elementary. Few have been cloned and, for several, their involvement in cancer has been deduced indirectly. There have been numerous reviews on the subject of onco-suppressor genes recently [22–24], so here we will focus on recent developments in the field.

3.1 Suppression of tumorigenicity

One of the earliest demonstrations of genetic control of tumorigenicity came from studies on cell hybrids. When normal and malignant cells of the same species were fused, the resulting hybrids, while retaining many of the features of transformed cells in culture, were non-malignant *in vivo* [25,26]. This suggested that the transformed phenotype and malignancy were under separate genetic control. The phenomenon, known as suppression of tumorigenicity, seems to be fairly general in that it occurs irrespective of the lineage of the normal or malignant parent (reviewed in ref. [27]). It is thought that chromosomes from the normal parent carry genes (onco-suppressor genes) whose expression complements a defect of the tumour cell and thereby suppresses the malignancy. Similar studies have shown that phenotypes for immortilazation [28], morphological transformation [29] and metastasis [30] can also be suppressed in appropriate hybrids.

Attempts to determine the genetic basis of suppression of tumorigenicity have centred on the observation that, on continued culture of non-malignant hybrids between normal and malignant cells, rare segregants sometimes arise that have acquired the ability to form tumours *in vivo* [27]. Random chromosome loss is a well-known feature of hybrid cell culture, so attempts have been made to correlate the loss of particular chromosomes with the acquisition of malignancy. In intraspecies hybrids this is not easy as chromosomes are frequently lost very rapidly and it may be difficult to determine the parental origin of chromosomes. Nevertheless, in 'stable' hybrids between malignant HeLa cervical carcinoma cells × normal human fibroblasts, loss of normal human chromosome 11 was implicated in re-expression of malignancy [31–33]. Support for involvement of chromosome 11 has come from studies where a single chromosome 11 introduced into HeLa cells by microcell fusion caused suppression of malignancy [34].

In a study of hybrids formed between HeLa cells and normal human diploid fibroblasts, RFLP analyses showed that non-tumorigenic hybrids contained four copies of chromosome 11, whereas tumorigenic hybrids had lost both copies of chromosome 11 of the normal parent [35]. The fact that two copies of chromosome 11 were required to suppress the tumorigenic phenotype of the hybrids suggests that there may be a gene dosage effect involved in suppressing tumorigenicity.

In crosses between normal human fibroblasts and HT1080 fibrosarcoma cells, loss of the normal chromosome 1 and possibly chromosome 4 was deduced to be important for re-expression of malignancy [36]. The fact that different chromosomes are implicated in different hybrids suggests that the parental tumour cells carry different defects and that suppression of malignancy occurs by different mechanisms depending on the genetic damage and the lineage of the fusion partners. This explanation is consistent with results where certain combinations of tumour types are fused; the hybrids are initially non-tumorigenic, suggesting that there are several tumour suppressor complementation groups [37].

While much valuable information has come from the study of hybrids involving HeLa cells, little in known about the original events that caused tumour formation. Recent studies have focused on better-defined systems such as the tumorigenicity of hybrids formed between pairs of tumour cell lines carrying different activated *ras* oncogenes [38]. When EJ colon carcinoma cells carrying an activated H-*ras* gene were fused with SW480 cells carrying an activated K-*ras* gene, the progeny hybrids expressed both oncogenes and were non-tumorigenic. This suggested firstly that an activated oncogene is not sufficient for tumorigenesis and secondly that complementation of the defects in each parental line had occurred in the hybrid.

By contrast, in crosses between HT1080 cells with an activated N-*ras* gene and EJ or SW480 or SW1271 cells which carry an activated N-*ras* gene, the hybrids expressed activated *ras* genes from both parents and were tumorigenic [38]. The N-*ras* gene in HT1080 is known to be important for tumorigenesis [39] so this result may indicate that the presence of activated N-*ras* in the same cell as other activated *ras* genes precludes suppression.

3.2 Phenotypic reversion in the presence of activated oncogenes

The study of phenotypic reversion of transformed cells has implicated genes responsible for controlling tumorigenesis and may provide a useful method for isolating onco-suppressor genes.

Reversion of virally transformed (K-*ras*) cells to phenotypic normality has been shown to occur by two different mechanisms. In the first, the transforming gene is lost or inactivated [40] and the cells can be retransformed by introduction of the original transforming gene. In the second, however, the transforming protein continues to be expressed and cells are resistant to transformation by the same gene [41]. Flat reverents selected from mutagenized, transformed cells containing two copies of the K-*ras* gene continued to express high levels of the K-*ras* protein, but cloned less efficiently in agarose and were non-tumorigenic. These revertants seem to carry dominant mutations in cellular genes that cause suppression of malignant characteristics. However, the putative onco-suppressor genes have not yet been isolated.

The technique of DNA-mediated gene transfer, which has been used very successfully to search for oncogenes [3], has now been applied to search for genes that suppress the phenotype of transformed cells [42–44]. It has been shown that morphologically non-transformed revertants with reduced tumorigenicity can be isolated from K-*ras* transformed cell by transfection with a cDNA expression library from normal human cells [44]. Revertants were produced which still expressed K-*ras*, were resistant to transformation by superinfecting K-*ras* and gave rise to non-transforming hybrids when fused to normal or K-*ras* transformed cells. Furthermore, in one revertant line an mRNA was detected that was absent in the original transformed parent. These properties suggested that a cDNA from the human library had contributed to the reversal of the transformed phenotype. This cDNA designated K*rev*1 has been cloned and sequenced [45]. It has the capacity to encode a protein with a molecular weight of 21 kDa, but the predicted amino sequence does not correspond to any known protein. However, it is identical to the deduced sequence of the human *rap*1 protein [46] and unexpectedly shares considerable homology to the guanine and phosphoryl binding domains of the K-*ras* protein itself. The gene seems to be conserved in several species and is expressed in many rat tissues, so K*rev*-1 may be involved in negative regulation of growth in many tissues.

Since K*rev*-1 DNA is not expressed in other K-*ras* revertants isolated by Noda and his colleagues, there are probably other onco-suppressor genes present in the human cDNA library [44]. Recovery and characterization of these may give further insights into mechanisms of growth control and transformation.

3.3 Gene loss or inactivation in cancer

Recessive mutations at specific locations have been implicated in a variety of human tumours. The most extensively studied of these malignancies are retinoblastoma and Wilms' tumour.

Retinoblastoma is a rare aggressive cancer of childhood that affects either one or both eyes. Following some early epidemiological studies, Knudson developed an influential two-hit theory which proposed that the development of the malignancy required two mutational events [47]. He recognized that the disease could have a hereditary susceptibility and proposed that, in bilateral retinoblastoma, a mutant gene was passed in the germline such that all cells in the body contained the mutation. A second mutation in the other allele of the same gene in the embryonic retina would lead to tumour formation. For the unilateral or sporadic form of the disease which has no familial predisposition, he suggested that somatic mutations affecting both alleles occurred in the same retinoblast. This theory has been substantiated by the studies of Cavenee *et al.* [48].

The chromosomal location of the retinoblastoma susceptibility gene was identified as 13q14 by cytogenetic analysis of retinoblastoma patients with a constitutional deletion [49]. Those with no microscopically visible deletion were shown by RFLP analysis of linked markers to have defects in the same chromosomal region [50]. Information on the genetic events giving rise to retinoblastoma has come from cytogenetic and RFLP analyses of tumours. Loss of the normal allele at the Rb-1 locus is a common feature of these tumours. Loss of the normal chromosome 13 [51] has also been observed often accompanied by duplication of the other so that the karyotype appears normal [52].

Individuals with hereditary susceptibility to retinoblastoma who have either never contracted the disease or been cured, are at a higher than normal risk of contracting certain other non-ocular primary cancers, the most common of which is osteosarcoma [53,54]. Molecular genetic analyses of osteosarcoma cell lines or primary bone tumours from osteosarcoma patients with no history of retinoblastoma showed loss of heterozygosity of the Rb-1 locus [55,56]. This is taken as evidence in support of the proposal [52] that inactivation of the same locus is important in both malignancies. It seems that homozygous loss of the Rb-1 gene can result in retinoblastoma if it occurs in a retinoblast and in oeseosarcoma if it occurs in bone tissue.

A cDNA that maps to 13q14 has been isolated from normal human fibroblasts and tentatively identified as the retinoblastoma gene on the basis of the fact that the gene was absent or altered in Rb patients who showed a cytogenetic abnormality in the putative Rb locus and RNA was altered in patients lacking visible deletion [57–59]. When the cloned cDNA was introduced into retinoblastoma or osteosarcoma cells that had inactivated endogeneous Rb genes, the neoplastic phenotype was suppressed [60]. This established, firstly, that the DNA did indeed encode the Rb protein and, secondly, that inactivation of this protein was an important step in the disease.

In all cases studied to date, loss of the Rb-1 gene or inability to produce functional Rb protein is correlated with increased cell proliferation and oncogenesis, leading to the suggestion that Rb-1 protein plays a critical role in limiting proliferation of normal cells.

Wilms' tumours is a cancer of the kidney that occurs in hereditary and sporadic forms [61]. The familial form is often associated with aniridia, genital abnormalities and mental retardation. Cytogenetic analysis of both hereditary and sporadic forms of the disease have suggested that 11p13 is the location of a gene predisposing to the disease [62,63]. In patients with no visible deletion, RFLP analyses confirmed the presence of deletions in the short arm of chromosome 11 [64,65]. When the short arm of chromosome 11 was introduced into a Wilms' tumour cell line by microcell fusion, the malignant phenotype was reversed, confirming that the Wilms' locus lies on this part of chromosome 11 [66]. Chromosome 11 introduction did not suppress transformation characteristics, so transformation and malignancy phenotypes seem to be under separate genetic control. The Wilms' tumour gene may regulate a late step in the progression to the neoplastic state rather than an early event in the preneoplastic state.

Early RFLP analyses of tumours that lacked cytogenetically visible deletions used the closest DNA probes then available and these mapped to 11p15 [64,65]. Loss of heterozygosity was found and it was assumed that the region of hemizygosity would also include 11p13. However, new studies using a set of probes that span the region 11p13–11p15 show that deletions of this extent occur in only a small fraction of tumours [67]. In several Wilms' tumours, the homozygosity was confined to 11p15.5. A possible explanation for these results is that both 11p13 and 11p15.5 contain oncosuppressor genes, and only one is involved in Wilms' tumour development.

There are several genes 'near' the Wilms' locus that might also play a role in tumour development. These include the insulin-like growth factor II gene on 11p15.5 and the H-*ras* allele on 11p14.1 [67–69]. It is of interest that in some sporadic Wilms' tumours there is loss of an H-*ras* allele and it has been proposed that the hemizygosity may have had a role in tumour initiation [69]. It is important to establish the exact extent of the Wilms' locus: further analyses are needed to establish the smallest region of overlap of deletions in patients having Wilms' tumour.

As with the retinoblastoma gene, loss of heterozygosity can occur by unequal mitotic recombination, deletion, unbalanced translocation and loss of a chromosome sometimes accompanied by duplication of the sister chromosome [67].

Loss of a normal allele in recessive diseases is probably more complicated than originally thought. In heritable cases the chromosome retained in the tumour derives from the parent with the affected gene. Analyses of cytogenetically visible defects in osteosarcoma and Wilms' tumours has established that it is the paternal chromosome that is retained in the tumour [70]. Although such a bias would not be expected in sporadic forms of the disease, it has now been shown that in non-familial forms of both Wilms' tumour and osteosarcoma the initial mutation also originates in the paternal chromosome [71,72]. The mechanism by which this occurs is unknown. Perhaps paternal and maternal chromosomes have different sensitivities to mutation or there may be a nearby onco-suppressor gene that is differently expressed on the parental chromosomes [73]. Epigenetic factors such as chromosomal imprinting may play a role.

Family studies and cytogenetic analysis suggest that other paediatric cancers may also have a hereditary component. The susceptibility is dominantly inherited, but the cellular mechanism of oncogenesis is recessive. Rare constitutional deletions in cancer patients have led to tentative identification of the chromosomal location of some recessive genes (see Table 1). Apart from the Rb gene, none of the others has yet been cloned.

Several cancers seem to involve defects in the same chromosomal location, e.g. 17p in colorectal cancer and breast cancer [82,85] and 11p in breast cancer, Wilms' tumour, hepatoblastoma, rhabdomyosarcoma and bladder cancer [62,74,79,83]. These regions are candidates for the location of onco-suppressor genes, but at present there is no direct evidence that precisely the same DNA sequences are defective in these diseases. However, a central role in carcinogenesis for the retinoblastoma gene at 13q14 is suggested by the finding that in retinoblastoma, osteosarcoma, breast cancer and small cell lung carcinoma the Rb gene is defective. SCLC is an adult carcinoma associated with heavy smoking but with no hereditary susceptibility [89]. Presumably the retinoblastoma gene in lung tissue is a target for carcinogens in tobacco smoke.

Loss of chromosomal material at 9p21–22 seems to be characteristic of acute lymphomatous leukaemia (ALL), suggesting that an onco-suppressor gene at or near this region functions as a regulatory gene to control the orderly proliferation of lymphoid precursors [89]. Band 9p21–22 is also the location of a heritable fragile site [93,94] and although no constitutional deletions have been found in lymphomatous ALL patients, there are the intriguing possibilities that the sites actually coincide or that the presence of a fragile site predisposes to loss of an onco-suppressor gene in this disease.

Table 1 — Genetic loss associated with human tumours

Tumour	Chromosome region	Reference
Retinoblastoma	13q14	[48]
Osteosarcoma	13q14	[56]
Wilms' tumour	11p13	[62]
Hepatoblastoma	11p13	[74]
Rhabdomyosarcoma	11p13	[74]
Small cell lung carcinoma	3p13	[75]
	13q14	[76]
Neuroblastoma	1p32	[77]
Renal cell carcinoma	3p	[78]
Bladder carcinoma	11p	[79]
Uveal melanoma	2	[80]
Colorectal carcinoma	5q	[81]
	17	[82]
Breast carcinoma	11p14	[83]
	13	[84]
	17p	[85]
Bilateral acoustic neuroma	22q	[86]
Meningioma	22	[87]
Myelodisplastia and acute myeloid leukaemia	5p and 7	[88]
Lymphomatous ALL	9p(21–22)	[89]
Follicular non-Hodgkin's lymphoma	6p and 13q32	[90]
Multiple endocrine neoplasia syndromes		
Type 1	11	[91]
Type 2	1	[92]

4. OTHER CANDIDATE ONCO-SUPPRESSOR GENES

The gene encoding the cellular protein p53 is another candidate onco-suppressor gene. Such a role in oncogenesis was suggested by the discovery that several mouse erythroleukaemia cell lines carry deletions of the p53 gene [95]. p53 seems to affect proliferation of primary cells and the mutant rather than the wild-type protein is oncogenic [96]. During SV40 infection of cells, a proportion of SV40 T antigen is bound to p53 protein (for review see [97]). Recent studies to establish the significance of this binding have shown that wild-type, but not high concentrations of mutant, protein inhibit initiation of viral DNA replication *in vitro* [98]. Although the function of p53 in the cell has not been established, it may be involved in control of cellular DNA replication.

Vascularization or angiogenesis is necessary for the unrestrained growth of tumours and without a blood supply tumours remain as nodules [99]. There is a clear correlation between the presence of factor that stimulate tumour growth and angiogenesis [100]. A previously unknown inhibitor of angiogenesis, a glycoprotein of $M_r \approx 140$ kDa, has been detected in the conditioned medium of BHK cells [101]. Through analyses of inhibitory activity in conditioned medium of transformed cells, revertants and segregating hybrids, the protein is believed to be produced by cells containing an active onco-suppressor gene. Although the inhibitor is abundant in conditioned medium, it does not correspond to currently known proteins, so its identity is unknown.

Tissue inhibitor of metalloproteinase (TIMP) is a secreted glycoprotein which inhibits a group of enzymes that degrade extracellular matrix and basement membrane components [102]. TIMP is encoded by a putative onco-suppressor gene since there is an inverse correlation between TIMP levels and the invasive potential of murine and human cells, implying an important role for TIMP in control of tumour invasion [102,103]. Recently, it has been shown that, when TIMP levels are reduced in mouse 3T3 cells by induction of anti-sense RNA, cells became more tumorigenic and metastatic than normal [104]. These results indicate that the presence of TIMP suppresses oncogenicity — at least in mouse 3T3 cells.

Studies on transformation of rat cells by Rous sarcoma virus suggest that the integrated provirus may be a useful indicator for genes that regulate transcription [105]. Rous sarcoma virus is an avian retrovirus that readily transforms chick cells in culture, but only occasionally transforms rat cells — possibly because transcription tends to be prevented. Rat cells do not lack positive regulatory factors and do not contain an excess of negative factors; rather, the lack of transcription seems to be caused by the site of integration of the virus. Trans-acting factors may also play a role in down regulating transcription and it has been suggested that loss of function of these genes may predispose to neoplasia.

5. RELATIONSHIP BETWEEN ONCOGENES AND ONCO-SUPPRESSOR GENES

It used to be thought that oncogenes were characterized by the fact that activation (mutation) caused expression of the gene, thus contributing to tumour formation, whereas oncogene suppressor genes were characterized by the fact that their inactivation or absence was important. It is now clear that the situation is more complicated and the same gene can act as an oncogene or an onco-suppressor gene depending on the cell type. Activated *ras* can act to transform fibroblasts, to cause PC12 phenochromocytoma cells to differentiate [106,107] and to induce growth arrest in several types of cell [108,109]. The *fos* oncogene, β transforming growth and tumour necrosis factors can also stimulate or inhibit cell proliferation depending on the target cell (for review see ref. [5]).

HT1080 cells contain an activated N-*ras* gene. As discussed above, suppression of tumorigenicity is likely to be mediated by chromosome 1 suggesting that an onco-suppressor gene is located on chromosome 1. This is the same chromosome to which N-*ras* itself maps [110] so this raises the possibility that the normal proto-oncogene suppresses the mutant allele. There is as yet no experimental evidence to support this proposal, but it is of interest that introduction of a normal *ras* gene to cells transformed by H-*ras* suppresses the tumorigenicity [111].

Studies on carcinogenesis in mouse skin, have shown that, whereas benign papillomas contain both normal and activated *ras* genes, progression to malignant carcinoma is frequently accompanied by loss of the normal gene [112]. In fresh breast cancer tissue one of the H-*ras* 1 alleles has frequently been deleted [83]. It is not known whether the remaining allele is mutant.

One interpretation of these results is that there are two classes of recessive mutation involved in tumorigenesis. The first is exemplified by retinoblastoma where mutation leads to loss of function of the gene. The second, exemplified by loss of a *ras* gene [69,83], may allow uncovering of an activating mutation in the other allele. However, proof of such an example has not yet been demonstated. It would be necessary to show that only the *ras* gene had been lost as there might be larger deletions that included onco-suppressor genes.

Comings [113] suggested that dominant 'suppressor' or 'regulatory' genes (onco-suppressor genes) controlled the expression of oncogenes. Loss of both copies of the regulatory gene might relieve repression, causing constitutive expression of the transforming gene, and tumour development would ensue. When it was found that retinoblastoma cells expressed elevated levels of N-*myc*, it was suggested that the Rb gene product controlled its expression [114]. This now seems unlikely since the levels of N-*myc*, while higher than in normal retinoblasts, are about the same level as found in foetal tissue.

A different approach to studying the relationship between oncogene and onco-suppressor activity has been to study by cell fusion the suppression of tumorigenicity in cells known to contain an activated oncogene. The EJ (human bladder carcinoma) cell line carries an activated H-*ras* gene that is thought to contribute to the tumorigenicity of the cell. Fusion with normal human fibroblasts produces hybrids that are non-tumorigenic yet still express the H-*ras* oncoprotein at high level. This result suggests that the onco-suppressor must act at a level after translation of the *ras* protein [38].

6. INTERACTION BETWEEN ONCOGENE AND ONCO-SUPPRESSOR GENE PRODUCTS

When the retinoblastoma gene was cloned, it was hoped that determination of its sequence would shed light on its role in the cell. However, the sequence did not correspond to any known protein and was not well conserved through evolution [59]. There was limited homology to DNA binding proteins and this was consistent with the finding that the Rb protein binds to DNA cellulose columns. Insights into the relationship of the Rb-1 gene product to transforming activity have come from an unexpected source — the study of DNA tumour viruses.

The human adenovirus 5 EIA protein is synthesized early in viral infection and together with the protein EIB is required for transformation of cells. EIA was known to bind to three host proteins of molecular weights 105, 110 and 300 kDa [115,116] and recent characterization of the 105 kDa protein has shown that it is the product of the retinoblastoma gene [117]. The regions of EIA that are required for transformation are precisely those that interact with the Rb-1 product [118]. Interaction with EIA protein is thought to sequester p105–Rb such that functional p105-Rb is denied to the cell. Although this association is necessary for cellular transformation, binding of the host 110 and 300 kDa proteins also seems to be required [119]. Retinoblastoma cells which lack the 105 kDa Rb-1 protein also lack the 107 kDa polypeptide, supporting the suggestion that its absence may also play a role in the development of retinoblastoma [119].

SV40 T antigen, an obligatory participant in cellular transformation by SV40 virus, also forms a specific complex with an un- or underphosphorylated form of the Rb-1 gene product [120]. Although neither SV40 nor adenovirus 5 is thought to cause cancer in man, binding p105-Rb to transforming proteins of certain viruses may be important in the genesis of certain human cancers. Human papiloma virus 16 is found in about 50% of human cervical carcinomas and the E7 gene product is required for transformation of rodent cells [121,122]. It has now been shown that the E7 protein binds to p105-Rb in *in vitro* mixing assays [123], leading to the suggestion that adenovirus, SV40 virus and papilloma virus 16 may share a common mechanism in transformation that involves binding of p105-Rb.

A computer search has shown that those regions of SV40 large T antigen and adenovirus EIA protein which are now known to be involved in Rb-1 binding share primary and secondary structure elements not only with each other, but also with certain regions of the v-*myc* and c-*myc* oncoproteins [124]. There is speculation that the *myc* proteins which are located in the neuclus may also bind p105-Rb-1. However there is no experimental evidence to support this prediction.

7. CLINICAL APPLICATIONS

RFLP analyses have already been used in antenatal diagnosis to identify people at risk for retinoblastoma [125,126]. Now that the extent of the Rb-1 gene is known [127] and cloned probes are available, the accuracy of diagnosis should be improved. Similar applications should become available when the genes predisposing to other recessive malignancies are cloned.

The use of recombinant technology to produce proteins which control cell growth may provide a powerful weapon against malignant cells. TIMP, angiogenesis inhibitory factor and other 'biological response modifiers' such as interferons and tumour necrosis factor [128,129] produced in this way might be useful therapeutically to inhibit tumour growth.

Although the therapeutic applications are at present distant, the recent results on the reversion of the tumorigenic phenotype of recipient cells by introducing the retinoblastoma [60] or K*rev*-1 [45] gene suggest that such gene products may eventually be of use in repressing malignancy.

8. CONCLUSIONS

Transformation of cells is in many instances the result of aberrant gene expression. It has been known for a long time that the overexpression of a normal proto-oncogene or expression of a mutated oncogene can contribute to malignancy and it is now becoming apparent that lack of expression of an onco-suppressor gene or functional inactivation of its product may play a similarly important role.

The study of onco-suppressor genes is experimentally difficult because we lack selection systems for their detection and we are ignorant of their identity and function. One of the most promising approaches to isolating more onco-suppressor genes is the revertant induction assay of Noda and colleagues whereby introduction of cDNA libraries into transformed cells is followed by recovery of revertants and characterization of the cDNA in revertants [45]. Application of the technique to reversion of transformants induced by different oncogenes may yield insights into mechanisms of growth control and transformation.

The interaction of the oncoproteins encoded by DNA transforming viruses and the Rb-1 protein suggests both a mechanism by which Rb-1 is functionally denied to the cell and a means by which other onco-suppressor proteins might by identifed. If sequestration of negative regulators of cell growth by oncoproteins is a common means by which they exert their transforming action, then investigation of interactions with cellular proteins may reveal the identity of other onco-suppressor proteins.

It is a major goal to identify and characterize more genes and their products so that we can learn more about the control of cell growth and proliferation and perhaps about the mechanism of carcinogenesis.

ACKNOWLEDGEMENTS

We gratefully acknowledge the support of the Medical Research Council, the Cancer Research Campaign of Great Britain and the National Hellenic Research Foundation who fund our current research.

REFERENCES

[1] Cairns, J. H. (1978). In: C. I. Davern (ed.) *Cancer: Science and Society*, Freeman, San Francisco, CA

[2] Boveri, T. (1929). *On the Problem of the Origin of Malignant Tumours* Williams and Wilkins, Baltimore, MD

[3] Varmus, H. (1984). The molecular genetics of cellular oncogenes. *Annu. Rev. Genet.* **18** 553–612

[4] Bishop, J. M. (1987). The molecular genetics of cancer. *Science* **235** 305–311

[5] Spandidos, D. A. & Anderson, M. L. M. (1989). Oncogenes and onco-suppressor genes; their involvement in cancer. *J. Pathol.* **157** 1–10

[6] Haluska, F. G., Tsujimoto, T. & Croce, C. M. (1987). Oncogene activation by chromosome translocation in human malignancy. *Annu. Rev. Genet.* **21** 321–345

[7] Brodeur, G. M., Seegar, R. C., Schwab, M., Varmus, H. E. & Bishop, J. M. (1984). Amplification of N-*myc* in untreated human neuroblastomas correlates with advanced disease stage. *Science* **224** 1121–1124

[8] Spandidos, D. A. (1987). Oncogene activation in malignant trasformation: a study of the H-*ras* in human breast cancer. *Anticancer Res.* **7** 991–996

[9] Wright, C., Angus, B., Nicholson, S. *et al.* (1989). Expression of c-*erb*B-2 oncoprotein: a prognostic indicator in human breast cancer. *Cancer Res.* **49** 2087–2090

[10] Shtivelman, E., Lifshitz, B., Gale, R. P. & Canaani, E. (1985). Fused transcript of *abl* and *bcr* genes in chronic myelogenous leukaemia. *Nature (London)* **315** 550–554

[11] Konopka, J. B., Watanabe, S. M. & Witte, O. N. (1984). An alteration of the human c-*abl* protein in K562 leukaemia cells unmasks associated tyrosine kinase activity. *Cell* **37** 1035–1042

[12] Barbacid, M. (1987). ras genes. *Annu. Rev. Biochem.* **56** 779–827

[13] Bos, J. L., Fearon, E. R., Hamilton, S. R., *et al.* (1987). Prevalence of *ras* gene mutations in human colorectal cancers. *Nature (London)* **327** 293–297

[14] Forrester, K., Almoguera, C., Han, K., Grizzle, W. E. & Perucho, M. (1987). Detection of high incidence K-*ras* oncogenes during human colon tumorigenesis. *Nature (London)* **327** 298–303

[15] Bos, J. L., Verlaan-deVries, M., van der Eb, A. J., *et al.* (1987). Mutations in N-*ras* predominance in acute myeloid leukaemia. *Blood* **69** 1237–1241

[16] Foulds, L. (1954). The experimental study of tumor progression. A review. *Cancer Res.* **14** 327–339

[17] Slamon, D. J., Dekernion, J. B., Verma, I. M. & Cline, M. J. (1984). Expression of cellular oncogenes in human malignancies. *Science* **224** 256–262

[18] Spandidos, D. A., Lamothe, A. & Field, J. N. (1985). Multiple trascriptional activation of cellular oncogenes in human head and neck tumors. *Anticancer Res.* **5** 221–224

[19] Sandberg, A. A. (1983). A chromosomal hypothesis of oncogenesis. *Cancer Genet. Cytogenet.* **8** 277–285

[20] Knudson, A. G. (1983). Hereditary cancers in man. *Cancer Invest.* **1** 187–193

[21] Todaro, G. J. (1986). Tumor growth factors. In: J. D. Fortner and J. E. Rhoades (eds) *Accomplishments in Cancer Research*, Lippincott, Philadelphia, PA

[22] Klein, G. (1987). The approaching era of the tumour suppressor genes. *Science* **238** 1539–1545

[23] Anderson, M. L. M. & Spandidos, D. A. (1988). Onco-suppressor genes and their invovlement in cancer. *Anticancer Res.* **8** 873–880

[24] Green, A. R. (1988), Recessive mechanisms of malignancy, *Br. J. Cancer* **58** 115–121

[25] Stanbridge, E. J. (1976). Suppression of malignancy in human cells. *Nature (London)* **260** 17–20

[26] Stanbridge, E. J. & Wilkinson, J. (1978). Analysis of malignancy in human cells: malignant and transformed properties are under separate genetic control. *Proc. Natl. Acad. Sci. USA* **75** 1466–1469

[27] Stanbridge, E. J. (1987). Genetic regulation of tumorigenic expression in somatic cell hybrids. *Adv. Viral Oncol.* **6** 83–101

[28] Huschtscha, L. I. & Holliday, R. (1983). Limited and unlimited growth of SV40-transformed cells from human diploid MRC-5 fibroblasts. *J. Cell Sci.* **63** 77–79

[29] Marshall, C. J. (1980). Suppression of the transformed phenotype with retention of the viral *src* gene in cell hybrids between Rous sarcoma virus-transformed rat cells and untransformed mouse cells. *Exp. Cell Res.* **127** 373–384

[30] Layton, M. G. & Franks, L. M. (1986). Selective suppression of metastasis, but not tumorigenicity of a mouse lung carcinoma by cell hybridisation. *Int. J. Cancer* **37** 723–730

[31] Stanbridge, E. J., Flandermeyer, R. R., Daniels, D. W. & Nelson-Rees, W. A. (1981). Specific chromosome loss associated with the expression of tumorigenicity in human cell hybrids. *Somatic Cell Genet.* **7** 699–712
[32] Kaelbling, M. & Klinger, H. P. (1986). Suppression of tumorigenicity in somatic cell hybrids III. Cosegregation of human chromosome 11 of a normal cell and suppression of tumorigenicity in intra species hybrids of normal diploid × malignant cells. *Cytogenet. Cell Genet.* **41** 65–70
[33] Srivatsan, E. S., Benedict, W. F. & Stanbridge, E. J. (1986). Implication of chromosome 11 in the suppression of neoplastic expression in human cell hybrids. *Cancer Res.* **46** 6174–6179
[34] Saxon, P. J., Srivatsan, E. S. & Stanbridge, E. J. (1986). Introduction of human chromosome 11 via microcell transfer controls tumorigenic expression of HeLa cells. *EMBO J.* **5** 3461–3466
[35] Klinger, H. P. & Kaelbling, M. (1986). Suppression of tumorigenicity in somatic cell hybrids. *Cytogenet. Cell Genet.* **42** 225–235
[36] Benedict, W. F., Weissman, B. E., Mark, C. & Stanbridge, E. J. (1984). Tumorigenicity of human HT1080 fibrosarcoma × normal fibroblasts is chromosome dose dependent. *Cancer Res.* **44** 3471–3475
[37] Weissmann, B. E. & Stanbridge, E. J. (1983). Complementation of the tumorigenic phenotype in human cell hybrids. *J. Natl. Cancer Inst.* **70** 667–672
[38] Geiser, A. G., Anderson, M. J. & Stanbridge, E. J. (1989). Suppression of tumorigenicity in human cell hybrids derived from cell lines expressing different activated *ras* oncogenes. *Cancer Res.* **49** 1572–1577
[39] Paterson, H., Reeves, B., Brown, R., *et al.* (1987). Activated N-*ras* controls the transformed phenotype of HT1080 human fibrosarcoma cells. *Cell* **51** 803–812
[40] Bassin, R. H. & Noda, M. (1987). Oncogene inhibition by cellular genes. *Adv. Viral Oncol.* **6** 103–127
[41] Noda, M., Selinger, Z., Scolnick, E. M. & Bassin, R. H. (1983). Flat revertants isolated from Kirsten sarcoma virus-transformed cells are resistant to the action of specific oncogenes. *Proc. Natl. Acad. Sci. USA* **80** 5602–5606
[42] Baker, R. F. (1983). DNA-mediated alteration of the reversion frequency of transformed NIH/3T3 cells. *Proc. Natl. Acad. Sci. USA* **80** 1174–1178
[43] Schaefer, R., Iyer, J., Item, E. & Nirkko, A. C. (1988). Partial reversion of the transformed phenotype in HRAS-transfected tumorigenic cells by transfer of a human gene. *Proc. Natl. Acad. Sci. USA* **85** 1590–1594
[44] Noda, M., Kitayama, H., Matsuzaki, T., *et al.* (1989). Detection of genes with a potential for suppressing the transformed phenotype associated with activated *ras* genes. *Proc. Natl. Acad. Sci. USA* **86**, 162–166
[45] Kitayama, N., Sugimoto, Y., Matsuzaki, T., *et al.* (1989). A *ras*-related gene with transformation suppressor activity. *Cell* **56** 77–84
[46] Pizon, V., Chardin, P., Lerosey, I., Olofsson, B. & Tavington, A. (1988). Human cDNAs *rap*1 and *rap*2 homologous to the Drosophila gene D *ras*3 encode proteins closely related to *ras* in the 'effector" region. *Oncogene* **3** 201–204
[47] Knudson, A. G. (1971). Mutation and cancer: statistical study of retinoblastoma. *Proc. Natl. Acad. Sci. USA* **68** 820–823

[48] Cavenee, W. K., Dryja, T. P., Phillips, R. A., Benedict, W. F., Godbout, R., Gallie, B. L., Murphree, A. L., Strong, L. C. & White, R. L. (1983). Expression of recessive alleles by chromosomal mechanisms in retinoblastoma. *Nature (London)* **305** 779–784
[49] Sparkes, R. S., Sparkes, M. N., Wilson, M. G. *et al.* (1980). Regional assignment of genes for human esterase D and retinoblastoma to chromosome band 13q14. *Science* **208** 1042–1044
[50] Sparkes, R. S., Murphree, A. L., Lingua, R. W. *et al.* (1983). Gene for hereditary retinoblastoma assigned to human chromosome 13 by linkage to esterase D. *Science* **219** 971–973
[51] Benedict, W. F., Murphree, A. L., Banerjee, A., Spina, C. A., Sparkes, M. C. & Sparkes, R. S. (1983). Patient with 13 chromosome deletion: evidence that the retinoblastoma gene is a recessive cancer gene. *Science* **219** 973–975
[52] Murphree, A. L. & Benedict, W. F. (1984). Retinoblastoma: clues to human oncogenes. *Science* **223** 1028–1033
[53] Abramson, D. H., Ellsworth, R. M., Kitchin, F. D. & Tung, G. (1984). Second non-ocular tumors in retinoblastoma survivors. Are they radiation induced? *Ophthalmology* **91** 1351–1355
[54] Draper, G. J., Sanders, B. M. & Kingston, J. E. (1986). Second primary neoplasms in patients with retinoblastoma. *Br. J. Cancer* **53** 661–671
[55] Dryja, T. P., Rapaport, J. M., Epstein, J., *et al.* (1986). Chromosome 13 homozygosity in osteosarcoma without retinoblastoma. *Am. J. Hum. Genet.* **38** 59–66
[56] Hansen, M. F., Koufos, A., Gallie, B. L. *et al.* (1985). Osteosarcoma and retinoblastoma: a shared chromosomal mechanism revealing recessive predisposition. *Proc. Natl. Acad. Sci. USA* **82** 6216–6220
[57] Friend, S. H., Bernards, R., Rojelj, S., Weinberg, R. A., Rapaport, J. M., Albert, D. M., & Dryja, T. P. (1986). A human DNA segment with the properties of the gene that predisposes to retinoblastoma and osteosarcoma. *Nature (London)* **323** 643–646
[58] Fung, Y.-K. T., Murphree, A. L., Tang, A., Qian, J., Hinrichs, S. H. & Benedict, W. F. (1987). Structural evidence for the authenticity of the human retinoblastoma gene. *Science* **236** 1657–1661
[59] Lee, H.-W., Brookstein, R., Hong, F., Young, L.-J., Shew, J. Y. & Lee, E. Y.-H. P. (1987). Human retinoblastoma susceptibility gene; cloning, identification and sequence. *Science* **235** 1395–1399
[60] Huang, H.-J. S., Yee, J.-K., Shew, J.-Y., Chen, P.-L., Bookstein, R., Friedmann, T., Lee, E. Y.-H. P. & Lee, W.-H. (1988). Suppression of the neoplastic phenotype by replacement of the RB gene in human cancer cells. *Science* **242** 1563–1566
[61] Knudson, A. G. & Strong, L. C. (1972). Mutation and cancer: a model for Wilms' tumor of the kidney. *J. Natl. Cancer Inst.* **48** 313–324
[62] Riccardi, V. M., Sujansky, E., Smith, A. C. & Francke, U. (1978). Chromosomal imbalance in the aniridia–Wilms' tumor association: 11p interstitial deletion. *Pediatrics* **61** 604–610
[63] Slater, R. M. (1986). The cytogenetics of Wilms' tumor. *Cancer Genet. Cytogenet.* **19** 37–41

[64] Fearson, E. R., Vogelstein, B. & Feinberg, A. P. (1984). Somatic deletion and duplication of genes on chromosome 11 in Wilms' tumors. *Nature (London)* **309** 176–178

[65] Koufos, A., Hansen, M. F., Lampkin, B. C., Workman, M. L., Copeland, N. G., Jenkins, N. A. & Cavenee, W. K. (1984). Loss of alleles at loci on human chromosome 11 during genesis of Wilms' tumor. *Nature (London)* **309** 170–172

[66] Weissman, B. E., Sazon, P. J., Pasquale, S. R., Jones, G. R., Geiser, A. G. & Stanbridge, E. J. (1987). Introduction of a normal human chromosome 11 into a Wilms' tumor cell line controls its tumorigeneic epxression. *Science* **236** 175–180

[67] Mannens, M., Slater, R. M., Heyting, C., *et al.* (1988). Molecular nature of genetic changes resulting in loss of heterozygosity of chromosome 11 in Wilms' tumours. *Hum. Genet.* **81** 41–48

[68] Porteous, D. J., Bickmore, W., Christie, S., *et al.* (1987). HRAS1-selected chromosome transfer generates markers that colocalise aniridia- and genito-urinary displasia-associated translocated breakpoints and the Wilms' tumor gene within band 11p13. *Proc. Natl. Acad. Sci. USA* **84** 5355–5359

[69] Reeve, A. E., Houseaux, P. J., Gardner, R. J. M., Chewings, W. E., Grindlay, R. M. & Millow, L. J. (1984). Loss of a Harvey *ras* allele in sporadic Wilms' tumor. *Nature (London)* **309** 174–176

[70] Toguchida, J., Ishizaka, K., Sasaki, M. S., *et al.* (1988). Chromosomal reorganisation for the expression of recessive mutation of retinoblastoma susceptibility gene in the development of osteosarcoma. *Cancer Res.* **48** 3939–3943

[71] Schroeder, W. T., Chao, L. Y., Dao, D. D., *et al.* (1987). Nonrandom loss of maternal chromosome 11 alleles in Wilms' tumors. *Am. J. Hum. Genet.* **40** 413–420

[72] Toguchida, J., Ishizaka, K., Sasaki, M. S., *et al.* (1989). Preferential mutation of the paternally derived RB gene as the initial event in sporadic osteosarcoma. *Nature (London)* **338** 156–158

[73] Wilkins, R. J. (1988). Genomic imprinting in carcinogenesis. *Lancet* **I'** 329–330

[74] Koufos, A., Hansen, M. F., Copeland, N. G., *et al.* (1985). Loss of heterozygosity in three embryonal tumours suggests a common pathogenic mechanism. *Nature (London)* **316** 330–334

[75] Whang-Peng, J., Bunn, P. A., Jr., Kao-shan, C. S., Lee, E. C., Carney, D. N., Gazdar, A. & Minna, J. A. (1982). A non-random chromosomal abnormality, del 3p (14–23) in human small cell lung cancer (SCLC). *Cancer Genet. Cytogenet.* **6** 119–134

[76] Harbour, J. W., Lar, S.-L., Whang-Peng, J., Gazdar, A. F., Minna, J. D. & Kaye, F. J. (1988). Abnormalities in structure and expression of the human retinoblastoma gene in SCLC. *Science* **241** 353–357

[77] Brodeur, G., Green, A. A., Hays, F. A., Williams, K. J., Williams, D. L. & Tsiakos, A. A. (1981). Cytogenetic features of human neuroblastomas and cell lines. *Cancer Res.* **41** 4678–4686

[78] Pathak, S., Strong, L. C., Ferrell, R. E. & Trindade, A. (1982). Familial renal

carcinoma with a 3:11 chromosome translocation limited to tumor cells. *Science* **217** 939–941

[79] Fearon, R., Feinberg, A. P., Hamilton, S. H. & Vogenstein, B. (1985). Loss of genes on the short arm of chromosome 11 in bladder cancer. *Nature (London)* **318** 377–380

[80] Mukai, S., & Dryja, T. P. (1986). Loss of alleles at polymorphic loci on chromosome 2 in uveal melanoma. *Cancer Genet. Cytogenet.* **22** 45–53

[81] Solomon, E., Voss, R., Hall, V., Bodmer, W. F., Jass, J. R., Jeffreys, A. J., Lucibello, F. C., Patel, & Ridee, S. H. (1987). Chromosome 5 allele loss in human colorectal carcinomas. *(London)* **328** 616–619

[82] Fearon, E. R., Hamilton, S. H. & Vogelstein, B. (1987). Clonal analysis of human colorectal tumours. *Science* **238** 193–197

[83] Ali, A. U., Lidereau, R., Theillet, C. & Callahan, R. (1987). Reduction to homozygosity of genes on chromosomes 11 in human breast neoplasia. *Science* **238** 185–188

[84] Lundberg, C., Skoog, L., Cavenee, W. K. & Nordenskjold. M. (1987). Loss of heterozygozity in human ductal breast tumours indicates a recessive mutation on chromosome 13. *Proc. Natl. Acad. Sci. USA* **84** 2372–2376

[85] Mackay, J., Steel, C. M., Elder, P. A., Forrest, A. P. M. & Evans, H. J. (1988). Allele loss on short arm of chromosome 17 in breast cancers. *Lancet* ii 1384–1385

[86] Seizinger, B. R., Matruza, R. L. & Gusella, J. F. (1986). Loss of genes on chromosome 22 in tumorogenesis of human acoustic neuroma. *Nature (London)* **322** 644–647

[87] Seizinger, B. R., de la Monte, S., Atkins, L., Gusella, J. F. & Martuza, R. L. (1987). Molecular genetic approach to human meningioma: loss of genes of chromosome 22. *Proc. Natl. Acad. Sci. USA* **84** 5419–5423

[88] Yunis, J. J. (1983). The chromosomal basis of human neoplasia. *Science* **221** 227–236

[89] Chilcote, R. R., Brown, E. & Rowley, L. D. (1985). Lymphoblastic leukaemia with lymphomatous features associated with abnormalities of the short arm of chromosome 9. *N. Engl. J. Med.* **313** 286–291

[90] Yunis, J. J., Frizzera, G., Oken, M. M., McKenna, J., Theologides, A. & Arneson, M. (1987). Multiple recurrent defects in follicular lymphoma. A possible model for cancer. *N. Engl. J. Med.* **316** 79–84

[91] Larsson, C., Skogseid, B., Oberg, K., Nakamura, Y. & Nordenskjold, M. (1988). Multiple endocrine neoplasia type 1 gene maps to chromosome 11 and is lost in insulinoma. *Nature (London)* **332** 85–87

[92] Mathew, C. G. P., Smith, B. A., Thorpe, K., Wong, Z., Royle, N. J., Jeffreys, A. J. & Ponder, B. A. J. (1987). Deletion of genes on chromosome 1 in endocrine neoplasia. *Nature (London)* **328** 524–528

[93] Sutherland, G. R., Jacky, P. B., Baker, E. & Manuel, A. (1983). Heritable fragile sites on human chromosomes. X. New folate-sensitive fragile sites: 6p23, 9p21, 9q32, and 11q23. *Am. J. Hum. Genet.* **35** 432–437

[94] Guichaoua, M., Mattei, M. G., Mattei, J. F. & Giraud, F. (1982). Aspects genetiques des sites fragiles autosomiques: a propos de 40 cas. *J. Genet. Hum.* **30** 183–198

[95] Mowat, M., Cheng, A., Kimura, N., Bernstein, A. & Benchimol, S. (1985). Rearrangement of the cellular p53 gene in erythroleukaemic cells transformed by Friend virus. *Nature (London)* **314** 633–636

[96] Hinds, P., Finlay, C. & Levine, A. J. (1989). Mutation is required to activate the p53 gene for cooperation with the *ras* oncogene and transformation. *J. Virol.* **63** 739–746

[97] Oren, M. (1985). The p53 cellular tumor antigen: gene structure, expression and protein properties. *Biochem Biophys. Acta* **823** 67–78

[98] Wang, E. H., Friedman, P. N. & Prives, C. (1989). The murine p53 protein blocks replication of SV40 DNA *in vitro* by inhibiting the initiation functions of SV40 large T antigen. *Cell* **57** 379–392

[99] Folkman, J. (1974). Tumor angiogenesis factor. *Cancer Res.* **34** 2109–2113

[100] Folkman, J., Watson, K., Ingber, D. & Hanahan, D. (1989). Induction of angiogenesis during the transition from hyperplasia to neoplasia. *Nature (London)* **339** 58–61

[101] Rastinejad, F., Polverini, P. J. & Bouck, N. P. (1989). Regulation of the activity of a new inhibitor of angiogenesis by a cancer suppressor gene. *Cell* **56** 345–355

[102] Edwards, D. R., Waterhouse, P., Holman, M. L. & Denhardt, D. T. (1986). A growth-responsive gene (16C8) in normal mouse fibroblasts homologous to a human collagenase inhibitor with erythroid-potentiating activity: evidence for inducible and constitutive transcripts. *Nucleic Acids Res.* **14** 8863–8878

[103] Hicks, N. J., Ward, R. V. & Reynolds, J. J. (1984). Fibrosarcoma model derived from mouse embryo cells: growth properties and secretion of collagenase and metalloproteinase inhibitor (TIMP) by tumor cell lines. *Int. J. Cancer* **33** 835–844

[104] Khokha, R., Waterhouse, P. *et al.* (1989) Antisense RNA-induced reduction in murine TIMP levels confers oncogenicity on Swiss 3T3 cells. *Science* **243** 947–950

[105] Wyke, J. A., Akroyd, J., Gillespie, D. A. F., Green, A. R. & Poole, C. (1989). Proviral position effects; possible probes for genes that suppress transcription. In: *Genetic Analyses of Tumor Suppression*, Ciba Foundation Symposium **142**, Wiley, Chichester, pp. 117–130

[106] Noda, M., Ko, M., Ogura, A., Liu, D.-G., Amano, T. & Ikawa, Y. (1985). Sarcoma viruses carrying *ras* oncogenes induce differentiation-associated properties in a neuronal cell line. *Nature (London)* **318** 73–75

[107] Spandidos, D. A. (1989). The effect of exogenous human *ras* and *myc* oncogenes in morphological differentiation of the rat pheochromocytoma PC12 cells. *Int. J. Dev. Neurosci.* **7** 1–4

[108] Hirakawa, T. & Ruley, H. E. (1988). Rescue of cells from *ras* oncogene-induced growth arrest by a second, complementing oncogene. *Proc. Natl. Acad. Sci. USA* **85** 1519–1523

[109] Ridley, A. J., Paterson, H. F., Noble, M. & Land, H. (1988). *ras*-mediated cell cycle arrest is altered by nuclear oncogenes to induce Schwann cell transformation. *EMBO J.* **7** 1635–1645

[110] Hall, A., Marshall, C. J., Spurr, N. K. & Weiss, R. A. (1983). Identification of

transformation gene in two human sarcoma cell lines as a new member of the *ras* gene family located on chromosome 1. *Nature (London)* **303** 396–400

[111] Spandidos, D. A. & Wilkie, N. M. (1988). The normal human H-*ras*-1 gene can act as an onco-suppressor. *Br. J. Cancer* **58** Suppl IX 67–71

[112] Quintanilla, M., Brown, K., Ramsden, M. & Balmain, A. (1986). Carcinogen-specific mutation and amplification of Ha-*ras* during mouse skin carcinogenesis. *Nature (London)* **322** 78–80

[113] Comings, D. E. (1973). A general theory of carcinogenesis. *Proc. Natl. Acad. Sci. USA* **70** 3324–3328

[114] Lee, W.-H., Murphree, A. L. & Benedict, W. F. (1984). Expression and amplification of the N-*myc* gene in primary retinoblastoma. *Nature (London)* **309** 458–460

[115] Yee, S.-P. & Branton, P. E. (1985). Detection of cellular proteins associated with human adenovirus type 5 early region E1A polypeptides. *Virology* **147** 142–153

[116] Harlow, E., Whyte, P., Franza, B. R., Jr., & Schley, C. (1986). Association of adenovirus early-region 1A proteins with cellular polypeptides. *Mol. Cell Biol.* **6** 1579–1589

[117] Whyte, P., Buchkovich, J. J., Horowitz, J. M., Friend, S. H., Raybuck, M., Weinberg, R. A. & Harlow, E. (1988). Association between an oncogene and an anti-oncogene: the adenovirus E1A proteins bind to the retinoblastoma gene product. *Nature (London)* **334** 124–129

[118] Whyte, P., Williamson, N. M. & Harlow, E. (1989). Cellular targets for transformation by the adenovirus E1A proteins. *Cell* **56** 67–75

[119] Egan, C., Bayley, S. T. & Branton, P. E. (1989). Binding of the *Rb* 1 protein to EIA products is required for adenovirus transformation. *Oncogene* **4** 383–388

[120] DeCaprio, J. A., Ludlow, J. W., Figge, J. *et al.*, (1988). SV40 large T antigen forms a specific complex with the product of the retinoblastoma susceptibility gene. *Cell* **54** 275–283

[121] Phelps, W. C., Yee, C. L., Munger, C. K. & Howley, P. M. (1988). The human papillomavirus type 16 E7 encodes transactivation and transformation functions similar to those of adenovirus E1A. *Cell* **53** 538–547

[122] Yasumoto, S., Burkhardt, A. L., Doniger, J. & DiPaolo, J. A. (1986). Human papilloma virus type 16 DNA-induced malignant transformation of NIH 3T3 cells. *J. Virol.* **57** 572–577

[123] Dyson, N., Howley, P. M., Munger, K. & Harlow, E. (1989). The human papilloma virus-16 E7 oncoprotein is able to bind to the retinoblastoma gene product. *Science* **243** 934–937

[124] Figge, J., Webster, T., Smith, T. F. & Paucha, E. (1988). Prediction of similar transforming regions in simian virus 40 large T, adenovirus, E1A and *myc* oncoproteins. *J. Virol.* **62** 1814–1818

[125] Cavenee, W. K., Murphree, A. K. L., Shull, M. M., *et al.* (1986). Prediction of familial predisposition to retinoblastoma. *N. Engl. J. Med.* **314** 1201–1207

[126] Wiggs, J., Nordenskjold, M., Yandell, D., *et al.* (1988). Prediction of the risk of hereditary retinoblastoma using DNA polymorphisms within the retinoblastoma gene. *N. Engl. J. Med.* **318** 151–157

[127] T'Ang, A., Wu, K.-J., Hashimoto, T., *et al.* (1989). Genomic organisation of the human retinoblastoma gene. *Oncogene* **4** 401–407
[128] Esteban, M. & Paez, E. (1985). Antiviral and antiproliferative properties of interferons: mechanism of actions. *Prog. Med. Virol.* **32** 159–173
[129] Pennica, D., Nedwin, G. E. *et al.* (1984). Human tumor necrosis factor: precursor structure, expression and homology to lymphotoxin. *Nature (London)* **312** 724–729

5

Ras oncogenes in human cancers

Nick Lemoine
ICRF Molecular Oncology Group, Hammersmith Hospital, Du Cane Road, London W12 0HS, UK

1. INTRODUCTION

The *ras* family of oncogenes was first implicated in human tumorigenesis in 1981 when DNA derived from human tumours was analysed by transfection assay [1–4] and found to contain dominantly transforming oncogenes related to those of the Harvey and Kirsten strains of rat sarcoma virus [5,6]. Since that time, an enormous amount of research has shown that they are frequently activated in a wide range of neoplasms, both human and experimental, and that the activated alleles differ from their normal proto-oncogene counterparts by specific point mutations that produce profound changes in the biological and biochemical activities of the encoded proteins. Furthermore, it is now appreciated that, in addition to their involvement in many malignancies, *ras* proteins play a key role in the regulation of proliferation and differentiation in normal cells [7–10].

Several excellent reviews of *ras* genes and their role in human and experimental tumours have appeared in the last couple of years (see [11–14]) and so this chapter will concentrate on the most recent and exciting findings in this field.

2. STRUCTURE AND FUNCTION OF THE *ras* PROTO-ONCOGENES

The three functional members of the *ras* gene family (N-*ras* located on the short arm of chromosome 1 at position 1p22–p32, Ki-*ras*2 on the short arm of chromosome 11 at position 11p15.1–15.5, and Ha-*ras*1 on the short arm of chromosome 12 at position 12p12.1-pter) are similar in that they each encode a 21 kDa protein in four exons (Ki-*ras*2 has two alternative fourth exons allowing production of a p21 protein in 189 or 188 amino acids). Their promoter regions lack characteristic TATA and CCAAT boxes, but are highly G+C rich [11, 15], possess several CCGCCC motifs [15–17] for potential interaction with Sp1 proteins (resembling the promoters of so-called housekeeping genes such as the genes of dihydrofolate reductase and the EGF receptor), and lie immediately upstream of a 5' non-coding exon. However, the intron sequences of each of these genes is very different, so that while Ha-*ras*1 is contained within 4.5 kb of genomic DNA [4,18–21] and N-*ras* within 13 kb [22], Ki-*ras*2 occupies about 50 kb [23–25]. There are two human *ras* pseudogenes (which lack introns and are not transcribed), Ki-*ras*1 at 6p21–p23 and Ha-*ras*2 on the X chromosome.

The amino acid sequence of the p21 proteins can be divided into four domains. The 85 amino acids at the amino terminus are identical in each member of the family and the adjacent more central sequence of 80 amino acids shows minor differences, while the carboxy terminal domain varies widely between the members of the *ras* family. The fourth domain comprises a characteristic motif of Cys186–A–A–X–COOH at the end of the tail ('A' representing an aliphatic amino acid). This tail is important as localization of the *ras* p21 protein depends on farnesylation of the cysteine residue at position 186 [26–28] which is then attached to the lipid bilayer [29, 30]. Post-translational modification of the *ras* protein is required prior to the critical acylation event. The first is clipping of the terminal tripeptide to expose the cysteine residue at position 186, which is associated with an increase in electrophoretic mobility of the protein. If the cysteine residue is replaced by serine, then removal of the tripeptide does not occur. Analysis of mutant *ras* p21s with deletions of residues between positions 181 and 185 has shown that these are critical to the efficient farnesylation of cysteine 186 (Hancock, J. & Marshall, C. J., personal communication). Repeated cycles of palmitylation and depalmitylation of upstream cysteine residues are thought to occur since the half-life of the incorporated palmitic acid is much shorter than that of the *ras* protein [31].

2.1 Three-dimensional structure of *ras* p21

An important recent development has been the description of the three-dimensional crystal structure of a *ras* p21 protein by Wittinghofer *et al.* [32], which proved to be very similar to a model put forward to McCormick *et al.* [33] based on sequence homology with the bacterial elongation factor EF-Tu, with a six-stranded β sheet structure bordered by four α helices with nine connecting loops (Fig. 1). X-ray crystallography of bacterially expressed Ha-*ras* protein (lacking 23 amino acids at the flexible carboxy terminus) complexed with GTP has allowed analysis of structure to a resolution of 2.7 Å. The GTP molecule lies in a pocket bounded by four loops, one of which carries amino acids 12 and 13 close to the phosphate groups of GTP and another bears amino acids 116–119 close to the guanine base. A loop running next to the phosphate-binding loop carries amino acids 59, 61 and 63. Significantly, substitution at any of these amino acid positions is capable of activating the *ras* protein to an oncogenic form.

The function of *ras* proteins is thought to involve interaction with guanine nucleotides in a manner similar to that of G proteins such as retinal transducin, G_s, G_i and G_o [34–37]. Although it has been known for some time that *ras* proteins can bind GDP and GTP and exhibit GTPase activity [38–43], it is only recently that direct evidence has been produced to support a model in which *ras* protein is activated by exchange of bound GDP for bound GTP [44–46]. The GTPase activity of purified normal p21 proteins is rather poor, and probably inadequate to maintain p21 in the

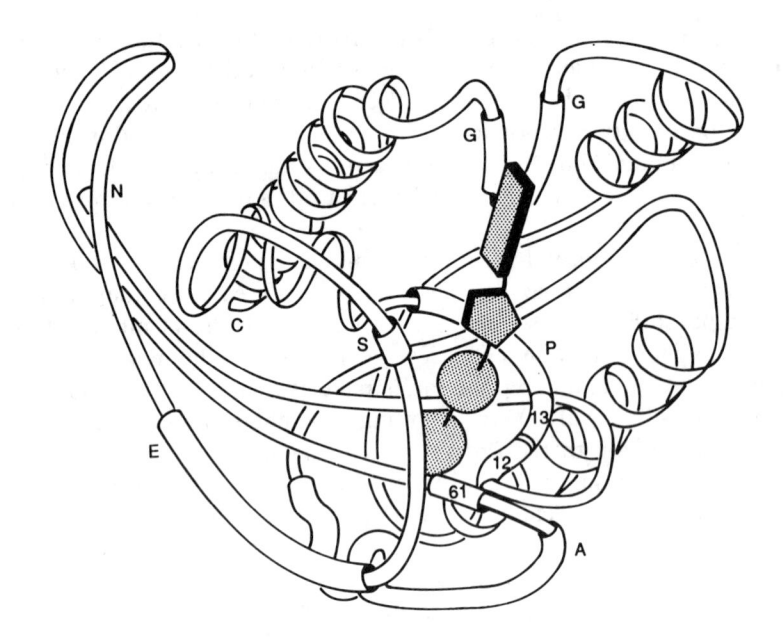

Fig. 1 — The three-dimensional backbone structure of normal Harvey *ras* p21 protein complexed to GDP, determined from X-ray analysis of crystals of the amino terminal 171 residues of this protein: N, N-terminus of protein; C, C-terminus of protein; G, region binding guanine base; S, region binding ribose sugar; P, region binding phosphates; Y13-259, region binding neutralizing antibody; E, effector region determined by *in vitro* mutagenesis. Figure based on Pai *et al.*, Tong *et al.* [32].

inactive GDP-bound form considering the likely exchange rates of guanine nucleotides from cellular pools in which GTP far exceeds GDP concentration. However, the normal p21 proteins interact with a cytoplasmic protein known as GTPase activating protein or GAP, which increases the speed of conversion of *ras*–GTP to *ras*–GDP by at least 100-fold. The isolation of this protein by McCormick and coworkers is a most exciting discovery, which has helped to fill the void that existed while the molecules upstream and downstream of *ras* p21 in the putative signal transduction pathway were unknown. In addition to its effect of down-regulating the stimulus produced by active *ras*–GTP complex by stimulating hydrolysis to GDP, GAP may also be the effector molecule for *ras* action.

2.2 *Ras* GTPase activating protein (GAP)

The 120 kDa GAP protein has been purified and cloned from both bovine brain and human placenta and the sequences were shown to be almost identical in these two species [47,48]. However, in human placenta (and a number of cell lines) a second form of GAP mRNA is produced by alternative splicing that encodes a 100 kDa protein without the highly hydrophobic amino terminal domain of the larger form. This region is not essential for catalytic activity, but it seems likely that it is involved in membrane attachment during interaction with the membrane-bound *ras* p21 protein; however, since most GAP activity in the cell types analysed appears to be present in the cytosolic fraction, this membrane association may only be transient.

In support of the effector role of GAP in *ras* signal transmission, two groups have carried out studies on the interaction of mutated *ras* proteins with cellular extracts or partially purified GAP [49,50]. Sigal et al. [51] identified amino acids 32–42 as the putative effector binding region of *ras* p21, and it was shown that most amino acid substitutions in this region abolished both transforming activity and stimulation by GAP. Only substitution of Ser39 by Ala and of Tyr40 by Phe failed to influence either of these parameters. Deletions or substitutions in regions known to be non-essential for transforming activity [52] did not reduce the stimulatory effects of GAP: deletion of the variable domain of p21, where the difference in Ha-, N- and Ki-*ras* proteins occur, did not affect GAP function. The results support a model proposed by McCormick [53] in which different *ras* proteins are associated with different upstream stimuli (specific growth factor receptors for example) and pass signals onto GAP for processing through a common downstream pathway.

Analysis of the amino acid sequence of GAP has shown close similarity of two regions to sequences in the non-receptor tyrosine kinases, phospholipase C-148 and the *crk* oncogene [47]. This finding may be related to the postulated effector action of GAP, in which it proposed that GAP transmits the received signal to another molecule (possibly a cytosolic serine kinase, as proposed by Yu et al. [54]) while associated with an active *ras*–GTP complex at the plasma membrane [53].

2.3 Downstream events from *ras* p21 in signal transduction

The use of microinjected neutralizing antibody has shown that *ras* p21 is important for response to a variety of growth factor stimuli [7], but the connections in the pathway have remained elusive. Stacey and coworkers [54,55] have used microinjection of neutralizing antibody to show that *ras* p21 activity is required for the transformation induced by oncogenes related to growth factors or their receptors (such as *fes*, *fms*, *src* and *sis*), but not for transformation induced by oncogenes related to cytsoplasmic kinases (*mos*, *raf*).

Over the last few years several groups have studied the relationship between *ras* p21 and phosphoinositide turnover which leads to the generation of 1,2-diacylglycerol (DAG) and inositol phosphates. The evidence from different systems is conflicting and this field remains highly controversial. Transformation by *ras* p21 does produce elevated levels of diacylglycerol [56–62], but has been shown to be associated with increased inositol phosphate production only under certain conditions [56–58,61]. The different results observed in different systems may be partly explained by the difficulty in separating the direct effects of *ras* p21 itself and any indirect effects it may evoke. In the search for an alternative source of DAG, several possibilities have been raised. Lacal et al. [59] found elevated steady state levels of phosphocholine and phosphoethanolamine in Ha-*ras*-transformed NIH3T3 cells, and suggested that these might be derived from increased hydrolysis of phosphatidylcholine and phosphatidylethanolamine by phospholipase C which could give rise to elevated diacylglycerol. However, Macara [63] has produced evidence that the elevated phosphocholine results instead from induction of the enzyme choline kinase, rather than phospholipase C, in fibroblasts stably transformed by *ras* oncogenes.

The complex relationship between *ras* p21 and various lipid molecules is compounded by the recent findings of Tsai *et al.* [64] who studied the effect of various lipids on the stimulation of *ras* GTPase activity by GAP *in vitro*. While diacylglycerols had no effect, GAP activity could be profoundly inhibited by phosphatidylinositol phosphates and arachidonic acid derivatives. If this occurred *in vivo*, then signals passed through *ras* p21 would be influenced by the levels of inositol phosphates and other lipids even if these were not directly involved in the signal transduction pathway.

It is possible that *ras* p21 acts via two different downstream pathways. An elegant approach to the analysis of the early direct effects of *ras* p21 introduction into quiescent cells was presented by Morris *et al.* [65] who used the scrape-loading technique to show that oncogenic p21 causes rapid activation of protein kinase C (within five minutes of introduction of protein) in the absence of any change in inositol phosphates. When cells pre-treated with TPA to abolish protein kinase C activity are scrape-loaded with oncogenic *ras* p21, they no longer show a stimulation of DNA synthesis [65], but are apparently still able to undergo characteristic morphological transformation (Lloyd, A. & Hall, A., personal communication), suggesting that this is achieved via an alternative pathway.

The effects on cellular proliferation and differentiation induced by *ras* protein are associated with changes in the level of transcription of various genes, which are likely to be controlled by specific transcription factors. The identity of the factors influenced by *ras* oncogenes is currently the subject of intense investigation and Imler *et al.* [66] have identified a *ras*-responsive element (RRE) in the polyoma enhancer A domain which appears to mediate stimulation not only by Ha-*ras* but also by phorbol ester and serum. The sequence of the RRE is strikingly similar to that recognized by the *jun*/AP-1 transcription factor, and so it may be that a final link in *ras* signal transduction to the nucleus involves this transcription factor. It is even possible that the same mechanism might control the transcription of *ras* genes themselves, since the human Ha-*ras* promoter has four motifs which resemble the consensus sequences of TPA-inducible and AP-1/*jun*-binding elements [67].

3. STRUCTURE AND DYSFUNCTION OF THE *RAS* ONCOGENES

3.1 Mutant *ras* p21

In vitro mutagenesis studies have demonstrated that it is possible to activate *ras* protein to an oncogenic form by amino acid substitutions at positions 12, 13, 56, 61, 63, 116 and 119 [51,68,69], but only positions 12, 13 and 61 have been found affected in any of the human tumours studied. Substitution of the normal position 12 glycine residue with any other amino acid except proline causes oncogenic activation [70], and a similar situation is observed for the glycine residue at position 13 although a few substitutions, such as Ser13, may not be transforming [68,71]. Both of these residues are part of the phosphoryl binding loop of the p21 protein [32], while position 61 is on the neighbouring loop. Missense mutations leading to the substitution of the usual 61 glutamine residue by any other amino acid except proline or glutamic acid produce a transforming protein. Recent work has shown that transforming substitutions can distort the normal structure with serious consequences for the functional activity of the protein. Tong et al. [72] have shown that substitution with valine at position 12 increases the radius of the phosphoryl binding loop such that two hydrogen bonds from the backbone NH groups of residues 12 and 13 to the β-phosphate are lost, with a likely change in the orientation of this phosphate that might account for the loss of GTPase activity of the mutant protein. Studies are now under way to determine the structure of different mutants including those with substitutions at position 61.

Mutant *ras* p21 proteins with activating amino acid substitutions at positions 12, 13 and 61 (and indeed with substitution of the normal alanine for threonine at position 59, as occurs in the viral Kirsten and Harvey *ras* p21s) have reduced intrinsic GTPase activity [41–43,73,74] and are unable to respond to the stimulatory effect of GAP [49,50]. This would prevent the deactivation of the active *ras*–GTP complex, and so lead to a prolonged stimulation of the signalling pathway.

Mutant *ras* proteins with substitutions at positions 116 and 119 (created by *in vitro* mutagenesis and shown to be transforming) are, in contrast, still able to hydrolyse GTP and indeed to respond to the GTPase stimulation induced by GAP [49,50]. It has been proposed that these mutations lead to a much reduced binding affinity for guanine nucleotides [51,69], so that GDP–GTP turnover is increased: since the intracellular concentration of GTP greatly exceeds that of GDP, this will favour the production of active *ras*–GTP complex.

3.2 Overexpression of *ras* p21

In vitro studies show that many of the features of transformation induced by mutant *ras* p21 can also be produced by overexpression of normal *ras* p21 [5,75–77]. Overexpression of normal *ras* p21 can result from amplification of a normal *ras* proto-oncogene, as occurs frequently in some human tumours with the *myc* gene family [78–80], and the c-*erbB*-2 proto-oncogene [81]. In human cervical cancer, *ras* gene amplification is reported to be a relatively frequent event [82] and it occurs occasionally in other tumours (Table 1). It appears to be an infrequent mechanism of activation of the *ras* gene family in most other human tumours (estimated at less than 1% of cases overall; ref. [84]).

Another possible mechanism for overexpression is a change in the regulatory sequences of a *ras* gene such as insertion of a strong promoter or enhancer element [94,95] or deletion of the first non-coding exon [96]. In human tumours, there is no evidence for any of these events occurring in an otherwise normal *ras* gene. However, it has recently been shown that a mutation in the last intron of the Ha-*ras* oncogene (which has a codon 12 mutation in the first coding exon) in the EJ/T24 human bladder carcinoma cell line produces enhanced expression of this oncogene [97]. Further studies of primary human tumour material are needed to test whether this finding reflects an artefact of long-term culture of a malignant cell line or whether mutations in this region could be involved in oncogenic transformation *in vivo*. Another possible mechanism for up-regulation was identified by Spandidos and Pintzas [98] who showed that a six base pair deletion in the first intron of the EJ/T24 Ha-*ras* gene could increase transcriptional activity. Other authors have reported both positive and negative regulatory elements in this intron [99].

Table 1 — Amplification of *ras* oncogenes in human tumours

	Reference
Ki-ras	
Ovarian carcinoma, 1 case	[83]
Ovarian carcinoma, 1 case	[84]
Ovarian carcinoma, 3/37 cases	[85]
Urinary tract cancers, 1/21 cases	[86]
Lung carcinoma, 1 case	[76]
Stomach carcinoma	
1 case	[87]
1/7 cases	[88]
Lu-65 lung giant cell line	[89]
Ha-ras	
Bladder carcinoma, 1 case	[90]
Thyroid adenoma, 1/4 cases	[91]
Thyroid cancer, 2/11 cases	[92]
N-ras	
MCF-7 breast cancer cell line	[93]

3.3 *Ras* allele loss

Deletion of one *ras* allele is observed in a significant proportion of tumour cell lines, but uncommonly in primary human tumour material [100]. Interestingly, the tumours in which *ras* allele loss has been observed are types which very rarely possess mutant *ras* oncogenes, such as breast carcinomas, ovarian carcinomas and urinary tract tumours [84,101–106]. One major exception to this rule is pancreatic adenocarcinomas, in which Ki-*ras* mutations are ubiquitous (see below) and in which loss of the normal allele is frequently observed [107]. Deletions usually affect Ha-*ras* — located on chromosome 11 which is suspected to be the location of a tumour-suppressor or anti-oncogene [108] — and it is possible that *ras* allele loss is merely incidental in these cases.

3.4 Ha-*ras* restriction fragment length polymorphism (RFLP) and predisposition to cancers

The possibility that some Ha-*ras* alleles are associated with increased risk of developing cancer was raised by a study performed by Krontiris *et al.* [102]. As a result of genetic variability in the number of tandemly repeated sequences (VTRs) downstream of the human Ha-*ras* gene, up to 20 polymorphic alleles (distinguished by RFLP of *Bam*HI- and *Taq*1-digested DNA) exist although four of these (so-called 'common' alleles) account for the majority in the population. Krontiris *et al.* [102] reported a significant increase in the frequency of 'rare' alleles in DNA extracted from leukocytes of a mixed group of cancer patients, and in patients with myelodysplastic syndrome. This stimulated several groups to analyse larger numbers of patients with particular neoplasms. Our own group [109] and Ceccherini-Nelli et al. [110] found no increased frequency of 'rare' alleles in patients with colorectal cancer, and Sutherland et al. [111], Gerhard et al. [112] and Radice et al. [113] reported negative findings for melanoma patients. Similar results are recorded in patients with lung cancer [103], urothelial cancer [114], cervical cancer [115], myelodysplasia [116] and breast cancer [103]. An excess of 'rare' alleles reported in breast cancer patients by Lidereau et al. [117] may be attributed to differences in electrophoretic mobility between tumour DNA (used in this study) and leukocyte DNA (used in other studies). The statistical significance of increased frequencies of 'rare' alleles in acute myeloid leukaemia [116,118] is largely due to the remarkably low prevalence of 'rare' alleles in the normal populations analysed by these authors. In summary, these studies have not confirmed any convincing association between 'rare' alleles and increased cancer risk. However, the data [103,109,110,113,115] do support a genuine positive association between possession of the larger two 'common' alleles in germline DNA and subsequent development of cancers. It is possible that this is related to an increase in the known enhancer action of the downstream VTR elements [102,119].

4. METHODS USED TO DETECT ACTIVATED *RAS* ONCOGENES

Until recently, the most widely used techniques for the detection of activated *ras* oncogenes were the biological assays of focus induction [1–3] and nude mouse tumorigenesis [68,120] following transfection of genomic DNA into NIH3T3 cells. These assays are very familiar and, although they have been successfully applied to a very large number of samples, both have drawbacks. Both require high quality DNA (not always available from human tumour material) to ensure transfection of intact genes: Bos et al. [121] estimate that only one-third of copies of the 45 kb Ki-*ras* gene are intact even in DNA of average size 100 kb. The assays are labour intensive and their results are produced only after a delay of some weeks. There is a risk that the process of transfection itself may produce artefactual mutations [122–124] which could activate *ras* alleles, or indeed generate artificial oncogenes by recombination of unrelated sequences (such as the *trk* oncogene, which is the result of recombination between a tyrosine kinase sequence and a tropomyosin gene [125]).

Direct analysis of specific codons for activating point mutations is a more attractive approach to the analysis of *ras* oncogenes, and recently several new techniques have been developed for this purpose. The most powerful of these is *in vitro* gene amplification by polymerase chain reaction which now allows up to 10^6-fold amplification of a target sequence [126]. In outline, DNA is denatured and allowed to reanneal in the presence of an excess of specific paired primers complementary to sequences (on opposite strands) upstream and downstream of the codon of interest; these primers are extended by incubation with a thermostable DNA polymerase and the cycle is repeated up to 50 times to generate multiple copies of the interprimer region. This technique can be successfully applied to DNA in thin sections from formalin-fixed, paraffin-embedded tissue [127], which is obviously ideal for the analysis of archival material.

The amplified product of the polymerase chain reaction applied to codons 12, 13 and 61 of the *ras* genes can be analysed by various methods. The most widely used is differential hybridization to sequence-specific oligonucleotide probes which was originally devised for *ras* genes by Bos and his colleagues [87,128,129]. By hybridization under stringent conditions of replicate dot- or slot-blots to individual oligonucleotides complementary to either the wild-type sequence or that sequence with a point mutation at a single base, the identity of mutant *ras* alleles can be determined by the higher thermal stability of fully matched hybrids.

A novel strategy developed by Kumar and Barbacid [130] allows the reliable detection of point mutations at the single-cell level. The PCR technique is used to generate multiple copies of DNA sequence that are hybridized under very stringent conditions to oligonucleotide probes in liquid phase and then run on polyacrylamide gels to demonstrate 'probe shift' (as the hemiduplex of the perfectly matched probe and target sequence will run more slowly than the free probe). Restriction fragment length polymorphisms within the amplified DNA can be detected with very high sensitivity by digestion of the PCR product before liquid phase hybridization to labelled oligonucleotide probe and separation on a polyacrylamide gel. The authors demonstrate that this RFLP assay can detect a single heterozygous mutant allele (in which a G-to-A transition in codon 12 of the rat Ha-*ras* oncogene results in the loss of a *Mn*II site) in a 10^5-fold excess of normal cells, compared with the maximum sensitivity of the standard hybridization technique of 1 in 10^3 cells. This will obviously facilitate the study of microscopic human neoplastic lesions (even those comprising a singe cell in a background of normal cells), and might provide for the highly sensitive detection of *ras*-positive relapses. Where no naturally occurring restriction site exists (such as the *Msp*I and *Nae*I sites which are lost in G-to-T transversion at Ha-*ras* codon 12, and the *Sac*I site created by G-to-C transversion at the first G of Ki-*ras* codon 12), in some cases it is possible to engineer novel RFLPs in the human *ras* genes by combining a mismatched sequence in the PCR primer with a mutated codon [239].

Another method to detect *ras* mutations, which can be applied to cellular RNA or PCR-amplified DNA, is RNAase A mismatch cleavage analysis developed by Perucho and coworkers [131,132]. The method depends on the ability of RNAase A to recognize and cleave single-base mismatches in RNA:RNA and RNA:DNA heteroduplexes. Total cellular RNA [133] or PCR-amplified genomic DNA [134] is hybridized to labelled antisense RNA, and then digested with RNAase A to yield fragments of the labelled RNA probe (whose size is determined by the position of mismatches) that can be resolved by electrophoresis. The position of the mutation (but not its identity) can therefore be defined; another drawback is that some mismatches (such as the pair G–U) are refractory to RNAase A digestion. However, the technique has an advantage over the alternative procedure of oligonucleotide probing because it allows a quantitative estimate to be made of the relative levels of expression of wild-type and normal alleles.

Mutations of *ras* genes can also be detected by analysis of the p21 proteins by immunoprecipitation followed by high resolution 2-D gel electrophoresis [135]. Early studies had shown that certain amino acid substitutions could cause changes in the electrophoretic mobility of *ras* p21 in 1-D SDS–polyacrylamide gels [68, 136–138], but this was a relatively crude tool. The high resolution technique permits the accurate distinction of individual mutant Ha- and N-*ras* proteins (but not Ki-*ras* proteins since these are more basic and do not focus well in this system), and allows estimation of the relative amounts of wild-type and mutant proteins present.

5. FREQUENCY OF *RAS* ONCOGENE ACTIVATION IN TUMOUR MATERIAL

The positive reports of activated *ras* oncogenes by many authors using these techniques are summarized in Tables 2 and 3. Although activated *ras* alleles have been detected in many human tumour cell lines (Table 2), because many genetic changes may develop *in vitro* it cannot be assumed that these were present in the parent tumours from which the lines were derived (as emphasized in ref. [13]). Table 4 shows the amino acid substitutions resulting from single point mutations reported at positions 12, 13 and 61 of each *ras* oncogene. The only report so far of an amino acid substitution resulting from a double mutation is that in the H-466B breast carcinoma line where glycine is replaced by phenylalanine because of two adjacent point mutations at Ki-*ras* codon 12 [208].

The studies of clinical material presented in table 3 merit some discussion here. The development of rapid diagnostic techniques easily applicable to archival material (outlined above) has led to accelerated accumulation of data on the prevalence of *ras* mutations in a range of human tumours, and will allow retrospective study of less common tumour types.

The highest prevalence of activated *ras* oncogenes in a human tumour type so far recorded is in carcinoma of the exocrine pancreas, where more than 95% were found to have mutations at codon 12 of Ki-*ras* [134]. Unpublished work by the same authors (Perucho, M., personal communication) shows that it is feasible to detect such mutations by RNAase A mismatch cleavage analysis after PCR amplification of DNA in needle biopsies, and that no false positives occurred (no *ras* mutations were found in benign pathological lesions). If this specificity for the identification of malignant cells can be validated by further studies, then it might be a very useful aid in the differential diagnosis of pancreatic cancer which is often problematic, particularly in fine needle specimens containing minute amounts of tissue.

A high prevalence of *ras* oncogene mutations, around 40%, has also been demonstrated in colonic adenocarcinomas [121,133], the majority affecting codon 12 of Ki-*ras*. In some cases, two different *ras* oncogenes were activated in the same tumour. More recently, similar mutations have been found in colonic adenomas [142]: while 58% of large adenomas (more than 1 cm in diameter) possessed mutated *ras* alleles, *ras* mutations were found in only 9% of adenomas smaller than 1 cm. This suggests that activation of *ras* oncogenes is not an initiating event for the development of this particular neoplasm, although it may occur relatively early, before progression to malignancy.

We have identified activated *ras* oncogenes at high prevalence (about 50%) in both benign and malignant tumours of the thyroid gland [146,147]. Neoplasia of the thyroid follicular epithelium may develop along two alternative pathways (each with a low risk of progression):

Table 2 — *Ras* oncogenes in human cell lines

		Assay used	Reference
Ki-ras			
Urinary tract neoplasms	A1698	FI	[140]
	A1163	SSO	[202]
	639 V	DMTA	[198]
Lung	LX-1	FI	[193]
	A2182	FI	[140]
	A427	FI	[140]
	Calu-1	FI	[24]
	SK-LU-1	FI	[203]
	PR310	FI	[25]
	PR317	FI	[25]
	LU-65	FI	[89]
	(Ki-*ras* amplification also present)		
	LU-12-JCK	FI	[204]
	A549	SSO	[202]
	A427	SSO	[202]
	Calu-4	DMTA	[198]
Colon	SW480	FI	[205]
	SK-CO-1	FI	[203]
	A2233	FI	[140]
	KMS-4	FI	[206]
	SW620	FI, SSO	[87]
	SW1116	FI, SSO	[87]
	SW1398	FI, SSO	[87]
	PC/AA	PCR–SSO	[143]
	PC/JW	PCR–SSO	[143]
	S/AN	PCR–SSO	[143]
Breast	MDA-MB231	NMT, PCR–SSO	[207]
	H-466B	FI	[208]
Pancreas	T3M-4	FI	[210]
	All65/Panc 1	NMT	[209]
	All65	SSO	[202]
Gall bladder	A1604	FI	[140]
Osteosarcoma	OHA	FI	[211]
AML	Rc2a	NMT, PCR–SSO	[180]
	(concurrent N12+K12 mutations)		
Intermediate T leukemia	CCRF-CEM	FI	[175]
N-ras			
Lung	SW-1271	FI	[212]
Colon	7060	FI	[213]
Neuroblastoma	SK-N-SH	FI	[24]
		FI	[203]
Fibrosarcoma	HT-1080	FI	[22]
		FI	[139]
	SHAC	FI, SEQ	[214]
Rhabdomyosarcoma	RD301	FI	[22]
Melanomas	XP44RO (mel)	NMT, PCR–SSO	[216]
	SK-Mel-93	FI	[201]
	SK-Mel-119	FI	[201]
	SK-Mel-147	FI	[201]
	Mel-Swift	FI	[215]
Teratocarcinoma	PA-1	FI	[217]
Acute promyelocytic leukaemia	HL-60	FI	[174]
CML	1M9	FI	[175]
AML	Rc2a	NMT, PCR, SSO	[180]
	(concurrent N12+K12 mutations)		
	KG-1	FI	[177]
	THP-1	NMT, SSO	[168]
	KG-1	NMT, SSO	[168]

(*Continued next page*)

Table 2 (*Continued*)

		Assay used	Reference
ALL	RPMI 8402	FI	[218]
	T-ALL-1	FI	[218]
	p-12	FI	[218]
	MOLT-3	FI	[175]
	MOLT-4	FI	[175]
Burkitt's lymphoma	AW Ramos	FI	[174]
H-ras			
Urinary tract neoplasms	EJ/T24	FI	[1]
		FI	[2]
		FI	[3]
		FI	[4]
		FI	[193]
		FI	[194]
		FI	[195]
Breast	HS578T	FI	[158]
	SK-BR-3	AI	[196]
Lung	HS242	FI	[197]
	KNS 62	DMTA	[198]
Stomach	BGC-823	FI	[199]
Melanoma	SK2	FI	[200]
	SK-MEL-146	FI	[201]
Neuroblastoma	LA-N-1	AI	[196]
Choriocarcinoma	GCC-SV	NMT	

Abbreviations: FI, focus induction assay; NMT, nude mouse tumorigenesis assay; PCR, polymerase chain reaction amplification; SSO, sequence-specific oligonucleotide probing; WB, Western blot; *Msp*I/*Hpa*II, diagnosis, of point mutation by loss of restriction site for these enzymes; RNAseMMC, RNAse A mismatch cleavage analysis; SEQ, nucleotide sequencing; Dirseq, direct sequencing; MDS, myelodysplastic syndromes; MPS, myeloproliferative syndromes; AML, acute myeloid leukaemia; ALL, acute lymphoblastic leukaemia; CML, chronic myeloid leukaemia; DMTA, defined medium transformation assay.

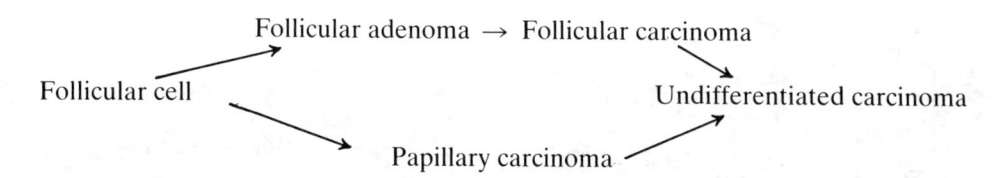

```
                    Follicular adenoma → Follicular carcinoma
                   ↗                                          ↘
    Follicular cell                                             Undifferentiated carcinoma
                   ↘                                          ↗
                    Papillary carcinoma
```

Analysis by transfection assays and by oligonucleotide hybridization assay shows that *ras* mutations occur in 50% of cellular (micro-) follicular adenomas and in 53% of follicular carcinomas. Activated alleles of all three members of the *ras* family (Ha-, Ki- and N-*ras*) have been found in these tumour types, in contrast to any other human solid tumour type. The predominant mutation was A-to-G transition at codon 61 of Ha- or N-*ras* leading to substitution of the normal glutamine by aspartic acid.

Activated *ras* oncogenes are found in a similar proportion (60%) of undifferentiated thyroid cancers, but in this group codon 12 mutations, particularly transversions, predominated. This suggests that the undifferentiated cancers develop from an uncommon subgroup of differentiated follicular tumours, in which codon 12 mutations are much less frequent. Only 20% of papillary thyroid cancers possess mutant *ras* alleles, and we and others have shown that non-*ras* dominant transform-

Table 3 — Activated ras oncogenes in human tumours

Tumour type	Number	Number of mutant alleles N	Ki	Ha	Percentage of cases with mutant alleles	Assay used	Reference
Colon cancers	1	1	—	—		FI	[139]
	2	—	2	—		FI	[140]
	27	—	10	—	37%	PCR–SSO	[121]
	66	—	26	—	39%	RNAaseMMC	[133]
	30	1	5	—	20%	FI	[133]
	16	—	4	—	25%	SSO	[141]
	92	—	38	—	47%	PCR/SSO	[142]
Colon adenomas	80	5	24	—	31%	PCR–SSO	[142]
	75	1	5	—	7%	PCR–SSO	[143]
	3	—	2	—	66%	RNAaseMMC	[144]
Thyroid cancers	1	—	—	1		FI	[145]
	6	1	2	1	67%	FI	[91]
	15	2	2	2	40%	NMT, PCR–SSO	[146]
	20	2	2	6	50%	NMT, PCR–SSO	[147]
Thyroid adenomas							
Microfollicular	16	4	1	3	50%	NMT, PCR–SSO	[147]
Macrofollicular	8	—	—	—	0%	NMT, PCR–SSO	[147]
Pancreatic cancers	2	—	2	—	95%	FI	[4]
	22	—	21	—		RNAaseMMC	[134]
	30	—	28	—	93%	PCR–SSO	[107]
Lung cancers	5	—	1	—	20%	FI	[100,140]
	19	—	2	—	11%	FI	[148]
	39	—	5	—	13%	PCR–SSO	[149]
	27	—	2	—	7%	SSO	[141]
	43	—	9	1	23%	PCR–SSO	[150]
Skin squamous cancers	7	—	—	1	14%	NMT	[151]
Xeroderma pigmentosum cancers	2	—	—	2		NMT, PCR–SSO	[152]
Keratoacanthoma	10	—	—	1	10%	NMT, SSO	[153]
	13	1	—	—	7%	NMT	[154]
Melanomas	37	7	—	—	19%	PCR–SSO	(Bos, J. L., personal communication)
	45	2	11	6	38%	PCR–SSO	(Hughes, D., personal communication)

Urothelial cancers	38	—	—	2	5%	FI	[86]
	15	—	—	1	7%	FI	[155]
	7	—	—	1	14%	WB	[88]
	24	—	—	4	17%	PCR–SSO	[156]
Renal cancers	16	—	—	2	13%	FI	[105]
Prostate cancers	8	—	1	—	13%	FI	[157]
Breast cancers	16	—	—	—	0%	FI	[158]
	34	—	—	—	0%	PCR–SSO	(Bos, J. L., personal communication)
Stomach cancers	24	—	—	2	8%	SSO	[159]
	8	—	1	—	13%	SSO	[141]
	49	—	4	—	8%	PCR–SSO	[160]
	1	1	1	—		FI, SSO	[87]
	1	1	—	—		NMT	[161]
	1	—	—	—		NMT, SSO	[162]
	26	—	—	—	0%	FI	[163]
	7	—	—	—	0%	FI	[164]
Ovarian cancers	5	—	1	—	20%	FI	[165]
	37	—	—	—	0%	PCR–SSO	[85]
Cervical cancers	30	—	—	—	0%	PCR–SSO	[13]
	76	—	—	8	11%	MspI/HpaII	[115]
Liver cancers	1	1	—	—		FI	[139]
	10	3	—	—	30%	FI	[166]
Gall bladder cancers	5	—	—	—	0%	RNAase MMC	[134]
Neuroblastoma	25	—	—	—	0%	PCR–SSO	[167]
Seminoma	1	—	1	—		FI	[148]
Rhabdomyosarcoma	2	—	1	—		FI	[140]
Glioblastoma	30	—	—	—		PCR–SSO	[13]
MPS	19	1	—	—	5%	PCR–SSO	[168]

MDS	8	3	—	—	38%	NMT	[169]
	1	—	1	—		NMT, PCR–SSO	[170]
	50	14	6	2	44%	NMT, PCR–SSO	[171]
	15	1	—	—	7%	PCR–SSO	[168]
	2	—	—	1		NMT, PCR–SSO	[172]
	34	1	2	—	9%	PCR–SSO	[173]
AML	1	—	—	—		FI	[174]
	2	1	—	—		FI	[175]
	1	—	—	—		FI	[176]
	5	4	—	—	80%	NMT	[71]
	3	—	—	—		FI	[177]
	6	3	2	—	50%	FI	[178]
	37	5	1	—	19%	PCR–SSO	[179]
	2	1	1	—		NMT, PCR–SSO	[180]
	6	4	—	—	83%	NMT	[181]
	3	3	—	—		NMT, SSO	[182]
	9	4	—	—		NMT, SSO	[168]
	18	5	—	1	28%	SSO	[183]
	9	2	—	—	22%	PCR–SSO	[184]
	52	14	—	—	19%	PCR–SSO	[185]
CML	2	—	1	—	10%	FI	[177]
	10	1	—	—	0%	FI	[186]
	26	—	—	—	33%	PCR–SSO	[168]
	12	1	—	3	0%	NMT, SSO	[187]
	25	—	—	—		PCR–SSO	[13]
	1	1	—	—	17%	PCR–DirSeq	[188]
	6	—	—	1		NMT, PCR–SSO	[172]
ALL	19	2	—	—	5%	PCR–SSO	[189]
	8	1	—	—	13%	FI	[186]
	4	4	—	—		NMT, SSO	[182]
	33	6	—	—	18%	PCR–SSO	[190]
Hodgkin's lymphoma	3	2	—	—		FI	[191]
NHL	1	1	—	—		NMT, SSO	[192]
	88	—	—	—	0%	PCR–SSO	[190]

Table 4 — Point mutations reported in human tumour material

Ki-ras 12 (GGT=glycine)

AGT serine	CGT arginine	TGT cysteine	GAT aspartic acid	GCT alanine	GTT valine
Colonic tumours [121, 141, 142]	Lung adenocarcinoma [100]	Colonic tumours [121, 141–143] Lung adenocarcinoma [149, 150]	Colonic tumours [121, 141–143] Lung adenocarcinoma [149, 150]	Colonic tumours [142] Lung adenocarcinoma [141, 150]	Colonic tumours [121, 142, 143] Lung adenocarcinoma [141]
Thyroid tumours [146]		Pancreas cancer [107]		Breast cancer [141]	Pancreas cancer [107] Breast cancer [160]
Stomach cancer [87]	Pancreas cancer [107] Hodgkin's disease [238] MDS [171]		MDS [171] AML [173, 179] Colon cancer line PC/JW [143]		AML [173, 179] Colon cancer line SW480, SW620, SW1116 SW1398, PC/AA [20, 87, 129, 143]
	Colon cancer line S/AN [143]	Colon cancer line KMS4 [206]			Breast cancer line PE6001 [160] Osteosarcoma line OHA [211]
Lung cancer line A549 [202]	Lung cancer line A2182 [100] Bladder cancer line A1698 [100]	Lung cancer line PR371 [25]; LU65 [89]	Lung cancer line A427 [202] Bladder cancer line A1163 [202] Pancreas cancer line Panc1 [161]		

Ki-ras 13 (GGC=glycine)

AGC serine	CGC arginine	TGC cysteine	GAC aspartic acid	GCC alanine	GTC valine
			Colonic tumours [142] Thyroid tumours [146] Breast cancer [160] MDS [170] AML [170] Breast cancer line MDA.MB231 [160, 207]		

N-ras 12 (GGT=glycine)

AGT serine	CGT arginine	TGT cysteine	GAT aspartic acid	GCT alanine	GTT valine
AML [185] MDS [168] CML [187] ALL [190]		AML [185] CML [186] ALL [186] Colonic tumours [121, 142] Thyroid tumours [147] Leukaemia line MOLT4 [129]	AML [168, 183, 185] MDS [171] MPS [168] ALL [190] Thyroid tumours [147] Leukaemia line THP-1 [168] Teratocarcinoma line PA-1 [217]	AML [185] MDS [171] ALL [190]	AML [168] MDS [171] Leukaemia line Rc2a [180]; KG1 [168]

N-ras 13 (GGT=glycine)

AGT serine	CGT arginine	TGT cysteine	GAT aspartic acid	GCT alanine	GTT valine
	AML [181] MDS [69] Colonic tumours [142] Stomach cancer [162]	AML [219] NHL [192]	AML [71, 179, 185] ALL [190] Colonic tumours [142]	MDS [171]	AML [71, 179]

Ha-ras 12 (GGC=glycine)

AGC serine	CGC arginine	TGC cysteine	GAC aspartic acid	GCC alanine	GTC valine
		Thyroid tumours [147]	Thyroid tumours [147] Breast cancer line HS578 [158]		Thyroid tumours [147] Lung adenocarcinoma [150] Breast cancer [119] AML [168] MDS [171, 172] CML blast crisis [172] Bladder cancer line EJ/T24 [20, 21, 136] Stomach cancer line BGC823 [199]

Ras oncogenes in human cancers

Ha-ras 13 (GGT=glycine)

AGT serine	CGT arginine	TGT cysteine	GAT aspartic acid	GCT alanine	GTT valine
		Bladder cancer [156]			

Ki-ras 61 (CAA=glutamine)

AAA lysine	GAA glutamic acid	CCA proline	CGA arginine	CTA leucine	CAC, CAT histidine
					Colonic tumours [121, 142] AML [168] Lung cancer line PR310 [15, 25] Pancreas cancer line T3M-4 [210]

N-ras 61 (CAA=glutamine)

AAA lysine	GAA glutamic acid	CCA proline	CGA arginine	CTA leucine	CAC, CAT histidine
AML [183, 185] Melanoma line MelSwift [215] Neuroblastoma line SK-N-SH [220] Fibrosarcoma line HT1080 [129, 221]			Thyroid tumours [146, 147] AML [168, 179] Bladder cancer [87] Stomach cancer [161] Melanoma line [154] Colonic cancer line SW1271 [212] Bladder cancer line HT1197 [128]	Thyroid tumours [147] Lung adenocarcinoma [150] CML line HL60 [129]	MDS [171] Xeroderma pigmentosum cancer [152] Melanoma line XP44RO(mel) [216]

Ha-ras 61 (CAG=glutamine)

AAG lysine	GAG glutamic acid	CCG proline	CGG arginine	CTG leucine	CAC, CAT histidine
Thyroid tumours [147]			Thyroid tumours [146, 147] Urothelial cancer [86] MDS [171]	Thyroid tumours [147] Urothelial cancer [86] Keratoacanthoma [153] Lung cancer line HS242 [197] Melanoma line SK2 [200]	

ing genes, such as the PTC oncogene [222], are more often involved in this tumour type.

The data from studies of *ras* activation in lung cancers have been rather surprising. As shown in Table 3, the overall prevalence of *ras* mutations in all types of lung cancer combined is relatively low, but Rodenhuis *et al.* [149, 150] have shown that the mutations are confined to one histological type, namely adenocarcinomas. The data also supported a strong correlation between smoking and the presence of *ras* mutations [150]. It is odd that smoking should apparently produce *ras* mutations in adenocarcinomas specifically, since the evidence indicting smoking as an aetiological agent for this histological type is much less strong than for squamous cell or small cell types [223]. However, it is intriguing that smoking is also a risk factor for carcinoma in the pancreas [224], in which *ras* oncogenes are activated at very high frequency.

Amongst haemopoietic malignancies, acute myeloid leukaemia (AML) shows the highest frequency of activated *ras* oncogenes (around 30%), with N-*ras* codon 12 as the favoured site of mutation (Tables 3 and 4). Furthermore, a similar prevalence of *ras* mutations has been found in patients with myelodysplastic syndromes that may progress to AML (Table 3). This implicates *ras* oncogene activation at an early stage of neoplastic development as for colonic and thyroid tumours mentioned above. Many of the mutations are G-to-A transitions, which has led to speculation that a particular carcinogen, such as one of the alkylating agents which are particularly associated with such base changes, may be involved in initiation of leukaemogenesis. However, a study that directly addresses this possibility found only one *ras* mutation (which has a G-to-T transversion) in nine cases of AML, and in none of four cases of myelodysplastic syndrome, developing after treatment with alkylating agents for other malignancies [219]. Follow-up of patients who relapse after treatment of *ras*-positive AML has revealed that in some cases the relapse leukaemic DNA lacks *ras* mutations [185, 225]. This suggests that activation of *ras* oncogenes is not the initiating event in leukaemia, and the fact that some presentation cases possess several different *ras* mutations indicates that multiple subclones may evolve in parallel.

Some tumours possess activated *ras* oncogenes in only a small minority of cases, or not at all. Examples include ovarian carcinomas, breast carcinomas, gastric carcinomas, and oesophageal carcinomas [226]. It is interesting to speculate on the alternative genetic events that produce initiation and progression in these tumour types, and the reasons for the difference between the groups showing high and low prevalence of *ras* oncogenes.

It has been argued on the basis of results from carcinogen-induced animal tumours (see ref. [14] for review) that *ras* oncogene activation is involved in tumour initiation, since mutations are present in the earliest lesions detectable. However, the mere presence of the mutations at this time (and presumably as a result of interaction with the carcinogen only at the time of its administration) does not necessarily imply that they alone are responsible for the initiation of tumorigenesis. Other genetic events might ordinarily occur before activation of *ras* oncogenes in tumours developing spontaneously *in vivo*, but could be produced simultaneously by the carcinogen, thereby telescoping the usual programme of molecular events. The experimental models of tumorigenesis have been very rewarding for the study of *ras* oncogene activation, but it might be misleading to extrapolate these directly to human neoplasia.

6. FUTURE PROSPECTS

In order to investigate the exact role of *ras* oncogenes in human tumorigenesis, it will be necessary to reconstruct the combination of molecular events which initiate and contribute to the progression of individual human neoplasms. This combination is likely to differ between various systems, and in some cases may not involve the *ras* family at all. *In vitro* gene transfer studies, mainly using rodent cells, have contributed greatly to our understanding of the biological and biochemical properties of *ras* genes, and the more recent development of human cell culture systems suitable for gene transfer by transfection of retroviral infection will extend this. One of the major frustrations in understanding *ras* function is the lack of evidence to link any particular growth factor receptor directly to *ras* p21 in a signal transduction pathway. In NIH3T3 fibroblasts expressing an inducible N-*ras* proto-oncogene, bombesin was able to stimulate a rise in inositol phosphates (not associated with change in receptor number or affinity) while EGF and PDGF could not [227]. Parries *et al.* [228] found that Ki-MuSV-transformed NIH3T3 cells showed an elevated level of bradykinin-stimulated phosphoinositide metabolism which was associated with an increase in high affinity binding of this growth factor (there were also increased sensitivities to bombesin and IGF-I). We have been investigating the action of *ras* oncogenes in human primary thyroid follicular cells and in cell lines, and their interaction with other oncogenes and growth factors, using such techniques. One particular area of interest is the relationship of *ras* p21 to the action of insulin-like growth factor I (IGF-I). Our group has shown that benign thyroid follicular tumours (which frequently possess activated *ras* oncogenes) are often capable of autocrine production of IGF-I so that they are independent of this growth factor for growth *in vitro* [229]. This is particularly interesting in view of the evidence implicating *ras* p21 as specific mediator of insulin action in *Xenopus* oocytes [230], and data in mammalian cells that suggest that GDP-bound *ras* p21 may influence the tyrosine kinase activity of the insulin receptor by inhibition of receptor autophosphorylation [231]. Preliminary data suggests that *ras* p21 is involved in the miotogenic response to IGF-I in Rat-1 fibroblasts [240].

The *ras* proto-oncogenes belong to a larger supergene family, which includes in mammals the R-*ras* [232], *ral* [233], *rap*1 and *rap*2 [234] and BRL-*ras* [235] genes. In common with the *ras* genes, the function of these related sequences is not yet known but very recently Kitayama *et al.* [236] have published the fascinating discovery that a *ras*-related gene is able to suppress the transformed phenotype in cells transformed by v-Ki-*ras*. This gene, which they term *Krev*-1, has a predicted amino acid sequence identical to that of the independently isolated *rap*2 human cDNA. The mechanism by which it induces reversion is unknown, but the fact that it retains the sequence of the so-called *ras* effector domain suggests that it could compete for the binding site of the effector molecule of *ras* p21 (probably GAP; see above). However the elegant experiments of Paterson *et al.* [237] showed that the transformed phenotype of the human fibrosarcoma line HT1080 (which is heterozygous for an N61 lysine mutation) is dependent on the absolute level of expression of the mutant *ras* p21, and was not influenced by the level of normal ras p21 expression (i.e. the ratio of mutant to normal p21 appeared to be unimportant). This implies that the *Krev*-1 product is unlikely to act by a mechanism other than simple 'dilution' of mutant *ras* p21, perhaps by transduction of signals in an inhibitory pathway. The possible role of this gene, and indeed those of the other *ras*-related sequences, is an intriguing problem.

ACKNOWLEDGEMENTS

I am grateful to Drs Helen Hurst and Bill Gullick for their constructive criticism of this manuscript. I thank Dr David Wynford-Thomas and Fiona Wyllie for their collaboration and helpful comments. The studies were supported by grants from the Cancer Research Campaign of Great Britain and the Welsh Office.

REFERENCES

[1] Shih, C., Padhy, L. C., Murray, M. & Weinberg, R. A. (1981) Tranmsforming genes of carcinomas and neuroblastomas introduced into mouse fibroblasts. *Nature (London)* **290** 261–264

[2] Krontiris, T. G. & Cooper, G. M. (1981). Transforming activity in human tumor DNAs. *Proc. Natl. Acad. Sci. USA* **78** 1181–1184

[3] Perucho, M., Goldfarb, M., Shimuzu, K., Lama, C., Fogh, J. & Wigler, M. (1981). Human tumor-derived cell lines contain common and different transforming genes. *Cell* **27** 467–476

[4] Pulciani, S., Santos, E., Lauver, A. V., Long, L. K., Robbins, K. C. & Barbacid, M. (1982). Oncogenes in human tumor cell lines: molecular cloning of a transforming gene from human bladder carcinoma cells. *Proc. Natl. Acad. Sci. USA* **79** 2845–2849

[5] DeFeo, D., Gonda, M. A., Young, H. A., Chang, E. H., Lowy, D. R., Scolnick, E. M. & Ellis, R. W. (1981). Analysis of two divergent rat genomic clones homologous to the transforming gene of Harvey murine sarcoma virus. *Proc. Natl. Acad. Sci. USA* **78** 3328–3332

[6] Ellis, R. W., DeFeo, D., Shih, T. Y., Gonda, M. A. & Young, H. A. (1981). The p21 *src* genes of Harvey and Kirsten sarcoma viruses originate from divergent members of a family of normal vertebrate genes. *Nature (London)* **292** 506–511

[7] Mulcahy, L. S., Smith M. R. & Stacey D. W. (1985). Requirement for *ras* proto-oncogene function during serum-stimulated growth of NIH 3T3 cells. *Nature (London)* **313** 241–243

[8] Noda, M., Ko, M., Ogura, A., Liu, D. G., Amano, T., Takano, T. & Ikawa, Y. (1985). Sarcoma viruses carrying *ras* oncogenes induce differentiation-associated properties in a neuronal cell line. *Nature (London)* **318** 73–75

[9] Bar-Sagi, D. & Feramisco, J. R. (1985). Microinjection of the *ras* oncogene protein into PC12 cells induces morphological differentiation. *Cell* **42** 841–848

[10] Guerrero, I., Wong, H., Pellicer, A.,.& Burstein, D. (1986). Activated N-*ras* gene induces neuronal differentiation of PC12 rat pheochromocytoma cells. *J. Cell Physiol.* **129** 71–76

[11] Barbacid, M. (1987). *Ras* genes. *Ann. Rev. Biochem.* **56** 779–827

[12] Guerrero, I. & Pellicer, A. (1987). Mutational activation of oncogenes in animal model systems of carcinogenesis. *Mutat. Res.* **185** 293–308

[13] Bos, J. L. (1988) The *ras* gene family and human carcinogenesis. *Mutat. Res.* **195** 255–271

[14] Balmain, A. & Brown, K. (1988). Oncogene activation in chemical carcinogenesis. *Adv. Cancer Res.* **51** 147–182

[15] Yamamoto, F. & Perucho, M. (1988). Characterization of the human c-K-*ras* gene promoter. *Oncogene Res.* **3** 125–138

[16] Ishii, S., Merlino, G. T. & Pastan, I. (1985). Promoter region of the human Harvey *ras* protooncogene: similarity to the EGF receptor protooncogene promoter. *Science* **230** 1378–1381

[17] Ishii, S., Kadonga, J., Tjian, R., Brady, J. N., Merlino, G. & Pastan, I. (1986). Binding of Sp1 transcription factor by the human Harvey *ras* 1 protooncogene promoter. *Science* **232** 1410–1413

[18] Goldfarb, M., Shimuzu, K., Perucho, M. & Wigler, M. (1982). Isolation and preliminary characterization of a human transforming gene from T24 bladder carcinoma cells. *Nature (London)* **296** 404–409

[19] Shih, C. & Weinberg, R. A. (1982). Isolation of transforming sequence from a human bladder carcinoma cell line. *Cell* **29** 161–169

[20] Capon, D. J., Chen, E. Y., Levinson, A. D., Seeburg, P. H. & Goeddel, D. V. (1983). Complete nucleotide sequences of the T24 human bladder carcinoma oncogene and its normal homologue. *Nature (London)* **302** 33–37

[21] Reddy, E. P. (1983). Nucleotide sequence analysis of the T24 human bladder carcinoma oncogene. *Science* **220** 1061–1063

[22] Hall, A., Marshall, C. J., Spurr, N. & Weiss, R. A. (1983). The transforming gene in two human sarcoma lines is a new member of the *ras* family located on chromosome one. *Nature (London)* **304** 135–139

[23] McGrath, J. P., Capon, D. J., Smith, D. H., Chen, E. Y., Seeburg, P. H., Goeddel, D. V. & Levinson, A. D. (1983). Structure and organization of the human Ki-*ras* poto-oncogene and a related processed pseudogene. *Nature (London)* **304** 501–506

[24] Shimuzu, K., Birnbaum, D., Ruley, M. A., Fasano, O., Suard, Y., Edlund, L., Taparowsky, E., Goldfarb, M. & Wigler, M. (1983). Structure of the Ki-*ras* gene of the human lung carcinoma cell line Calu-1. *Nature (London)* **304** 497–500

[25] Nakano, H., Yamamoto, F., Neville, C., Evans, D., Mizuno, T. & Perucho, M. (1984). Isolation of transforming sequences of two human lung carcinomas: structural and functional analysis of the activated c-K-*ras* oncogenes. *Proc. Natl. Acad. Sci. USA* **81** 71–75

[26] Sefton, B. M., Trowbridge, I. S., Cooper, J. A. & Scolnick, E. M. (1982). The transforming proteins of Rous sarcoma virus, Harvey sarcoma virus and Abelson virus contain tightly bound lipid. *Cell* **31** 465–474

[27] Chen, Z. Q., Ulsh, L. S., DuBois, G. & Shih, T. Y. (1985). Posttranslational processing of p21 *ras* proteins involves palmitylation of the C-terminal tetrapeptide containing cysteine-186. *J. Virol.* **56** 607–612

[28] Buss, J. E. & Sefton, B. M. (1986). Direct identification of palmitic acid as a lipid attached to p21 *ras. Mol. Cell Biol.* **6** 116–122

[29] Willingham, M. C., Pastan, I., Shih, T. Y. & Scolnick, E. M. (1980). Localisation of the src gene product of the Harvey strain of MSV to plasma membrane of transformed cells by electron microscopic immunocytochemistry. *Cell* **19** 1005–1014

[30] Willumsen, B. M., Christensen, A., Hubbert, N. L., Papageorge, A. G. & Lowly, D. R. (1984). The p21 ras C-terminus is required for transformation and membrane association. *Nature (London)* **310** 583–586

[31] Magee, A. I., Gutierrez, L., McKay, I. A., Marshall, C. J. & Hall, A. (1987). Dynamic fatty acylation of p21 N-*ras*. *EMBO J.* **6** 3353–3357

[32] (a) Pai, E. F., Kabsch, W., Krengel, U., Holmes, K. C., John, J., Wittinghofer, A. (1989). Structure of the guanine-nucleotide-binding domain of the Ha-*ras* oncogene product p21 in the triphosphate conformation. *Nature* **341** 209–214
(b) Tong, L., Milburn, M. V., De Vos, A. M., Kim, S. H. (1989). *Science* **245** 244

[33] McCormick, F., Clark, B. F., la Cour, T. F., Kjeldgaard, M., Norskov-Lauritsen, L. & Nyborg, J. (1985). A model for the tertiary structure of p21, the product of the *ras* oncogene. *Science* **230** 78–82

[34] Hurley, J. B., Simon, M. I., Teplow, D. B., Robishaw, J. D. & Gilman, A. G. (1984). Homologies between signal transducing G proteins and *ras* gene products. *Science* **226** 860–862

[35] Tanabe, T., Nukada, T., Nishikawa, Y., Sugimoto, K., Suzuki, H., Takahashi, H., Noda, M., Haga, T., Ichiyama, A., Kangawa, K., Minamino, N., Matsuo, H. & Numa, S., (1985). Primary structure of the alpha-subunit of transducin and its relationship to *ras* proteins. *Nature (London)* **315** 242–245

[36] Lochrie, M. A., Hurley, J. B. & Simon, M. I. (1985). Sequence of the alpha subunit of photoreceptor G protein: homologies between transducin, *ras*, and elongation factors. *Science* **228** 96–99

[37] Itoh, H., Kozasa, T., Nagata, S., Nakanura, S., Katada, T., Ui, M., Iwai, S., Ohtsuka, E., Kawasaki, H., Suzuki, K. & Kaziro, Y. (1986). Molecular cloning and sequence determination of cDNAs for alpha subunits of the guanine nucleotide-binding proteins Gs, Gi and Go from rat brain. *Proc. Natl. Acad. Sci. USA* **83** 3776–3780

[38] Scolnick, E. M., Papageorge, A. G. & Shih, T. Y. (1979). Guanine nucleotide binding activity as an assay for the src protein of rat-derived murine sarcoma viruses. *Proc. Natl. Acad. Sci. USA* **76** 5355–5359

[39] Shih, T. Y., Papageorge, A. G., Stokes, P. E., Weeks, M. O. & Scolnick, E. M. (1980). Guanine nucleotide-binding and autophosphorylation activities associated with p21 src protein of Harvey murine sarcoma virus. *Nature (London)* **287** 686–691

[40] Tamanoi, F., Walsh, M., Kataoka, T. & Wigler, M. (1984). A product of yeast RAS2 gene is a guanine nucleotide binding protein. *Proc. Natl. Acad. Sci. USA* **81** 6924–6928

[41] Temeles, G. L., Gibbs, J. B., D'Alonzo, J. S., Sigal, I. S. & Scolnick, E. M. (1985). Yeast and mammalian *ras* proteins have conserved biochemical properties. *Nature (London)* **313** 700–703

[42] Gibbs, J. B., Sigal, I. S., Poe, M. & Scolnick, E. M. (1984). Intrinsic GTPase activity distinguishes normal and oncogenic ras p21 molecules. *Proc. Natl. Acad. Sci. USA* **81** 5704–5708

[43] McGrath, J. P., Capon, D. J., Goeddel, D. V. & Levinson, A. D. (1984). Comparative biochemical properties of normal and activated human *ras* p21 protein. *Nature (London)* **310** 644–649
[44] Trahey, M. & McCormick, F. (1987). A cytoplasmic protein stimulates normal N-*ras* p21 GTPase, but does not affect oncogenic mutants. *Science* **238** 542–545
[45] Field, J., Broek, D., Kataoka, T. & Wigler, M. (1987). Guanine nucleotide activation of, and competition between, *ras* proteins from *Saccharomyces cerevisiae*. *Mol. Cell. Biol.* **7** 2128–2133
[46] Satoh, T., Nakamura, S. & Kaziro, Y. (1987). Induction of neurite formation in PC12 cells by microinjection of protoncogenic Ha-*ras* protein preincubated with guanosine-5'-O-(3-thiophosphate). *Mol. Cell. Biol.* **7** 4553–4556
[47] Vogel, U. S., Dixon, R. A. F., Schaber, M. D., Diehl, R. E., Marshall, M. S., Scolnick, E. M., Sigal, I. S. & Gibbs, J. B. (1988). Cloning of bovine GAP and its interaction with oncogenic *ras* p21. *Science* **235** 90–93
[48] Trahey, M., Wong, G., Halenbeck, R., Rubinfeld, B., Martin, G. A., Ladner, M., Long, C. M., Crosier, W. J., Watt, K., Koths, K. & McCormick, F. (1988). Molecular cloning of two types of GAP complementary DNA from human placenta. *Science* **242** 1697–1700
[49] Adari, H., Lowry, D. R., Willumsen, B. M., Der, C. J. & McCormick, F. (1988). Guanosine triphosphatase activating protein (GAP) interacts with the p21 *ras* effector binding domain. *Science* **240** 518–521
[50] Cales, C., Hancock, J. F., Marshall, C. J. & Hall, A. (1988). The cytoplasmic protein GAP is implicated as the target for regulation by the *ras* gene product. *Nature (London)* **332** 548–551
[51] Sigal, I. S., Gibbs, J. B., D'Alonzo, J. S., Temeles, G. L., Wolanski, B. S., Socher, S. H. & Scolnick, E. M. (1986). Mutant *ras*-encoded proteins with altered nucleotide binding exert dominant biological effects. *Proc. Natl. Acad. Sci. USA* **83** 952–956
[52] Willumsen, B. M., Papageorge, A. G., Kung, H. F., Bekesi, E., Robins, T., Johnsen, M., Vass, W. C. & Lowy, D. R. (1986). Mutational analysis of a *ras* catalytic domain. *Moll. Cell. Biol.* **6** 2646–2654
[53] McCormick, F. (1989). *Ras* GTPase activating protein: signal transmitter and signal terminator. *Cell* **56** 5–8
[54] Yu, C. L., Tsai, M. H. & Stacey, D. W. (1988). Cellular *ras* activity and phospholipid metabolism. *Cell* **52** 63–71
[55] Smith, M. R., Degudicibus, S. J. & Stacey, D. W. (1986). Requirement for c-*ras* proteins during viral oncogene transformation. *Nature (London)* **320** 540–543
[56] Berridge, M. J. & Irvine, R. F. (1984). Inositol triphosphate, a novel second messenger in cellular signal transduction. *Nature (London)* **312** 315–321
[57] Nishizuka, Y. (1984). The role of protein kinase C in cell surface signal transduction and tumor promotion. *Nature (London)* **308** 693–698
[58] Fleischmann, L. F., Chahwala, S. B. & Cantley, L. (1986). *Ras*-transformed cells: altered levels of phosphatidylinositol 4,5-biphosphate and catabolites. *Science* **231** 407–410
[59] Lacal, J. C., Moscat, J. & Aaronson, S. A., (1987). Novel source of 1,2-

diacylglycerol elevated in cells transformed by Ha-*ras* oncogene. *Nature (London)* **330** 269–271

[60] Wolfman, A. & Macara, I. G. (1987). Elevated levels of diacylglycerol and decreased phorbol ester sensitivity in *ras*-transformed fibroblasts. *Nature (London)* **325** 359–361

[61] Hancock, J. F., Marshall, C. J., McKay, I. A., Gardner, S., Houslay, M. D., Hall, A. & Wakelam, M. J. O. (1988). Mutant but not normal p21 *ras* elevates inositol phospholipid breakdown in two different cell systems. *Oncogene* **3** 187–193

[62] Seuwen, K., Lagarde, A. & Pouyssegur, J. (1988). Deregulation of hamster fibroblast proliferation by mutated *ras* oncogenes is not mediated by constitutive activation of phosphoinositide-specific phospholipase C. *EMBO J.* **7** 161–168

[63] Macara, I. G. (1989). Elevated phosphocholine concentration in *ras*-transformed NI3T3 cells arises from increased choline kinase activity, not from phosphatidylcholine breakdown. *Mol. Cell. Biol.* **9** 325–328

[64] Tsai, M. H., Yu, C. L., Wei, E. S. & Stacey, D. W. (1989). The effect of GTPase activating protein upon *ras* is inhibited by mitogenically responsive lipids. *Science* **243** 522–526

[65] Morris, J. D. H., Price, B., Lloyd, A. C., Self, A. J., Marshall, C. J. & Hall, A. (1989). Scrape-loading of Swiss 3T3 cells with *ras* protein rapidly activates protein kinase C in the absence of phosphoinositide hydrolysis. *Oncogene* **4** 27–31

[66] Imler, J. L., Schatz, C., Wasylyk, C., Chatton, B. & Wasylyk, B. (1988). A Harvey-*ras* responsive transcription element is also responsive to a tumour-promoter and to serum. *Nature (London)* **332** 275–278

[67] Spandidos, D. A., Nichols, R. A. B., Wilkie, N. M. & Pintzas, A. (1988). Phorbol ester-responsive H-*ras*1 gene promoter contains multiple TPA-inducible/AP-1-binding consensus sequence elements. *FEBS Lett.* **240** 191–195

[68] Fasano, O., Aldrich, T., Tamanoi, F., Taparowsky, E., Furth, M. & Wigler, M. (1984). Analysis of the transforming potential of the human H-*ras* gene by random mutagenesis. *Proc. Natl. Acad. Sci. USA* **81** 4008–4012

[69] Walter, M., Clark, S. G. & Levinson, A. D. (1986). The oncogenic activation of human p21 *ras* by a novel mechanism. *Science* **233** 649–652

[70] Seeburg, P. H., Colby, W. W., Capon, D. J., Goeddel, D. V. & Levinson, A. D. (1984). Biological properties of human c-Ha-*ras*1 genes mutated at codon 12. *Nature (London)* **312** 71–75

[71] Bos, J. L., Toksoz, D., Marshall, C. J., Verlaan-de Vries, M., Veeneman, G. H., van der Eb, A. J., van Boom, J. H., Janssen, J. W. G. & Steenvoorden, A. C. M. (1985). Amino-acid substitutions at codon 13 of the N-*ras* oncogene in human acute myeloid leukaemia. *Nature (London)* **315** 726–730

[72] Tong, L., de Vos, A. M., Milburn, M. V., Jancarik, J., Noguchi, S., Nishimura, S., Miura, K., Ohtsuka, E. & Kim, S. H. (1989). Structural differences between a *ras* oncogene protein and the normal protein. *Nature (London)* **337** 90–93

[73] Sweet, R. W., Yokoyama, S., Kamata, T., Feramisco, J. R., Rosenberg, M.

& Gross, M. (1984). The product of *ras* is a GTPase and the T24 oncogenic mutant is deficient in this activity. *Nature (London)* **311** 273–275

[74] Manne, V., Bekesi, E. & Kung, H. F. (1985). Ha-*ras* proteins exhibit GTPase activity: point mutations that activate H-*ras* gene product result in decreased GTPase activity. *Proc. Natl. Acad. Sci. USA* **82** 376–380

[75] Chang, E. H., Furth, M. E., Scolnick, E. M. & Lowy, D. R. (1982). Tumorigenic transformation of mammalian cells induced by a normal human gene homologous to the oncogene of Harvey murine sarcoma virus. *Nature (London)* **297** 479–483

[76] Pulciani, S., Santos, E., Long, L. K., Sorrentino, V. & Barbacid, M. (1985). *Ras* gene amplification and malignant transformation. *Mol. Cell. Biol.* **5** 2836–2841

[77] McKay, I. A., Marshall, C. J., Cales, C. & Hall, A. (1986). Transformation and stimulation of DNA synthesis in NIH-3T3 cells are a titratable function of normal p21 N-*ras* expression. *EMBO J.* **5** 2617–2621

[78] Little, C. D., Nau, M. N., Carney, D. N., Gazdar, A. F. & Minna, J. D. (1983). Amplification of the c-*myc* oncogene in human lung carcinoma cell lines. *Nature (London)* **306** 194–196

[79] Brodeur, G. M., Seegers, R. C., Schwab, M., Varmus, H. E. & Bishop, J. M. (1984). Amplification of the N-*myc* gene in untreated human neuroblastomas correlated with advanced disease stage. *Science* **224** 1221–1224

[80] Nasu, M. N., Brooks, B. J., Battey, J., Sansville, E., Gazdav, A. F., Kirsch, I. R., McBride, O. W., Bertness, V., Hollis, G. F. & Minna, J. D. (1985). L-*myc*, a new *myc*-related gene amplified and expressed in human small cell lung cancer. *Nature (London)* **318** 69–73

[81] Slamon, D. J., Clark, G. M., Wong, S. G., Levin, W. J., Ullrich, A. & McGuire, W. L. (1987). Human breast cancer: correlation of relapse and survival with amplification of the HER-2/*neu* oncogene. *Science* **235** 177–181

[82] Riou, G., Barrois, M., Tordjman, I., Dutronquay, V. & Orth, G. (1984). Presence of papillomavirus genomes and amplification of the c-*myc* and c-Ha-*ras* oncogenes in invasive cancers of the uterine cervix. *C.R. Acad. Sci. (III)* **299** 575–580

[83] Filmus, J. E. & Buick, R. N. (1985). Stability of c-K-*ras* amplification during progression in a patient with adenocarcinoma of the ovary. *Cancer Res.* **45** 4468–4472

[84] Yokota, J., Tsunetsugu-Yokota, Y., Battifora, H., Lefevre, C. & Cline, M. J. (1986). Alterations of *myc*, *myb*, and *ras*-Ha protooncogenes in cancers are frequent and show clinical correlation. *Science* **231** 261–265

[85] van 't Verr, L. J., Hermens, R., van den Berg-Bakker, L. A. M., Cheng, N. C., Fleuren, G. J., Bos, J. L., Cleton, F. J. & Schrier, P. I. (1988). *Ras* oncogene activation in human ovarian carcinoma. *Oncogene* **2** 157–165

[86] Fujita, J., Srivastava, S. K., Kraus, M. H., Rhim, J. S., Tronick, S. R. & Aaronson, S. A. (1985). Frequency of molecular alterations affecting *ras* protooncogenes in human urinary tract tumors. *Proc. Natl. Acad. Sci. USA* **82** 3849–3853

[87] Bos, J. L., Verlaan-de Vries, M., Marshall, C. J., Veeneman, G. H., van Boom, J. H. & van der Eb, A. J. (1986). A human gastric carcinoma contains a

single mutated and an amplified normal allele of the Ki-*ras* oncogene. *Nucleic Acids Res.* **14** 1209–1217
[88] Fujita, J., Nakayama, H., Onoue, H., Rhim, J. S., El-Bolkainy, M. N., El-Aaser, A. A. & Kitamura, Y. (1987). Frequency of active *ras* oncogenes in human bladder cancers associated with schistosomiasis. *Jpn. J. Cancer Res.* **78** 915–920
[89] Taya, Y., Hosogai, K., Hirohashi, S., Shimosato, Y., Tsuchiya, R., Tsuchida, N., Fushimi, M., Sekiya, T. & Nishimura, S. (1984). A novel combination of K-*ras* and *myc* amplification accompanied by point mutational activation of K-*ras* in a human lung cancer. *EMBO J.* **3** 2943–2946
[90] Hayashi, K., Kakizoe, T. & Sugimura, T. (1983). In vivo amplification and rearrangement of the c-Ha-*ras*-1 sequence in a human bladder carcinoma. *Jpn. J. Cancer Res.* **74** 798–801
[91] Suarez, H. G., Du Villard, J. A., Caillou, B., Schlumberger, M., Tubiana, M., Parmentier, C. & Monier, R. (1988). Detection of activated *ras* oncogenes in human thyroid carcinomas. *Oncogene* **2** 403–406
[92] Knyazev, P. G., Chafer, R., Willecke, K., Pluzhnikovas, G. F., Serova, O. M., Fedorov, S. N. & Seita, I. F. (1987) Activation of protooncogenes from the *ras* and *myc* families in human carcinoma of the breast and neuroblastoma. *Mol. Biol. (Moscow)* **20** 1236–1243
[93] Young, D., Waitches, G., Birchmeier, C., Fasano, O. & Wigler, M. (1986). Isolation and characterization of a new cellular oncogene encoding a protein with multiple transmembrane domains. *Cell* **45** 711–719
[94] George, D. L. Glick, B., Trusko, S. & Freeman, N. (1986). Enhanced c-Ki-*ras* expression associated with Friend virus integration in a bone marrow-derived mouse cell line. *Proc. Natl. Acad. Sci. USA* **83** 1651–1655
[95] Westaway, D., Papkoff, J., Moscovici, C. & Varmus, H. E. (1986). Identification of a proverbially activated c-Ha-*ras* oncogene in a avian nephroblastoma via a novel procedure: cDNA cloning of a chimaeric viral-host transcript. *EMBO J.* **5** 301–309
[96] Cichutek, K. & Duesberg, P. H. (1986). Harvey *ras* genes transform without mutant codons, apparently activated by truncation of a 5' exon (exon −1). *Proc. Natl. Acad. Sci. USA* **83** 2340–2344
[97] Cohen, J. B. & Levinson, A. D. (1988). A point mutation in the last intron responsible for increased expression and transforming activity of the c-Ha-*ras* oncogene. *Nature (London)* **334** 119–124
[98] Spandidos, D. A. & Pintzas, A. (1988). Transcriptional enhancer activity in the variable tandem repeat DNA sequence downstream of the human Ha-*ras*1 gene. *FEBS Lett.* **232** 269–274
[99] Hashimoto-Gotoh, T., Kikuno, R., Takahashi, M. & Honkawa, H. (1988). Possible role of the first intron of c-H-*ras* in gene expression: anti-cancer elements in oncogenes. *Anticancer Res.* **8** 851–860
[100] Santos, E., Martin-Zanca, D., Reddy, E. P., Pierotti, M. A., Della Porta, G. & Barbacid, M. (1984). Malignant activation of a K-*ras* oncogene in lung carcinoma but not in normal tissue of same patient. *Science* **223** 661–664
[101] Fearon, E. R., Feinberg, A. P., Hamilton, S. H. & Vogelstein, B. (1985). Loss

of genes on the short arm of chromosome 11 in bladder cancer. *Nature (London)* **318** 377–380

[102] [a] Krontiris, T. G., DiMartino, N. A., Colb, M. & Parkinson, D. R. (1985). Unique allelic restriction fragments of the human Ha-*ras* locus in leukocyte and tumor DNAs of cancer patients. *Nature (London)* **313** 369–373; [b] Krontiris, T. G., Nancy, A., DiMartino, N. A., Colb, M., Mitchison, H. D. & Parkinson, D. R. (1986). Human restriction fragment length polymorphisms and cancer risk assessment. *J. Cell. Biochem.* **30** 319–329

[103] Heighway, J., Thatcher, N., Cerny, T. & Hasleton, P. S. (1986). Genetic predisposition to human lung cancer. *Br. J. Cancer* **53** 453–457

[104] Theillet, C., Lidereau. R., Escot, C., Hutzell, P., Brunet, M., Gest, J., Sclom, J. & Callahan, R. (1986). Loss of a c-Ha-*ras*-1 allele and aggressive human primary breast carcinomas. *Cancer Res.* **46** 4776–4781

[105] Fujita, J., Kraus, M. H., Onoue, H., Srivastava, S. R., Ebi, Y., Kitamura, Y. & Rhim, J. S. (1988). Activated H-*ras* oncogenes in human kidney tumors. *Cancer Res.* **48** 5251–5255

[106] Lee, J. H., Kavanagh, J. J., Wharton, J. T., Wildrick, D. M. & Blick, M. (1989). Allele loss at the c-Ha-*ras*1 locus in human ovarian cancer. *Cancer Res.* **49** 1220–1222

[107] Smit, V. T. H. B. M., Boot, A. J. M., Smits, A. M. M., Fleuren, G. J., Cornelisse, C. J. & Bos, J. L. (1988). K-*ras* codon 12 mutations occur very frequently in pancreatic adenocarcinomas. *Nucleic Acids Res.* **16** 7773–7782

[108] Larsson, C., Skogseid, B., Oberg, K., Nakamura, Y. & Nordenskjold, M. (1988). Multiple endocrine neoplasia type 1 gene maps to chromosome 11 and is lost in insulinoma. *Nature (London)* **332** 85–87

[109] Wyllie, F. S., Wynford-Thomas, V., Lemoine, N. R., Williams, G. T., Williams, E. D. & Wynford-Thomas, D. (1988). Ha-*ras* restriction fragment length polymorphisms in colorectal cancer. *Br. J. Cancer* **57** 135–138

[110] Ceccherini-Nelli, L., De Re, V., Viel, A., Molaro, G., Zilli, L., Clemente, C. & Boiocchi, M. (1987) Ha-*ras*-1 restriction fragment length polymorphism and susceptibility to colon adenocarcinoma. *Br. J. Cancer* **56** 1–5

[111] Sutherland, C., Shaw, H. M., Roberts, C., Grace, J., Stewart, M. M., McCarthy, W. H. & Kefford, R. F. (1986). Harvey-*ras* oncogene restriction fragment alleles in familial melanoma kindreds. *Br. J. Cancer* **54** 787–790

[112] Gerhard, D. S., Dracopoli, N. C., Bale, S. J., Houghton, A. N., Watkins, P., Payne, C. E., Greene, M. H. & Housman, D. E. (1987). Evidence against Ha-*ras*-1 involvement in sporadic and familial melanoma. *Nature (London)* **325** 73–75

[113] Radice, P., Pierotti, M. A., Borrello, M. G., Illeni, M. T., Rovini, D. & Della Porta, G. (1987). H-*ras*1 protooncogene polymorphisms in human malignant melanoma: *Taq*I defined alleles significantly associated with the disease. *Oncogene* **2** 91–95

[114] Ishikawa, J., Maeda, S., Takahashi, R., Kamidono, S. & Sugiyama, T. (1987). Lack of correlation between rare Ha-*ras* alleles and urothelial cancer in Japan. *Int. J. Cancer* **40** 474–478

[115] Riou, G., Barrois, M., Sheng, Z. M., Duvillard, P. & Lhomme, C. (1988).

Somatic deletions and mutations of c-Ha-*ras* gene in human cervical cancers. *Oncogene* **3** 329–333

[116] Carter, G., Worwood, M. & Jacobs, A. (1988). The H-*ras* polymorphism in myelodysplasia and acute myeloid leukaemia. *Leukemia Res.* **12** 385–391

[117] Lidereau, R., Escot, C., Theillet, C., Champeme, M. H., Brunet, M., Gest, J. & Callahan, R. (1986). High frequency of rare alleles of the human c-Ha-*ras*-1 proto-oncogene in breast cancer patients. *J. Natl. Cancer Inst.* **77** 697–701

[118] Boehm, T. L. J., Hirth, H. P., Kornhuber, B. & Drahovsky, D. (1987). Oncogene amplifications, rearrangements, and restriction fragment length polymorphism in human leukaemia. *Eur. J. Cancer Clin. Oncol.* **23** 623–629

[119] Spandidos, D. A. & Holmes, L. (1987). Transcriptional enhancer activity in the variable tandem repeat DNA sequence downstream of the human Ha-*ras*1 gene. *FEBS Lett.* **218** 41–46

[120] Blair, D. G., Cooper, C. S., Oskarsson, M. K., Eader, L. A. & Vande Woude, G. F. (1982). New method for detecting cellular transforming genes. *Science* **281** 1122–1125

[121] Bos, J. L., Fearon, E. R., Hamilton, S. R., Varlaan-de Vries, M., van Boom, J. H., van der Eb, A. J. & Vogelstein, B. (1987). Prevalence of *ras* gene mutations in human colorectal cancers. *Nature (London)* **327** 293–297

[122] Calos, M. P., Lebkowski, J. S. & Botchan, M. R. (1983). High mutation frequency in DNA transfected into mammalian cells. *Proc. Natl. Acad. Sci. USA* **80** 3015–3019

[123] Miller, J. H., Lebkowski, J. S., Greisen, K. S. & Calos, M. P. (1984). Specificity of mutations induced in transfected DNA by mammalian cells. *EMBO J.* **3** 3117–3121

[124] Hauser, J., Levine, A. S. & Dixon, K. (1987). Unique patterns of point mutations arising after gene transfer into mammalian cells. *EMBO J.* **6** 63–67

[125] Martin-Zanca, D., Hughes, S. H. & Barbacid, M. (1986). A human oncogene formed by the fusion of truncated tropomyosin and protein tyrosine kinase sequences. *Nature (London)* **319** 743–748

[126] Saiki, R. K., Gelfand, D. H., Stoffel, S., Scharf, S. J., Higuchi, R., Horn, G. T., Mullis, K. B. & Erlich, H. A. (1988). Primer-directed enzymatic amplification of DNA with a thermostable DNA polymerase. *Science* **239** 487–491

[127] Shibata, D., Arnheim, N. & Martin, J. (1988). Detection of human papilloma virus in paraffin-embedded tissue using the polymerase chain reaction. *J. Exp. Med.* **167** 225–230

[128] Bos, J. L., Verlaan-de Vries, M., Jansen, A. M., Veeneman, G. H., van Boom, J. H. & van der Eb, A. J. (1984). Three different mutations in codon 61 of the human N-*ras* gene detected by synthetic oligonucleotide hybridization. *Nucleic Acids Res.* **12** 9155–9163

[129] Verlaan-de Vries, M., Bogaard, M. E., van den Elst, H., van Boom, J. H., van der Eb, A. J. & Bos, J. L. (1986). A dot-blot screening procedure for mutated *ras* oncogenes using synthetic oligodeoxynucleotides. *Gene* **50** 313–320

[130] Kumar, R. & Barbacid, M. (1988). Oncogene detection at the single cell level. *Oncogene* **3** 647–651

[131] Myers, R. M., Larin, Z. & Maniatis, (1985). Detection of single base base

substitutions by ribonuclease cleavage at mismatches in RNA:DNA duplexes. *Science* **230** 1242–1246
[132] Winter, E., Yamamoto, F., Almoguera, C. & Perucho, M. (1985). A method to detect and characterize point mutations in transcribed genes: amplification and overexpression of the mutant c-K-*ras* allele in human tumour cells. *Proc. Natl. Acad. Sci. USA* **82** 7575–7579
[133] Forrester, K., Almoguera, C., Han, K., Grizzle, W. E. & Perucho, M. (1987). Detection of high incidence of K-*ras* oncogenes during human colon tumorigenesis. *Nature (London)* **327** 298–303
[134] Almoguera, C., Shibata, D., Forrester, K., Martin, J., Arnheim, N. & Perucho, M. (1988). Most human carcinomas of the exocrine pancreas contain mutant c-K-*ras* genes. *Cell* **53** 549–554
[135] Shen, W. P., Aldrich, T. H., Venta-Perez, G., Franza, B. R. & Furth, M. E. (1987). Expression of normal and mutant *ras* proteins in human acute leukemia. *Oncogene* **1** 157–165
[136] Tabin, C. J., Bradley, S. M., Bargmann, C. I., Weinberg, R. A., Papageorge, A. G., Scolnick, E. M., Dhar, R., Lowy, D. R. & Chang, E. H. (1982). Mechanism of activation of a human oncogene. *Nature (London)* **300** 143–149
[137] Srivastava, S. K., Yuasa, Y., Reynold, S. H. & Aaronson, S. A. (1985). Effects of two major activating lesions on the structure and conformation of human *ras* oncogene products. *Proc. Natl. Acad. Sci. USA* **82** 38–42
[138] Der, C. J. & Cooper, G. M. (1986). Altered gene products are associated with activation of cellular *ras*K genes in human lung and colon carcinomas. *Cell* **32** 201–208
[139] Notario, V., Sukumar, S., Santos, E. & Barbacid, M. (1984). In: G. Vande Woude *et al.* (eds) *Cancer Cells 2: Oncogenes and Viral Oncogenes*, Cold Spring Harbor Laboratories, pp. 425–432
[140] Pulciani, S., Santos, E., Lauver, A. V., Long, L. K., Aaronson, S. A. & Barbacid, M. (1982) Oncogenes in solid human tumors. *Nature (London)* **300** 539–542
[141] Yanez, L., Greffen, J. & Valuenzuala, D. M. (1987) c-K-*ras* mutations in human carcinomas occur preferentially in codon 12. *Oncogene* **1** 315–318
[142] Vogelstein, B., Fearon, E. R., Hamilton, S. R., Kern, S. E., Preisinger, A. C., Leppert, M., Nakamura, Y., White, R., Smits, A. M. M. & Bos, J. L. (1988). Genetic alterations during colorectal-tumor development. *N. Engl. J. Med.* **319** 525–532
[143] Farr, C. J., Marshall, C. J., Easty, D. J., Wright, N. A., Powell, S. C. & Paraskeva, C. (1988). A study of *ras* gene mutations in colonic adenomas from familial polyposis coli patients. *Oncogene* **3** 673–678
[144] Lundy, J., Chen, J., Wang, P., Fromowitz, F., Schuss, A., Lynch, S., Brugge, J. & Viola, M. V. (1988). Phenotypic and genetic alterations in precancerous cells in the colon. *Anticancer Res.* **8** 1005–1014
[145] Suarez, H. G., Nargeux, P. C., Andeol, Y. & Sarasin, A. (1987). Multiple activated oncogenes in human tumors. *Oncogene Res.* **1** 201–207
[146] Lemoine, N. R., Mayall, E. S., Wyllie, F. S., Farr, C. J., Hughes, D., Padua,

R. A., Thurston, V., Williams, E. D. & Wynford-Thomas, D. (1988). Activated *ras* oncogenes in human thyroid cancers. *Cancer Res.* **48** 4459–4463
[147] Lemoine, N. R., Mayall, E. S., Wyllie, F. S., Williams, E. D., Goyns, M., Stringer, B. M. J. & Wynford-Thomas, D. (1989). High frequency of *ras* oncogene activation in all stages of human thyroid tumorigenesis. *Oncogene* (in press)
[148] Stanton, V. P. & Cooper, G. M. (1987). Activation of human (*ras*) transforming genes by deletion of normal amino-terminal coding sequences. *Mol. Cell. Biol.* **7** 1171–1179
[149] Rodenhuis, S., van de Wetering, M. L., Mooi, W. J., Evers, S. G., van Zandwijk, N. & Bos, J. L. (1987). Mutational activation of the K-*ras* oncogene, a possible pathogenetic factor in adenocarcinoma of the lung. *N. Engl. J. Med.* **317** 929–935
[150] Rodenhuis, S., Slebos, R. J. C., Boot, A. J. M., Evers, S. G., Mooi, W. J., Wagenaar, S. S., van Bodegom, P. C. & Bos, J. L. (1988). K-*ras* oncogene activation in adenocarcinoma of the lung: incidence and possible clinical significance. *Cancer Res.* **48** 5738–5741
[151] Ananthaswamy, H. N., Price, J. E., Goldberg, L. H. & Bales, E. S. (1988). Detection and identification of activated oncogenes in human skin cancers occurring on sun-exposed body sites. *Cancer Res.* **48** 3341–3346
[152] Saurez, H. G., Daya, Grosjean, L., Sclaifer, D., Nardeux, P., Renault, G., Bos, J. L. & Sarasin, A. (1989). Activated oncogenes in human skin tumors from a repair-deficient syndrome, xeroderma pigmentosum. *Cancer Res.* **49** 1223–1228
[153] Leon, J., Kamino, H., Steinberg, J. J. & Pellicer, A. (1988). H-*ras* activation in benign and self-regressing skin tumors (keratoacanthomas) in both humans and an animal model system. *Mol. Cell. Biol.* **8** 786–793
[154] Raybaud, F., Noguchi, T., Marics, I., Adelaide, J., Planche, J., Batoz, M., Aubert, C., de Lapeyriere, O. & Birnbaum, D. (1988). Detection of a low frequency of activated *ras* genes in human melanomas using a tumorigenicity assay. *Cancer Res.* **48** 950–953
[155] Malone, P. R., Visvanathan, K. V., Ponder, B. A., Shearer, R. J. & Summerhayes, I. C. (1985). Oncogenes and bladder cancer. *Br. J. Urol.* **57** 664–667
[156] Visvanathan, K. V., Pocock, R. D. & Summerhayes, I. C. (1988). Preferential and novel activation of H-*ras* in human bladder carcinomas. *Oncogene Res.* **3** 77–86
[157] Peehl, D. M., Wehner, N. & Stamey, T. (1987) Activated Ki-*ras* oncogenes in prostatic adenocarcinoma. *Prostate* **11** 281–289
[158] Kraus, M. H., Yuasa, Y. & Aaronson, S. A. (1984). A position 12-activated H-*ras* oncogene in all HS578T mammary carcinosarcoma cells but not in normal mammary cells of the same patient. *Proc. Natl. Acad. Sci. USA* **81** 5384–5388
[159] Spandidos, D. A. (1987). Oncogene activation in malignant transformation: a study of H-*ras* in human breast cancer. *Anticancer Res.* **7** 991–996
[160] Rochlitz, C. F., Scott, G. K., Dodson, J. M., Liu, E., Dollbaum, C., Smith, H. S. & Benz, C. C. (1989). Incidence of activating *ras* oncogene mutations

associated with primary and metastatic human breast cancer. *Cancer Res.* **49** 357–360

[161] O'Hara, B. M., Oskarsson, M., Tainsky, M. A. & Blair, D. G. (1986). Mechanism of activation of human *ras* genes cloned from a gastric adenocarcinoma and a pancreatic carcinoma cell line. *Cancer Res.* **46** 4695–4700

[162] Nishida, J., Kobayashi, Y., Hirai, H. & Takaku, F. (1987). A point mutation at codon 13 of the N-*ras* oncogene in a human stomach cancer. *Biochem. Biophys. Res. Commun.* **146** 247–252

[163] Sakato, H., Mori, M., Taira, M., Yoshida, T., Matsukawa, S., Shimuzu, K., Sekiguchi, M., Terada, M. & Sugimura, T. (1986). Transforming gene from human stomach cancers and a noncancerous portion of stomach mucosa. *Proc. Natl. Acad. Sci. USA* **83** 3997–4001

[164] Fujita, K., Ohuchi, N., Yao, T., Okumura, M., Fukushima, Y., Kanakura, V., Kitamura, Y. & Fujita, J. (1987). Frequent overexpression, but not activation by point mutation, of *ras* genes in primary human gastric cancers. *Gastroenterology* **93** 1339–1345

[165] Feig, L. A., Bast, R. C., Knapp, R. C. & Cooper, G. M. (1984). Somatic activation of *ras*K gene in a human ovarian carcinoma. *Science* **223** 698–700

[166] Gu, J. R., Hu, L. F., Cheng, Y. C. & Wan, D. F. (1986). Oncogenes in human primary hepatic cancer. *J. Cell Physiol. (Suppl.)* **4** 13–20

[167] Ballas, K., Lyons, J., Janssen, J. W. G. & Bartram, C. R. (1988). Incidence of *ras* gene mutations in neuroblastoma. *Eur. J. Pediatr.* **147** 313–314

[168] Janssen, J. W. G., Steenvoorden, A. C. M., Lyons, J., Anger, B., Bohlke, J. U., Bos, J. L., Seliger, H. & Bartram, C. R. (1987). *Ras* gene mutations in acute and chronic myelocytic leukemias, chronic myeloproliferative disorders and myelodysplastic syndromes. *Proc. Natl. Acad. Sci. USA* **84** 9228–9232

[169] Hirai, H., Kobayashi, Y., Mano, H., Hagiwara, K., Maru, Y., Omine, M., Mizoguchi, H., Nishida, J. & Takaku, F. (1987). A point mutation at codon 13 of the N-*ras* oncogene in myelodysplastic syndrome. *Nature (London)* **327** 430–432

[170] Liu, E., Hjelle, B., Morgan, R., Hecht, F. & Bishop, J. M. (1987). Mutations of the Kirsten-*ras* proto-oncogene in human preleukemia. *Nature (London)* **330** 186–188

[171] Padua, R. A., Carter, G., Hughes, D., Gow, J., Farr, C. J., Oscier, D., McCormick, F. & Jacobs, A. (1988). *Ras* mutations in myelodysplasia detected by amplification, oligonucleotide hybridisation, and transformation. *Leukemia* **2** 503–510

[172] Gow, J., Hughes, D., Farr, C. J., Hamblin, T., Oscier, D., Brown, R. & Padua, R. A. (1988). Activation of Ha-*ras* in human chronic granulocytic and chronic myelomonocytic leukemia. *Leukaemia Res.* **12** 805–810

[173] Lyons, J., Janssen, J. W. G., Bartram, C., Layton, M. & Mufti, G. J. (1988). Mutations of Ki-*ras* and N-*ras* oncogenes in myelodysplastic syndromes. *Blood* **71** 1707–1712

[174] Murray, M. J., Cunningham, J. M., Parada, L. F., Duatry, F., Lebowitz, P. & Weinberg, R. A. (1983). The HL-60 transforming sequence: a *ras* oncogene coexisting with altered *myc* genes in hematopoietic tumors. *Cell* **33** 749–757

[175] Eva, A., Tronick, S. R., Gol, R. A., Pierce, J. H. & Aaronson, S. A. (1983).

Transforming genes of human hematopoietic tumors: frequent detection of *ras*-related oncogenes whose activation appears to be independent of tumor phenotype. *Proc. Natl. Acad. Sci. USA* **80** 4926–4930

[176] Gambke, C., Hall, A. & Moroni, C. (1984). Activation of an N-*ras* gene in acute myeloblastic leukemia through somatic mutation in the first exon. *Proc. Natl. Acad. Sci. USA* **82** 879–882

[177] Janssen, J. W. G., Steenvoorden, A. C. M., Collard, J. G. & Nüsse, R. (1985). Oncogene activation in human myeloid leukemia. *Cancer Res.* **45** 3362–3367

[178] Needleman, S. W., Kraus, M. H., Srivastava, S. K., Levine, P. H. & Aaronson, S. A. (1986). High frequency of N-*ras* activation in acute myelogenous leukemia. *Blood* **67** 753–757

[179] Bos, J. L., Verlaan-de Vries, M., van der Eb, A. J., Janssen, J. W. G., Delwel, R., Lowenberg, B. & Colly, L. P. (1987). Mutations in N-*ras* predominate in acute myeloid leukemia. *Blood* **69** 1237–1241

[180] Janssen, J. W. G., Lyins, J., Steenvoorden, A. C. M., Seliger, H. & Bartram, C. R. (1987). Concurrent mutations in two different *ras* genes in acute myelocytic leukemias. *Nucleic Acids Res.* **15** 5669–5680

[181] Hirai, H., Nishida, J. & Takaku, F. (1987). Highly frequent detection of transforming genes in acute leukemias by transfection usin in vivo selection assays. *Biochim. Biophys. Res. Commun.* **147** 108–114

[182] Nishida, J., Hirai, H. & Takaku, F. (1987). Activation mechanism of the N-*ras* oncogene in human leukemias detected by synthetic oligonucleotide probes. *Biochem. Biophys. Res Commun.* **147** 870–875

[183] Senn, H. P., Tran-Thang, C., Wodnar-Filipowicz, A., Jiricny, J., Fopp, M., Gratwohl, A., Signer, E., Weber, W. & Moroni, C. (1988). Mutation analysis of the N-*ras* proto-oncogene in active and remission phase of human acute leukemias. *Int. J. Cancer* **41** 59–64

[184] Buschle, M., Janssen, J. W. G., Drexler, H., Lyons, J., Anger, B. & Bartram, C. R. (1988). Evidence for pluripotent stem cell origin of idiopathic myelofibrosis: clonal analysis of a case characterized by a N-*ras* gene mutation. *Leukemia* **2** 658–660

[185] Farr, C. J., Saiki, R. K., Ehrlich, H. A., McCormick, F. & Marshall, C. J. (1988). Analysis of *ras* gene mutations in acute myeloid leukemia using the polymerase chain reaction and oligonucleotide probes. *Proc. Natl. Acad. Sci. USA* **85** 1629–1633

[186] Hirai, H., Yanaka, S., Azuma, M., Anraku, Y., Kobayashi, Y., Fujisawa, M., Okabe, T., Urabe, A. & Takaku, F. (1986). Transforming genes in human leukemia cells. *Blood* **66** 1371–1378

[187] Liu, E., Hjelle, B. & Bishop, J. M. (1988). Transforming genes in chronic myelogenous leukemia. *Proc. Natl. Acad. Sci. USA* **85** 1952–1956

[188] Collins, S. J. (1988). Direct sequencing of amplified genomic fragments documents N-*ras* point mutations in myeloid leukemia. *Oncogene Res.* **3** 117–123

[189] Rodenhuis, S., Bos, J. L., Slater, R. M., Behrendt, H., van't Veer, M. & Smets, L. A. (1986). Absence of oncogene amplifications and occasional

activation of N-*ras* in lymphoblastic leukemia of childhood. *Blood* **67** 1698–1704

[190] Neri, A., Knowles, D. M., Greco, A., McCormick, F. & Dalla-Favera, R. (1988). Analysis of *ras* oncogene mutations in human lymphoid malignancies. *Proc. Natl. Acad. Sci. USA* **85** 9268–9272

[191] Sklar, M. D. & Kitchingham, G. R. (1985). Isolation of activated *ras* transforming genes from two patients with Hodgkin's disease. *Int. J. Radiat. Oncol. Biol. Phys.* **11** 49–55

[192] Wodnar-Filipowicz, A., Senn, H. P., Jiricny, J., Signer, E. & Moroni, C. (1987). Glycine–cysteine substitution at codon 13 of the N-*ras* proto-oncogene in a human T cell non-Hodgkin's lymphoma. *Oncogene* **1** 457–461

[193] Der, C. J., Krontiris, T. G. & Cooper, G. M. (1982). Transforming genes of human bladder and lung carcinoma cell lines are homologous to the *ras* genes of the Harvey and Kirsten sarcoma viruses. *Proc. Natl. Acad. Sci. USA* **79** 3637–3640

[194] Parada, L. F., Tabin, C. J., Shih, C. & Weinberg, R. A. (1982). Human EJ bladder carcinoma oncogene is homologue of Harvey sarcoma virus *ras* gene. *Nature (London)* **297** 474–478

[195] Santos, E., Tronick, S. R., Aaronson, S. A., Pulciani, S. & Barbacid, M. (1982). T24 human bladder carcinoma oncogene is an activated form of the normal human homologue of BALB- and Harvey-MSV transforming genes. *Nature (London)* **298** 343–347

[196] Knyazev, P. G., Fedorov, S. N., Serova, O. M., Pluzhnikova, G. F., Novikov, L. B., Kalinovsky, V. P. & Seitz, J. F. (1986). Molecular-genetic analysis of *myc* and c-Ha-*ras* proto-oncogene alterations in human carcinoma. *Haematol. Blood Transfus.* **31** 469–473

[197] Yuasa, Y., Srivastava, S. K., Dunn, C. Y., Rhim, J. S., Reddy, E. P. & Aaronson, S. A. (1983). Acquisition of transforming properties by alternative point mutations within c-*bas/has* human proto-oncogene. *Nature (London)* **303** 775–779

[198] Zhan, X., Culpepper, A., Reddy, M., Loveless, J. & Goldfarb, M. (1987). Human oncogenes detected by a defined medium culture assay. *Oncogene* **1** 369–376

[199] Deng, G., Lu, Y., Chen, S., Miao, J., Lu, G., Li, H., Cai, H., Xu, X., Zheng, E. & Liu, P. (1987). Activated c-Has-*ras* oncogene with a guanine to thymine transversion in a human stomach cancer cell line. *Cancer Res.* **47** 3195–3198

[200] Sekiya, T., Fushimi, M., Hori, H., Hirohashi, S., Nishimura, S. & Sugimura, T. (1984). Molecular cloning and the total nucleotide sequence of human c-Har-*ras*-1 gene activated in a melanoma from a Japanese patient. *Proc. Natl. Acad. Sci. USA* **81** 4771–4775

[201] Albino, A. P., LeStrange, R., Oliff, A. I., Furth, M. E. & Old, L. J. (1984). Transforming *ras* genes from human melanoma: a manifestation of tumour heterogeneity? *Nature (London)* **308** 69–72

[202] Valenzuela, D. M. & Groffen, J. (1986). Four human carcinoma cell lines with novel mutations in position 12 of c-Ki-*ras* oncogene. *Nucleic Acids Res.* **14** 843–852

[203] Shimuzu, K., Goldfarb, M., Guard, Y., Perucho, M., Lik, Y., Kamata, T., Feramisco, J., Stavnezer, E., Fogh, J. & Wigler, M. (1983). Three human transforming genes are related to the viral *ras* oncogenes. *Proc. Natl. Acad. Sci. USA* **80** 2112–2116

[204] Fukui, M., Yamamoto, T., Kawai, S., Maruo, K. & Toyoshima, K. (1985). Detection of a *raf*-related and two other transforming DNA sequences in human tumors maintained in nude mice. *Proc. Natl. Acad. Sci. USA* **82** 5954–5958

[205] McCoy, M. S., Toole, D. J., Smith, D. H., Chen, E. Y., Seeburg, P. H., Goeddel, D. V. & Levinson, A. D. (1983). Structure and organization of the human Ki-*ras* proto-oncogene and a related processed pseudogene. *Nature (London)* **310** 644–649

[206] Yuasa, Y., Oto, M., Sato, C., Miyaki, M., Iwana, T., Tonomura, A. & Namba, M. (1986). Colon carcinoma K-*ras* 2 oncogene of a familial polyposis coli patient. *Jpn. J. Cancer Res.* **77** 901–907

[207] Kozma, S. C., Bogaard, M. E., Buser, K., Saurer, S. M., Bos, J. L., Groner, B. & Hynes, N. E. (1987). The human c-Kirsten *ras* gene is activated by a novel mutation in codon 13 in the breast carcinoma cell line MDA-MB231. *Nucleic Acids Res.* **15** 5963–5971

[208] Prosperi, M. T., Even, J., Clavo, F., Lebeua, J. & Goubin, G. (1987). Two adjacent mutations at position 12 activate the K-*ras* oncogene of a human mammary tumor cell line. *Oncogene Res.* **1** 121–128

[209] Cooper, C. S., Blair, D. G., Oskarsson, M. K., Tainsky, M. A., Eader, L. A. & Vande Woude, G. F. (1984). Characterization of human transforming genes from chemically-transformed, teratocarcinoma, and pancreatic carcinoma cell lines. *Cancer Res.* **44** 1–10

[210] Hirai, H., Okabe, T., Anraku, Y., Fujisawa, M., Urabe, A. & Takaku, F. (1985). Activation of the c-Ki-*ras* oncogene in a human pancreas carcinoma. *Biochem. Biophys. Res. Commun.* **127** 168–174

[211] Nardeux, P. C., Daya-Grosjean, L., Landin, R. M., Andeol, Y. & Saurez, H. G. (1987). A c-*ras*-Ki oncogene is activated, amplified and overexpressed in a human osteosarcoma cell line. *Biochem. Biophys. Res. Commun.* **146** 395–402

[212] Yuasa, Y., Fol, R. A., Chang, A., Chiu, I. M., Reddy, E. P., Tronick, S. R. & Aaronson, S. A. (1984). Mechnaism of activation of an N-*ras* oncogene of SW-1271 human lung carcinoma cells. *Proc. Natl. Acad. Sci. USA* **81** 3670–3674

[213] Yuasa, Y., Reddy, E. P., Rhim, J. S., Tronick, S. R. & Aaronson, S. A. (1986).Activated N-*ras* in a human rectal carcinoma cell line associated with clonal homozygosity in *myb* locus-restriction fragment polymorphism. *Jpn. J. Cancer Res.* **77** 639–647

[214] Andeol, Y., Nardeux, P. C., Daya-Grosjean, L., Brison, O., Cebrian, J. & Saurez, H. (1988). Both N-*ras* and c-*myc* are activated in the SHAC human stomach fibrosarcoma cell line. *Int. J. Cancer* **41** 732–737

[215] Padua, R. A., Barrass, N. C. & Currie, G. A. (1985). Activation of N-*ras* in a human melanoma cell line. *Mol. Cell. Biol.* **5** 582–585

[216] Keizer, W., Mulder, M. P., Langeveld, J. C. M., Smit, E. M. E., Bos, J. L., Bootsma, D. & Hoeijmakers, J. H. J. (1989). Establishment and characteriza-

tion of a melanoma cell line from a xeroderma pigmentosum patient: activation of N-*ras* at a potential pyrimidine dimer site. *Cancer Res.* **49** 1229–1235
[217] Tainsky, M. A., Cooper, C. S., Giovanella, B. C. & Van De Woude, G. F. (1984). An activated *ras*N gene: detected in late but not early passage human PA1 teratocarcinoma cells. *Science* **223** 643–645
[218] Souyri, M. & Fleissner, E. (1983). Identification of transfection of transforming sequences in DNA of human T-cell leukemias. *Proc. Natl. Acad. Sci. USA* **80** 6676–6679
[219] Pedersen-Bjergaard, J., Janssen, J. W. G., Lyons, J., Philip, P. & Bartram, C. R. (1988). Point mutation of the *ras* oncogenes and chromosome aberrations in acute nonlymphocytic leukemia and preleukemia related to therapy with alkylating agents. *Cancer Res.* **48** 1812–1817
[220] Taparowsky, E., Shimuzu, K., Goldfarb, M. & Wigler, M. (1983). Structure and activation of the human N-*ras* gene. *Cell* **34** 581–586
[221] Brown, R., Marshall, C. J., Pennie, S. G. & Hall, A. (1984). Mechanism of activation of an N-*ras* gene in the human fibrosarcoma cell line HT1080. *EMBO J.* **3** 1321–1326
[222] Fusco, A., Grieco, M., Santoro, M., Berlingieri, M. T., Pilotti, S., Pierotti, M. A., Della Porta, G. & Vecchio, G. (1987). A new oncogene in human thyroid papillary carcinomas and their lymph-nodal metastases. *Nature (London)* **328** 170–172
[223] Robbins, S. L., Cotran, R. S. & Kumar, V. (1984). In: *Pathological Basis of Disease*, Saunders, Philadelphia, PA, pp. 705–766
[224] Wynder, E. L., Mabuchi, K., Maruchi, N. & Fortner, J. G. (1973). Epidemiology of cancer of the pancreas. *J. Natl. Cancer Inst.* **50** 6454–667
[225] Senn, H. P., Fopp, J. M., Schmid, L. & Moroni, C. (1988). Relapse cell population differs from acute onset clone as shown by absence of the initially activated N-*ras* oncogene in a patient with acute myelomonocytic leukemia. *Blood* **72** 931–935
[226] Hollstein, M. C., Smits, A. M., Galiana, C., Yamasaki, H., Bos, J. L., Mandard, A., Partensky, C. & Montesano, R. (1988). Amplification of epidermal growth factor receptor gene but no evidence of *ras* mutations in primary human esophageal cancers. *Cancer Res.* **48** 5119–5123
[227] Wakelam, M. J. O., Davies, S. A., Houslay, M. D., McKay, I., Marshall, C. J. & Hall, A. (1986). Normal p21 N-*ras* couples bombesin and other growth factors to inositol phosphate production. *Nature (London)* **323** 173–176
[228] Parries, G., Hoebel, R. & Racker, E. (1987). Opposing effects of a *ras* oncogene on growth factor-stimulated phosphoinositide hydrolysis: desensitization to platelet-derived growth factor and enhanced sensitivity to bradykinin. *Proc. Natl. Acad. Sci. USA* **84** 2648–2652
[229] Williams, D. W., Williams, E. D. & Wynford-Thomas, D. (1988). Loss of dependence on IGF-1 for proliferation of human thyroid adenoma cells. *Br. J. Cancer* **57** 535–539
[230] Korn, L. J., Siebel, C. W., McCormick, F. & Roth, R. A. (1987). *Ras* p21 as a potential mediator of insulin action in *Xenopus* oocytes. *Science* **236** 840–843
[231] O'Brien, R. M., Siddle, K., Houslay, M. D. & Hall, A. (1987). Interaction of

the human insulin receptor with the *ras* oncogene product p21. *FEBS Lett.* **217** 253–259
[232] Lowy, D. R., Capon, D. J., Delwart, E., Sakaguchi, A. Y., Naylor, S. L. & Goeddel, D. V. (1987). Structure of the human and murine R-*ras* genes, novel genes closely related to *ras* proto-oncogenes. *Cell* **48** 137–146
[233] Chardin, P. & Tavitian, A. (1986). The ral gene: a new *ras* related gene isolated by the use of a synthetic probe. *EMBO J.* **5** 2203–2208
[234] Pizon, V., Chardin, P., Lerosey, I., Olofsson, B. & Tavitian, A. (1988). Human cDNAs *rap*1 and *rap*2 homologous to the Drosophila gene Dras3 encode proteins closely related to *ras* in the "effector" region. *Oncogene* **3** 201–204
[235] Bucci, C., Frunzio, R., Chiarotti, L., Brown, A. L., Rechler, M. M. & Bruni, C. B. N. (1988). A new member of the *ras* gene superfamily identified in a rat liver cell line. *Nucleic Acids Res.* **16** 9979–9993
[236] Kitayama, H., Sugimoto, Y., Masuzaki, T., Ikawa, Y. & Noda, M. (1989). A *ras*-related gene with transformation suppressor activity. *Cell* **56** 77–84
[237] Paterson, H., Reeves, B., Brown, R., Hall, A., Furth, M., Bos, J. L., Jones, P. & Marshall, C. J. (1987). Activated N-*ras* controls the transformed phenotype of HT1080 human fibrosarcoma cells. *Cell* **51** 803–812
[238] Steenvorden, A. C. M., Janssen, J. W. G., Drexler, H. G., Lyons, J., Tesch, H., Binder, T., Jones, D. B. & Bartram, C. R. (1988). *Ras* mutations in Hodgkin's disease. *Leukemia* **2** 325–326
[239] Kumar, R., & Dunn, L. L. (1989) Designed diagnostic restriction fragment length polymorphisms for the detection of point mutations in *ras* oncogenes. *Oncogene Res.* **1** 235–241
[239] Burgering, B. M. T., Smijders, A. J., Maasens, J. A., van der Eb, A. J. & Bos, J. L. (1989) Possible involvement of normal p21 H-*ras* in the insulin/insulin-like growth factor 1 signal transduction pathway. *Mol. Cel. Biol.* **9** 4312–4322

6

Anti-cancer elements in oncogenes

Hidenori Honkawa*, Reiko Kikuno, Mikiko Takahashi, Ken-Ich Tezuka and
Tamotsu Hashimoto-Gotoh
Laboratory for Molecular Biology, Pharma Research Laboratories, Hoechst
Japan Ltd., 3-2 Minamidai 1, Kawagoe 350, Japan
*Present address: Clinical Research Planning Department, Pharma Clinical
Research Division, Hoechst Japan Ltd., 10–16 Akasaka 8, Minatoku Tokyo 107,
Japan.

1. INTRODUCTION

Oncogenes are a set of genes that are thought to be involved in or responsible for carcinogenesis. They were found first in retroviral genomes and then later in cellular chromosomes from tumour-derived cells as well as from normal cells [1]. Oncogenes derived from retrovirus genomes or tumour cells are often designated as activated types and those from normal cells as prototypes [2–8]. Two gene families in particular, the *ras* and *myc* genes, have been studied extensively from various sources including clinical samples in relation to both structure and activity [8–27].

In the case of *ras* oncogenes (H-*ras*, K-*ras* and N-*ras*), although extensive data from investigations of the transformation mechanism caused by structural mutation have accumulated, a convincing explanation does not yet seem to be available as to how the activated *ras* gene products, p21s, incite the cellular events of cell proliferation and carcinogenesis [15,27–38]. Thus, the regulatory mechanisms of the expression of *ras* oncogene appear to have become a focal topic for researchers recently [7–9,19,20,23,39–44]. However, little is known about the area. Therefore, it is very important to know what types of regulation mechanism exist, how the expression of c-H-*ras* oncogene is thus regulated and, most vital, the relationship between carcinogenesis and these regulatory mechanisms.

Two possible trigger events which convert prototype oncogenes to the activated type are currently being discussed among researchers in this field. For example, one argument is that the oncogene may be activated by overexpression of the gene resulting from gene truncation, chromosome translocation or gene amplification [14,17,44–50] or otherwise that some structural mutation in the coding region resulting in amino acid changes at certain key positions in proteins causes enzymatic activation of the oncogene [4,51–54]. The *myc* and *ras* oncogenes are argued to be typical examples of the former and the latter cases respectively [55].

Good evidence has accumulated showing in the case of *ras* oncogenes that an amino acid substitution at or around position 12 or 61 is involved in focus or tumour formation in *in vitro* and *in vivo* experiments and in clinical samples as well (for a review see ref. [55]). Nevertheless, some discrepancies between *in vitro* experiments and clinical samples [50] remain to be resolved. Specifically, it has been reported that accelerated expression of prototype H-*ras* without structural mutation can also induce focus formation in mouse fibroblast cells such as NIH3T3 [45,56,57]. It has also been shown that such transformed cells can subsequently form tumours in nude mice [45], although there have been no such reports from clinical samples.

In this article based on a study using c-H-*ras* oncogene, we propose that oncogenes whose overexpression may induce tumours have their own regulatory mechanism located in the preceding intron sequences upstream from the coding regions. These postulated regulatory sequences, named anti-cancer elements, may work to suppress gene expression at the level of protein synthesis by reducing the translation initiation frequency, probably by means of a complex secondary structure located in front of the translation initiation site. These sequences will work only when the elements remain in mature mRNA after unusual splicing as a result of gene truncation or chromosome translocation. In fact, we have found some sequences in intron −1, the intron located in the 5′-flanking region of c-H-*ras*, which are unusually highly conserved in mammalian genomes and show an inhibitory function on gene expression when the donor site of intron −1 is deleted.

2. LOCALIZATION OF PRINCIPAL PROMOTER REGION OF c–H-*ras* ONCOGENE

To elucidate the regulation of expression of the c-H-*ras* oncogene and the influence of the DNA sequences in the 5′-flanking region on transforming activity, the promoter sequence and the initiation site for transcription were determined. Various test deletion mutants in the 5′- and 3′-flanking regions of the c-H-*ras* oncogene originating from pEJ [4] are detailed in Fig. 1. These deletion mutants, cloned onto the vector pHSG367 [58], were introduced into mouse NIH3T3 cells by calcium-phosphate-mediated DNA transfection as described previously [23] with or without pHSG274 [59], which confers resistance to the antibiotic G418, and transforming activities were monitored. The results of transforming activities of these constructions in the focus assay are summarized in Fig. 1. Deletion of the 3′-flanking region downstream of the *Sph*I site, which contains the so-called variable tandem repeat region [9], did not influence transforming activity in this system. On the other hand, a deletion between the two *Sac*I sites in the 5′-flanking region drastically diminished transforming activity, suggesting that sequences important for transcription are located in this region (from −1458 to −616). This is consistent with observations with *neo* gene fusions and G418 transformation assay [23]. To localize the promoter region more precisely, we isolated many more deletions in this region.

The 5′ terminus of the essential sequences in the 5′ regulatory region was located at around position −1450 (Fig. 2). Transforming activity of the 5′ deletion mutants

Ch. 6] Anti-cancer elements in oncogenes

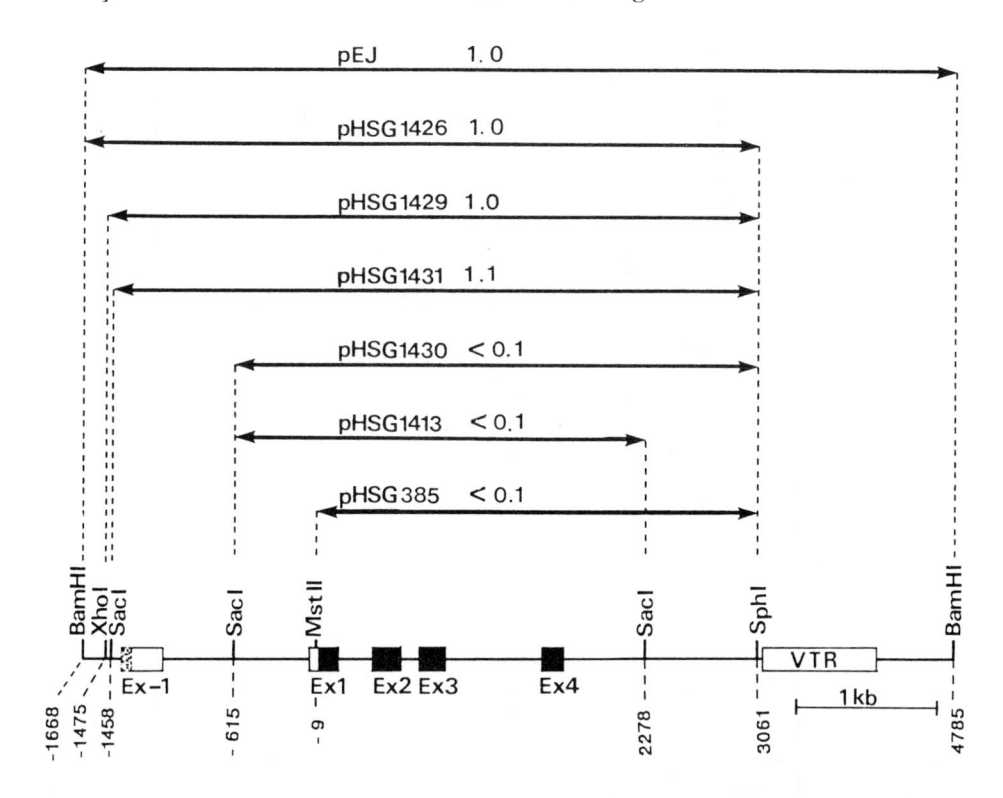

Fig. 1 — Construction and transforming activities of various deletion mutants. The structures of various test deletion mutants are presented. The structure of the entire c-H-*ras* oncogene is shown at the bottom of the figure for comparison; coding and non-coding exon regions are indicated by closed and open boxes respectively. The dotted box represents the principal promoter region. VTR represents a variable tandem repeat (see text). Restriction sites used for construction of these deletion mutants are also shown. All the deletion mutants in the 5'- and 3'-flanking regions of c-H-*ras* are shown by thick lines with arrows, which indicate the remaining parts of the genome. Relative transforming activities are expressed as a fraction of the mean activity obtained with pHSG1426 or pEJ. The plasmids pHSG1426 (*ras*) and pHSG274 (*neo*) transformed NIH3T3 cells at efficiencies of 7.3×10^4 and 1.6×10^5 transformants per 10^6 cells per pmol of DNA respectively under our experimental conditions, as described previously [23]. pHSG274 contained herpes simplex virus thymidine kinase promoter in front of the *neo* gene [59]. Nucleotide positions are defined in relation to the adenine residue of the initiation methionine codon, which is position 1.

started to decrease when the guanine nucleotide at position −1457 was deleted and was abolished completely when the deletion reached guanine at position −1307. It was never restored by further deletions.

In contrast to the 5' deletions, transforming activities of 3' deletions in this region were abolished at position −121, recovered at position −293, abolished again at position −422 and recovered again at position −1070 (Fig. 3). Thus there are two sets of positive (P) and negative (N) elements in the 5'-flanking region which are not necessarily required for expression of the transforming activity of the c-H-*ras* oncogene, but the absence of which influences the transforming activity negatively

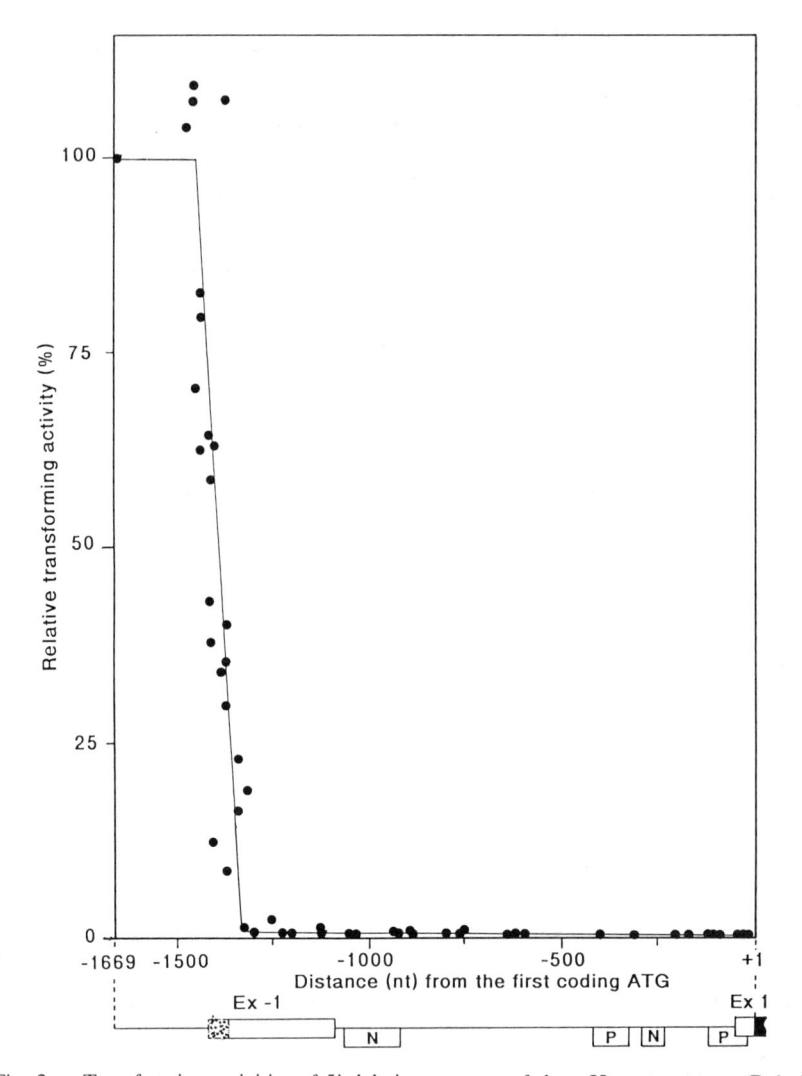

Fig. 2 — Transforming activities of 5′ deletion mutants of the c-H-*ras* oncogene. Relative transforming activity is expressed as a percentage of the mean activity obtained by pHSG1426. The abscissa represents the endpoint of deletions of each 5′ deletion mutant of the c-H-*ras* oncogene. The structure of the 5′-flanking region of the c-H-*ras* oncogene is schematically shown at the bottom of the figure for comparison; coding and non-coding exon regions are indicated by closed and open boxes respectively. Open boxes with N and P indicate negative and positive regions (see text). Dotted box represents the principal promoter region.

and positively respectively [23]. This does not, however, imply necessarily that these elements function as positive and negative regulatory factors, when they are present. P (positive) regions were located between positions −17 and −121 and positions −326 and −422, and N (negative) regions occurred between positions −232 and −293 and positions −922 and −1070. The 5′ deletion analysis indicated that the 5′

Anti-cancer elements in oncogenes

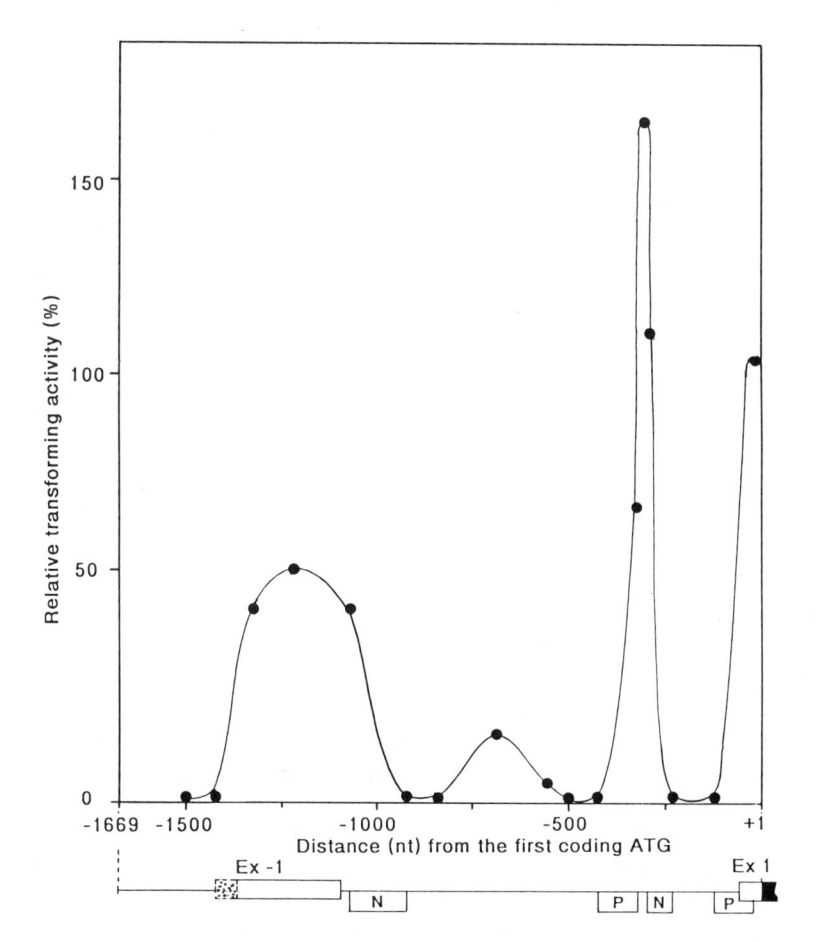

Fig. 3 — Transforming activities of 3′ deletion mutants in the 5′-flanking region of the c-H-*ras* oncogene. Relative transforming activity is expressed as a percentage of the mean activity obtained by pHSG1426. Each 3′ deletion mutant was placed in front of the coding region of the c-H-*ras* oncogene on pHSG385 which contains nucleotides from positions −11 to 3056. The abscissa represents the endpoint of deletions of each 3′ deletion mutant of the c-H-*ras* oncogene. The structure of the 5′-flanking region of the c-H-*ras* oncogene is shown at the bottom of the figure for comparison as in Fig. 2.

endpoint of the largest deletion mutant that gave an equivalent transforming activity to that of pEJ was at position −1458 (Fig. 2) [23].

Subsequently, we localized the promoter region by S1 endonuclease mapping and deletion analysis from both sides [23]. The results of S1 endonuclease mapping analysis indicated that the two major transcription start sites were located at positions −1371 and −1297. On the other hand, six different deletion mutants were isolated for focus formation assay, which retained nucleotides from position −1458 to position −1287, −1320, −1342, −1368 or −1420 and nucleotides from positions −1418 to −1368; these were designated p*ras*172, p*ras*139, p*ras*117, p*ras*91, p*ras*39 and p*ras*51 respectively. The various deletions were placed in front of the c-H-*ras* coding region in the c-H-*ras* promoter probe vector pHSG385 [44] and their transforming activity was examined [23]. All promoter sequences which contained the region from position −1418 to position −1368 gave efficient transforming activities at levels of 20–59% of that in pHSG1426. Rather surprisingly, smaller promoters such as p*ras*51, p*ras*91, p*ras*117 and p*ras*139 had somewhat higher transforming activities than p*ras*172 did, although the significance of these differences was not clear [23].

These results show that the principal promoter sequence of c-H-*ras* oncogene may be located in the relatively short stretch of DNA sequence, 51 bp or less, somewhere between −1418 and −1368 on the basis of focus-forming analysis of various deletion mutants in the 5′-flanking region [23,44]. A similar observation was reported by Trimble and Hozumi on the basis of CAT assay analysis [43]. The proposed principal promoter does not contain the consensus sequences, such as the CAAT and TATA box [60], which are usually observed in promoter regions directed by RNA polymerase II (pol II), but is instead very GC rich (78%). The typical pol II directed promoter containing the consensus sequences generates many unspecific mRNA start sites if the TATA sequence is deleted [61]. The TATA box sequence may therefore be responsible for a fixed unidirectional initiation site for RNA polymerase II [62]. In fact, S1 nuclease mapping analysis revealed the presence of two clusters of the mRNA start sites as briefly described in the preceding section which contain multiple start sites, one within and the other further downstream of the principal promoter. The upstream major start site was located on the adenine residue at position −1371 at the 3′ end of the principal promoter [23,44]. The location of this start site agrees well with the hypothesis that the exon sequence is generally well conserved when compared among different species such as human and rodents [44].

Although the length of 51 bp promoter is unusually short, it is possible that only one GC box and an appropriate DNA sequence to bind RNA polymerase II may be sufficient to induce transcription.

In contrast to our finding, Osborne *et al.* [63] have reported that, in the hamster HMG coenzyme A reductase gene, a so-called housekeeping gene, the promoter region is dispersed over 500 bp, extending to 300 bp upstream of transcription start sites.

On the other hand, Melton *et al.* [64] have reported that in the mouse HPRT gene, another housekeeping gene, the promoter is located within a 49 bp fragment containing a GC box [65]. These findings may be consistent with each other if the dispersed promoter contains several sets of subpromoters that increase the principal promoter activity to various degrees or that exhibit minor transcriptional activities themselves.

In conclusion, (1) the 51 bp promoter of c-H-*ras*, the location of which proposed to be between −1418 and −1368, may indeed represent the minimum requirement for initiation of transcription of the gene, and (2) there is a set of sequence elements which influence gene expression of c-H-*ras* in the 5′-flanking region between the promoter region and the coding region when sequence deletion is introduced from the 3′ side at nucleotide position −11 [44].

3. REGULATORY ELEMENTS IN INTRON −1

The positions of the intron −1 sequence together with its donor and acceptor sites have been reported by two independent groups (in ref. [19] and by Nomura *et al.*, personal communication cited in ref. [23]) to be located between nucleotide positions −1093 and −54. P and N regions found during deletion analysis from the 3' side in the 5' flanking region are therefore located mainly within the intron −1 sequence. However, since all the 3' deletions lack the acceptor site of intron −1, interpretation of the results would be hindered by the complex effect of possible splicing disorder. For example, the donor site of intron −1 could use the acceptor site of intron 1, and therefore coding exon 1 could be spliced out as observed by RNase protection assay [44]. To analyse further the influence of these regions on gene expression, we have therefore constructed various internal deletions between the promoter and the acceptor site by combining 3' and 5' deletions. The structures of the internal deletions are shown in Fig. 4.

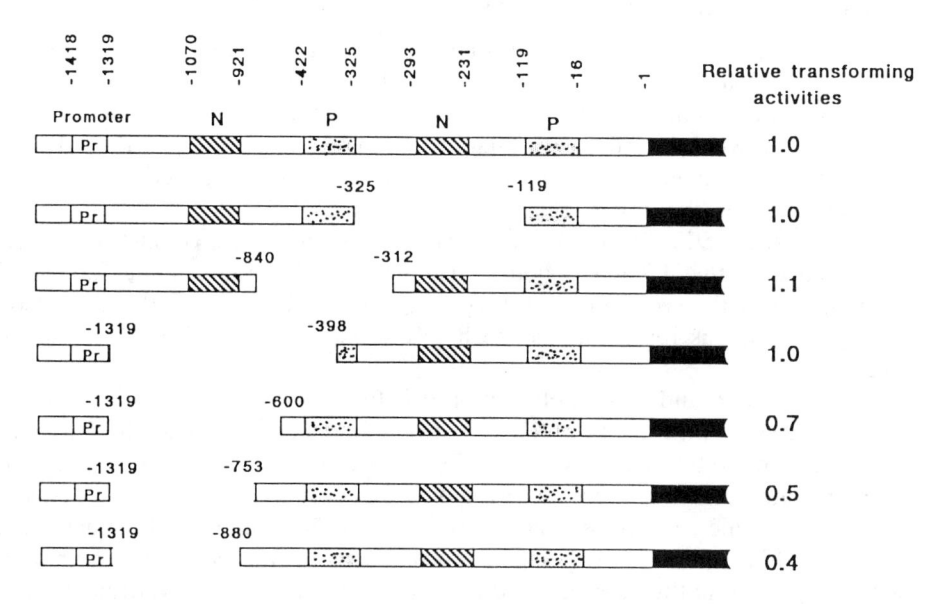

Fig. 4 — Structure and transforming activities of internal deletions. All internal deletion mutants were constructed by combining 3' and 5' deletions. The remaining sequences in each internal deletion mutant are shown by boxes. The relative transforming activities are scored compared with that of pHSG1426. The deletion endpoints of each mutant are shown above each box. Open boxes with Pr represent the principal promoter region. Hatched and dotted boxes represent negative and positive elements respectively in the 5'-flanking region (see text). Closed boxes represents the coding region of exon 1.

In cases where both the donor site and the acceptor site were retained and either the middle P region (from −840 to −312) or the middle N region (from −325 to −119) was removed, no influence was observed on transforming activity. However, when the donor site alone was deleted to various extents, or when the deletion originated from position −1318 in exon −1 to several positions in intron −1 such as −880, −753, −600 or −398, definite inhibitory effects on focus formation efficiency were observed depending on the sequence length retained. The relative transforming activities were 0.4, 0.5, 0.7 and 1.0 respectively, compared with pHSG1426 or pEJ which has a valine residue at amino acid position 12, one of the strongest mutations [15] (Fig. 4). By RNase protection assay, it was observed that the intron −1 sequence contained in each deletion was preserved in the mature mRNA and that there was not much difference in the amount of mRNA produced from among these samples with deletions. On the other hand, consistent reduction in the amount of H-*ras* protein p21 was observed in NIH3T3 cells transformed with these deletions by immunoprecipitation analysis using anti-H-*ras* p21 antibodies [66, in preparation]. Taking into account both these observations, it is very likely that some of the intron sequences between the upstream N and P regions can reduce the efficiency of translation initiation when they are preserved in mRNA.

4. ANTI-CANCER ELEMENT IN ONCOGENE

As described above, it is very likely that the first intron (intron −1) of c-H-*ras* contains a regulatory element (or elements) which affects gene expression post-transcriptionally in some deletion mutants. However, the intron −1 sequence must be present in mRNA in order to regulate the translational activity of the mRNA. Although intron sequences are normally spliced out from the transcript before the translational event, the splicing out of intron −1 must be disordered with respect to deletion mutants. If the intron −1 sequence has biological function in such a manner, the sequence in the deleted regions must therefore be under some functional constraint during the evolutionary history of oncogenes. This implies that the intron −1 sequence in this region should be highly conserved throughout mammalian evolution.

In fact, Kikuno and Hashimoto-Gotoh [67] found a highly conserved segment (−511 to −281) in intron −1 or c-H-*ras* between human and rat (Fig. 5). Previous studies of comparative analysis of DNA sequence homology have revealed that intron sequences, except for boundary regions, are in general highly diverged except those of organelle and ribosomal genes [68,69]. The value of the nucleotide difference K [70], defined as the number of mismatches per nucleotide site in a pair of aligned sequences in this segment, is 0.26, which is substantially lower than that of the remaining part of this region ($K=0.41$). This is also significantly low when compared with other introns studied previously [70]. K values of introns normally fall within a range between 0.35 and 0.51 when compared between human and rodents, which branched at the point of mammalian divergence about 75 million years ago. The rate of nucleotide substitution in this conserved region is significantly lower ((2.1×10^{-9}/site)/year) than those observed in introns ((3.7×10^{-9}–5.3×10^{-9}/site)/year) [67,70] and in silent positions in the coding exons ((5.3×10^{-9}–5.5×10^{-9}/site)/year) [71]. In cases where functional constraints should operate, for example on amino acid replacement positions, the rate falls within a range between (5×10^{-11}/site)/year and (2.7×10^{-9}/site)/year [71]. This clearly implies the conserved sequence observed in the intron −1 of c-H-*ras* has been under evolutionary constraint since the branching of mammalian species.

More strikingly, there is an internal segment from −338 to −289 in this conserved sequence which is especially highly conserved; $K=0.05$. These values can be explained if there is a common secondary structure(s) or a functional reading frame(s) between human and rat.

Ch. 6] Anti-cancer elements in oncogenes 127

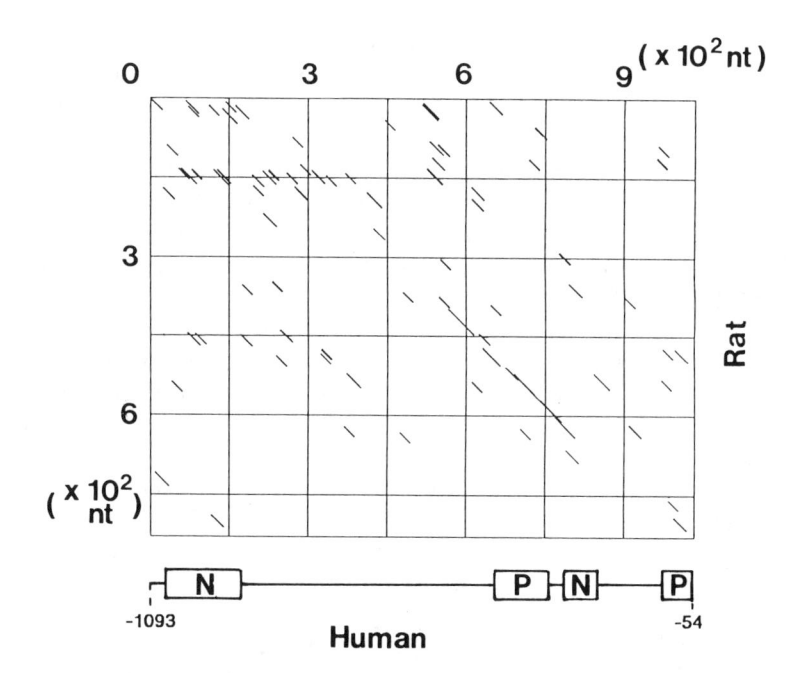

Fig. 5 — Homology matrix of c-H-*ras* intron −1 sequences between human and rat. The horizontal axis represents human sequences and the vertical axis represents the rat sequences. Each diagonal line indicates a segment of 20 nucleotides (nt) with sequence similarity of more than 70%. The boxes with N and P are as described in Fig. 2. The intron −1 sequences of human and rat are 1040 nucleotides and 831 nucleotides long respectively.

5. POSSIBLE ROLE OF THE FIRST INTRON OF c-H-*ras* IN GENE EXPRESSION

A human oncogene, c-*sis*, has been reported to contain a translational inhibitory element in the 5′-non-coding exon [72]. Its mRNA may form secondary structures within the untranslated region (stability of the structure $\Delta G = -100$ kcal/mol). If the conserved sequence in intron −1 of c-H-*ras* has such inhibitory activity, the expression of c-H-*ras* would be suppressed when the intron is present in the transcript by removal of the intron −1 donor sites. If this is the case, these elements would represent an anti-cancer element against oncogene activation by gene truncation or translocation anywhere between the donor site position and the 5′ end of the conserved region which corresponds roughly to the region between the first negative and positive elements (Fig. 5). These putative anti-cancer elements found in human and rodent c-H-*ras* genes are very GC rich (ca. 70%), and two possible secondary structures were found in this region with stabilities ranging from $\Delta G = -140$ kcal/mol to -130 kcal/mol as was observed in the case of c-*sis* [44].

More interestingly, nearly complete inverted repeats were found in the region from −480 to −452 and in the region from −25 to +5 including the first ATG [44] as reported previously [73,74]. Therefore, if the intron −1 donor site of c-H-*ras* is removed by a truncation or translocational event, formation of this secondary structure may serve also as an inhibitory element of translation or as an anti-cancer element. A stem structure of mRNA between the 5'-untranslated exon and the first coding exon of the c-*myc* gene has been proposed to inhibit efficient translation in normal cells, and some truncation or translocation event which removes the 5'-untranslated exon from c-*myc* gene may activate translation of mRNA by escaping the suppression mechanism [46]. Since the intron -1 of c-*myc* is also highly conserved ($K=0.29$), there is a possibility that the conserved region in intron -1 may also exhibit similar function of translation inhibition as proposed previously for c-H-*ras* [44,67].

On the other hand, it is of interest to note whether all introns of oncogenes are more conserved than those of non-oncogene, or whether intron -1 is particularly conserved. We have observed that oncogene introns are generally more conserved not only in intron -1 sequences but also in some plus introns [44,67]. However, the significance of this finding is not clear. In any case, our model predicts that there may be some rare clinical cases in which tumour cells contain normal *ras* oncogene whose expression has been elevated by some means such as gene recombination (Fig. 6).

ACKNOWLEDGEMENT

This chapter is dedicated to Professor Hilger on his 60th birthday and we thank him for his strong support of our research activity in Japan.

REFERENCES

[1] Bishop, J. M. (1983). Cellular oncogenes and retroviruses. *Annu. Rev. Biochem.* **52** 301

[2] Der, C. J., Krontiris, T. G. & Cooper, G. M. (1982). Transforming genes of human bladder and lung carcinoma cell lines are homologous to the *ras* genes of Harvey and Kirsten sarcoma viruses. *Proc. Natl. Acad. Sci. USA* **79** 3637–3640

[3] Pulciani, S., Santos, E., Lauver, A. V., Long, L. K., Robbins, K. C. & Barbacid, M. (1982). Oncogenes in human tumor cell lines: molecular cloning of a transforming gene from human bladder carcinoma cells. *Proc. Natl. Acad. Sci. USA* **79** 2845–2849

[4] Tabin, C. J., Bradley, S. M., Bargmann, C. I., Weinberg, R. A., Papageorge, A. G., Scolnick, E. M., Dhar, R., Lowy, D. R. & Chang, E. H. (1982). Mechanism of activation of a human oncogene. *Nature (London)* **300** 143–149

[5] Shimizu, K., Goldfarb, M., Perucho, M. & Wigler, M. (1983). Isolation and preliminary characterization of the transforming gene of a human neuroblastoma cell line. *Proc. Natl. Acad. Sci. USA* **80** 383–387

[6] Albino, A. P., Strange, R. L., Oliff, A. I., Furth, M. E. & Old, L. J. (1984). Transforming *ras* genes from human melanoma: a manifestation of tumour heterogeneity? *Nature (London)* **308** 69–72

[7] Damante, G., Filetti, S. & Rapoport, B. (1987). Nucleotide sequence and characterization of the 5'-flanking region of the rat Ha-*ras* protooncogene. *Proc. Natl. Acad. Sci. USA* **84** 774–778

[8] Hoffman, E. K., Trusko, S. P., Freeman, N. & George, D. L. (1987).

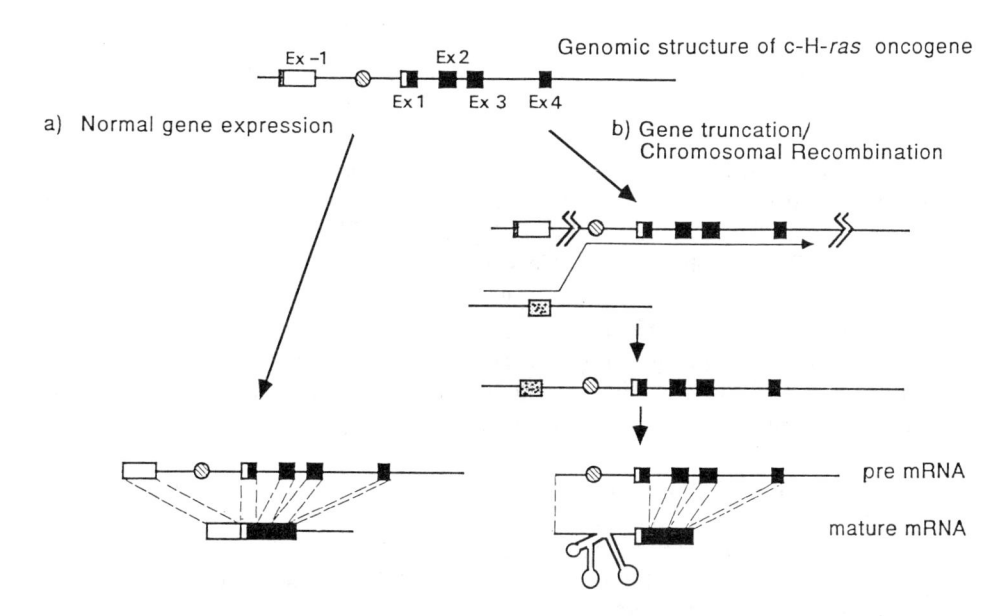

Fig. 6 — Model for anti-cancer elements. (a) During normal gene expression, the four introns (intron −1 to intron 3) are spliced out from the pre mRNA of the c-H-*ras* oncogene. Thus the mature mRNA does not contain the possible regulatory elements located in intron −1. (b) Gene truncation or chromosomal translocation events will change the genomic structure in the 5′-flanking region of c-H-*ras* oncogene. When the rearrangement occurs between the intron −1 splicing donor site and 5′ termini of the putative anti-cancer element resulting in possible transcriptional activation directed by a newly positioned promoter, the element will remain in the mature mRNA by splicing disorder of intron −1, which may reduce the expression at the translation level by forming stable secondary structure(s) and compensate for the wayward transcriptional activation, thus acting as a safety valve. Open and closed boxes indicate the non-coding and coding exons respectively. Dotted boxes show the promoter region. Putative anti-cancer elements are shown by hatched circles and the assumed stem structure is also shown at the bottom.

Structural and functional characterization of the promoter regions of the mouse c-Ki-*ras* gene *Mol. Cell. Biol.* **7** 2592–2596

[9] Capon, D. J., Chen, E. Y., Levinson, A. D., Seeburg, P. H. & Goeddel, D. V. (1983). Complete nucleotide sequences of the T24 human bladder carcinoma oncogene and its normal homologue. *Nature (London)* **302** 33–37

[10] Fasano, O., Taparowsky, E., Fiddes, J., Wigler, M. & Goldfarb, M. (1983). Sequence and structure of the coding region of the human H-*ras*-1 gene from T24 bladder carcinoma cells. *J. Mol. Appl. Genet.* **2** 173–180

[11] Kelly, K., Cochran, B. H., Stiles, C. D. & Leder, P. (1983). Cell-specific regulation of the c-*myc* gene by lymphocyte mitogens and platelet-derived growth factor. *Cell* **35** 603–610

[12] McGrath, J. P., Capon, D. J., Smith, D. H., Chen, E. Y., Seeburg, P. H., Goeddel, D. V. & Levinson, A. D. (1983). Structure and organization of the human Ki-*ras* proto-oncogene and related processed pseudogene. *Nature (London)* **304** 501–506

[13] Shimizu, K., Goldfarb, M., Suard, Y., Perucho, M., Li, Y., Kamata, T., Feramisco, J., Stavnezer, E., Fogh, J. & Wigler, M. H. (1983). Three human transforming genes are related to the viral *ras* oncogenes. *Proc. Natl. Acad. Sci. USA* **80** 2112–2116

[14] Brodeur, G. M., Seeger, R. C., Schwab, M., Varmus, H. E. & Bishop, J. M. (1984). Amplification of N-*myc* in untreated human neuroblastomas correlates with advanced disease stage. *Science* **224** 1121–1124

[15] Seeburg, P. H., Colby, W. W., Capon, D. J., Goeddel, D. V. & Levinson, A. D. (1984). Biological properties of human c-Ha-*ras*1 genes mutated at codon 12. *Nature (London)* **312** 71–75

[16] Sekiya, T., Fushimi, M., Hori, H., Hirohashi, S., Nishimura, S. & Sugimura, T. (1984). Molecular cloning and the total nucleotide sequence of the human c-Ha-*ras*-1 gene activated in a melanoma from a Japanese patient. *Proc. Natl. Acad. Sci. USA* **81** 4771–4775

[17] Darveau, A., Pelletier, J. & Sonenberg, N. (1985). Differential efficiencies of *in vitro* translation of mouse c-*myc* transcripts differing in the 5' untranslated region. *Proc. Natl. Acad. Sci. USA* **82** 2315–2319

[18] Hall, A. & Brown, R. (1985). Human N-*ras*: cDNA cloning and gene structure. *Nucleic Acids Res.* **13** 5255–5268

[19] Ishii, S., Merlino, G. T. & Pastan, I. (1985). Promoter region of the human Harvey *ras* proto-oncogene: similarity to the EGF receptor proto-oncogene promoter. *Science* **230** 1378–1381

[20] Ishii, S., Kadonaga, J. T., Tjian, R., Brady, J. N., Merlino, G. T. & Pastan, I. (1986). Binding of the sp1 transcription factor by the human Harvey *ras*1 proto-oncogene promoter. *Science* **232** 1410–1413

[21] Land, H., Chen, A. C., Morgenstern, J. P., Parada, L. F. & Weinberg, R. A. (1986). Behavior of *myc* and *ras* oncogenes in tranformation of rat embryo fibroblasts. *Mol. Cell. Biol.* **6** 1917–1925

[22] Trimble, W. S., Johnson, P. W., Hozumi, N. & Roder, J. C. (1986). Inducible cellular transformation by a metallothionein-*ras* hybrid oncogene leads to natural killer cell susceptibility. *Nature (London)* **321** 782–784

[23] Honkawa, H., Masahashi, W., Hashimoto, S. & Hashimoto-Gotoh, T. (1987). Identification of the principal promoter sequence of the c-H-*ras* transforming oncogene: deletion analysis of the 5'-flanking region by focus formation assay. *Mol. Cell. Biol.* **7** 2933–2940

[24] Sinn, E., Muller, W., Pattengale, P., Tepler, I., Wallace, R. & Leder, P. (1987). Coexpression of MMTV/v-Ha-*ras* and MMTV/c-*myc* genes in transgenic mice: synergistic action of oncogenes *in vivo*. *Cell* **49** 465–475

[25] Spandidos, D. A. (1987). Oncogene activation in malignant transformation: a study of H-*Ras* in human breast cancer. *Anticancer Res.* **7** 991–996

[26] De Vos, A. M., Tong, L., Milburn, M. V., Matias, P. M., Jancarik, J., Noguchi, S. & Nishimura, S. (1988). Three-dimensional structure of an oncogene protein: catalytic domain of human c-H-*ras* p21. *Science* **239** 888–893

[27] Leon, J., Kamino, H., Steinberg, J. J. & Pellicer, A. (1988). H-*ras* activation in benign and self-regressing skin tumors (keratoacanthomas) in both humans and an animal model system. *Mol. Cell. Biol.* **8** 786–793

[28] Alema, S., Casalbore, P., Agostini, E. & Tato, F. (1985). Differentiation of

PC12 phaeochromocytoma cells induced by v-*src* oncogene. *Nature (London)* **316** 557–559

[29] Bar-Sagi, D. & Feramisco, J. R. (1985). Microinjection of the *ras* oncogene protein into PC12 cells induces morphological differentiation. *Cell* **42** 841–848

[30] Gibbs, J. B., Sigal, I. S. & Scolnick, E. M. (1985). Biochemical properties of normal and oncogenic *ras* P21. *TIBS* **10** 350–353

[31] Manne, V., Bekesi, E. & Kung, H.-F. (1985). Ha-*ras* proteins exhibit GTPase activity: point mutations that activate Ha-*ras* gene products result in decreased GTPase activity. *Proc. Natl. Acad. Sci. USA* **82** 376–380

[32] Noda, M., Ko, M., Ogura, A., Liu, D.-G., Amano, T., Takano, T. & Ikawa, Y. (1985). Sarcoma viruses carrying *ras* oncogenes induce differentiation-associated properties in a neuronal cell line. *Nature (London)* **318** 73–75

[33] Srivastava, S. K., Yuasa, Y., Reynords, S. H. & Aaronson, S. A. (1985). Effects of two major activating lesions on the structure and conformation of human *ras* oncogene products. *Proc. Natl. Acad. Sci. USA* **82** 38–42

[34] Brown, K., Quintanilla, M., Ramsden, M., Kerr, I. B., Young, S. & Balmain, A. (1986). v-*ras* genes from Harvey and BALB murine sarcoma viruses can act as initiations of two-stage mouse skin carcinogenesis. *Cell* **46** 447–456

[35] Clanton, D. J., Hattori, S. & Shih, T. Y. (1986). Mutations of the *ras* gene product p21 that abolish guanine nucleotide binding. *Proc. Natl. Acad. Sci. USA* **83** 5076–5080

[36] Lacal, J. C., Srivastava, S. K., Anderson, P. S. & Aaronson, S. A. (1986). Ras p21 proteins with high or low GTPase activity can efficiently transform NIH/3T3 cells. *Cell* **44** 609–617

[37] Papageorge, A. G., Willumsen, B. M., Johnsen, M., Kung, H.-F., Stacey, D. W., Vass, W. C. & Lowy, D. R. (1986). A transforming *ras* gene can provide an essential function ordinarily supplied by an endogenous *ras* gene. *Mol. Cell. Biol.* **6** 1843–1846

[38] Tarpley, W. G., Hopkins, N. K. & Gorman, R. R. (1986). Reduced hormone-stimulated adenylate cyclase activity in NIH-3T3 cells expressing the EJ human bladder *ras* oncogene. *Proc. Natl. Acad. Sci. USA* **83** 3703–3707

[39] Spandidos, D. A. (1985). Mechanism of carcinogenesis: the role of oncogenes, transcriptional enhancers and growth factors. *Anticancer Res.* **5** 485–498

[40] Zarbl, H., Sukumar, S., Arthur, A. V., Martin-Zanca, D. & Barbacid, M. (1985). Direct mutagenesis of Ha-*ras*-1 oncogenes by *N*-nitroso-*N*-methylurea during initiation of mammary carcinogenesis in rats. *Nature (London)* **315** 382–385

[41] Spandidos, D. A. & Riggio, M. (1986). Promoter and enhancer like activity at the 5′-end of normal and T24 Ha-*ras*1 genes. *FEBS Lett.* **203** 169–174

[42] Leon, J., Guerrero, I. & Pellicer, A. (1987). Differential expression of the *ras* gene family in mice. *Mol. Cell. Biol.* **7** 1535–1540

[43] Trimble, W. S. & Hozumi, N. (1987). Deletion analysis of the c-Ha-*ras* oncogene promoter. *FEBS Lett.* **219** 70–74

[44] Hashimoto-Gotoh, T., Kikuno, R., Takahashi, M. & Honkawa, H. (1988). Possible role of the first intron of c-H-*ras* in gene expression: anti-cancer elements in oncogenes. *Anticancer Res.* **8** 851–860

[45] Chang, E. H., Furth, M. E., Scolnick, E. M. & Lowy, D. R. (1982).

Tumorigenic transformation of mammalian cells induced by a normal human gene homologous to the oncogene of Harvey murine sarcoma virus. *Nature (London)* **297** 479–483

[46] Saito, H., Hayday, A. C., Wiman, K., Hayward, W. S. & Tonogawa, S. (1983). Activation of the c-*myc* gene by translocation: a model for translational control. *Proc. Natl. Acad. Sci. USA* **80** 7476–7480

[47] Adams, J. M., Gerondakis, S., Webb, E., Corcoran, L. M. & Cory, S. (1983). Cellular *myc* oncogene is altered by chromosome translocation to an immunoglobulin locus in murine plasmacytomas and is rearranged similarly in human Burkitt lymphomas. *Proc. Natl. Acad. Sci. USA* **80** 1982–1986

[48] Schwab, M., Alitalo, K., Varmus, H. E. & Bishop, J. M. (1983). A cellular oncogene (c-Ki-*ras*) is amplified, overexpressed, and located within karyotypic abnormalities in mouse adrenocortical tumour cells. *Nature (London)* **303** 497–501

[49] George, D. L., Scott, A. F., Trusko, S., Glick, B., Ford, E. & Dorney, D. J. (1985). Structure and expression of amplified c-Ki-*ras* gene sequences in Y1 mouse adrenal tumor cells. *EMBO J.* **4** 1199–1203

[50] Cichutek, K. & Duesberg, P. H. (1986). Harvey *ras* genes transform without mutant codons, apparently activated by truncation of a 5' exon (exon −1). *Proc. Natl. Acad. Sci. USA* **83** 2340–2344

[51] Shimizu, K., Birnbaum, D., Ruley, M. A., Fasano, O., Suard, Y., Edlung, L., Taparowsky, E., Goldfarb, M. & Wigler, M. (1983). Structure of the Ki-*ras* gene of the human lung carcinoma cell line Calu-1. *Nature (London)* **304** 497–500

[52] Taparowsky, E., Shimizu, K., Goldfarb, M. & Wigler, M. (1983). Structure and activation of the human N-*ras* gene. *Cell* **34** 581–586

[53] Bos, J. L., Toksoz, D., Marshall, C. J., Vries, M. V., Veeneman, G. H., van der Eb, A. J., van Boom, J. H., Janssen, J. W. G. & Steenvoorden, A. C. M. (1985). Amino-acid substitutions at codon 13 of the N-*ras* oncogene in human acute myeloid leukaemia. *Nature (London)* **315** 726–730

[54] Der, C. J., Finkel, T. & Cooper, G. M. (1986). Biological and biochemical properties of human *ras*[H] genes mutated at codon 61. *Cell* **44** 167–176

[55] Nishimura, S. & Sekiya, T. (1987). Human cancer and cellular oncogenes. *Biochem. J.* **243** 313–327

[56] Pulciani, S., Santos, E., Long, L. K., Sorrentino, V. & Barbacid, M. (1985). *ras* gene amplification and malignant transformation. *Mol. Cell. Biol.* **5** 2836–2841

[57] Mekay, I. A., Marshall, C. J., Cales, C. & Hall, A. (1986). Transformation and stimulation of DNA synthesis in NIH-3T3 cells are a titratable function of normal p21^{N-ras} expression. *EMBO J.* **5** 2617–2621

[58] Hashimoto-Gotoh, T., Kume, A., Masahashi, W., Takeshita, S. & Fukuda, A. (1986). Improved vector, pHSG664, for direct streptomycin-resistance selection: cDNA cloning with G:C-tailing procedure and subcloning of double-digest DNA fragments. *Gene* **41** 125–128

[59] Brady, G., Jantzen, H. M., Bernard, H. U., Brown, R., Schutz, G. & Hashimoto-Gotoh, T. (1984). New cosmid vector developed for eukaryotic DNA cloning. *Gene* **27** 223–232

[60] Breathnach, R. & Chambon, P. (1981). Organization and expression of eukaryotic split genes coding for proteins. *Annu. Rev. Biochem.* **50** 349–383

[61] Grosschedl, R. & Birnstiel, M. L. (1980). Identification of regulatory sequences in the prelude sequences of an H2A histone gene by the study of specific deletion mutants *in vitro*. *Proc. Natl. Acad. Sci. USA* **77** 1432–1436

[62] Farnham, P. J., Abrams, J. M. & Schimke, R. T. (1985). Opposite-strand RNAs from the 5' flanking region of the mouse dihydrofolate reductase gene. *Proc. Natl. Acad. Sci. USA* **82** 3978–3982

[63] Osbone, T. F., Goldstein, J. L. & Brown, M. S. (1985). 5' end of HMG CoA reductase gene contains sequences responsible for cholesterol-mediated inhibition of transcription. *Cell* **42** 203–212

[64] Melton, D. W., McEwan, C., McKie, A. B. & Reid, A. M. (1986). Expression of the mouse HPRT gene: deletional analysis of the promoter region of an X-chromosome linked housekeeping gene. *Cell* **44** 319–328

[65] Byrne, B. J., Davis, M. S., Yamaguchi, J., Bergsma, D. J. & Subramanian, K. N. (1983). Definition of the simian virus 40 early promoter region and demonstration of a host range bias in the enhancement effect of the simian virus 40 72-base-pair repeat. *Proc. Natl. Acid. Sci. USA* **80** 721–725

[66] Ishihara, H., Nakagawa, H., Ono, K. & Fukuda, A. (1987). Antibodies against synthetic carboxy-terminal peptides distinguish H-*ras* and K-*ras* oncogene products p21. *J. Immunol. Methods* **103** 131–139

[67] Kikuno, R. & Hashimoto-Gotoh, T. (1989). Submitted for publication.

[68] Lomber, A., Hensgens, M., Bonen, L., deHaan, M. M., van den Horst, G. & Grivell, L. A. (1983). Two intron sequences in yeast mitochondrial COX1 gene: homology among ORF-containing introns and strain dependent variation in flanking exons. *Cell* **32** 379–389

[69] Michel, F. & Dujon, B. (1983). Conservation of RNA secondary structures in two intron families including mitochondrial-, chloroplast- and nuclear-encoded members. *EMBO J.* **2** 33–38

[70] Hayashida, H. & Miyata, T. (1983). Unusual evolutionary conservation and frequent DNA segment exchange in class I genes of the major histocompatibility complex. *Proc. Natl. Acad. Sci. USA* **80** 2671–2675

[71] Miyata, T., Hayashida, H., Kikuno, R., Toh, H. & Kawade, Y. (1985). Evolution of interferon genes. In: I. Grösser (ed) *Interferon*, Vol. 6, Academic Press, London, pp. 1–30

[72] Rao, C. D., Pech, M., Robbins, K. C. & Aaronson, S. A. (1988). The 5' untranslated sequence of the c-*sis*/platelet-derived growth factor 2 transcript is a potent translational inhibitor. *Mol. Cell. Biol.* **8** 284–292

[73] Kozak, M. (1986). Influences of mRNA secondary structure on initiation by eukaryotic ribosomes. *Proc. Natl. Acad. Sci. USA* **83** 2850–2854

[74] McPheeters, D. S., Christensen, A., Young, E. T., Stormo, G. & Gold, L. (1986). Translational regulation of expression of the bacteriophage T4 lysozyme gene *Nucleic Acids Res.* **14** 5813–5826

7

The oncogenic potential of fibroblast growth factor genes

Sagrario Ortega and **Kenneth A. Thomas**
Department of Biochemistry, Merck Institute for Therapeutic Research, Merck Sharp and Dohme Research Laboratories, P.O. Box 2000, Rahway, NJ 07065, USA

1. INTRODUCTION

Protein growth factors play a major role in the control of cell division and differentiation. These molecules are, in essence, hormones some of which act in a classical systemic endocrine manner. Most growth factors, however, appear to act in a local, or paracrine, fashion by non-systemic diffusion. Even blood-borne growth factors that are released at sites of injury by degranulation of platelets are presumably meant to provide paracrine mitogenic support to promote repair of damaged tissue. Autocrine stimulation might have normal physiological functions including support for rapid growth during embryonic development. In this case, however, the duration of the autocrine stimulation is presumably limited by undefined means, perhaps including secondary regulation of autocrine growth factor expression.

Aberrant endocrine, paracrine and autocrine control by growth factors can lead to a variety of pathologies. For example, elevated levels of circulating insulin-like growth factor I resulting from excess growth hormone can cause acromegaly whereas local non-cancerous hyperplasias such as psoriasis might be mediated by defective paracrine control. A particularly insidious lesion of growth factor regulation, however, is generation of unrestrained autocrine stimulation since a population of self-stimulatory cells has little, if any, feedback control on its persistent expansion.

In principle, oncogene-associated transformation of cells from normal to cancerous can be mediated not only by defects in the regulation of autocrine growth factor expression, but also by any donwstream lesion in the signal transduction cascade of events initiated by growth factor that results in a chronic stimulation mitogenic signal. Although constitutive induction of post-receptor mitogenic signals might account for the majority of tumours, clear examples exist of both naturally occurring and experimentally induced autocrine growth factor-dependent transformation.

The first such direct link in a naturally occurring tumour between protein growth factor expression and transformation was established with the finding of structural and functional similarities between the protein product of the simian sarcoma virus oncogene v-sis and the B chain of platelet-derived growth factor [1,2]. Recently, three oncogenes that encode proteins of the fibroblast growth factor (FGF) family have been found that provide additional perspective concerning the relevance of the autocrine expression of growth factors to cellular transformation [3]. In this chapter we will review the oncogenic potential of members of the FGF family.

2. ACIDIC AND BASIC FIBROBLAST GROWTH FACTORS

Although FGF-like activities have been recognized for 50 years [4,5], the purifications to apparent homogeneity of the first two FGFs, acidic FGF (aFGF) and basic FGF (bFGF), were only recently reported [6,7]. The 155 residue primary structures of these two prototypic FGFs (Fig. 1) are 55% identical as determined by direct amino acid sequencing [8–10] and, subsequently, as deduced from the corresponding cDNA sequences [11–16]. The genomic clones for both aFGF and bFGF each contain two introns located at equivalent sites flanked by coding regions [17,18]. Although a stop codon immediately upstream of the cDNA coding sequencing demonstrates that longer forms of aFGF are not expected to exist, no such upstream termination codon has been observed preceding the analogous bFGF coding sequence. In fact, larger forms of bFGF with masses of up to approximately 25 kDa have been identified which contain amino terminal extensions demonstrating the existence of alternative upstream translational initiation sites [15,19]. The sequence of a hepatoma cDNA has recently been described encoding 21 and 22.5 kDa bFGFs that are initiated from CUG start codons [20]. Curiously, neither aFGF nor bFGF, translated from any of the currently recognized start sites, contains hydrophobic leader sequences assumed to be required for active secretion from cells. These forms of FGF must contain unrecognized cryptic leader sequences, be secreted by unknown mechanisms or only be released from cells by leakage and lysis.

Both aFGF and bFGF are mitogenic for a wide range of targets including many, if not all, cells of mesodermal (fibroblasts, endothelial cells, myoblasts, chondrocytes, osteoblasts) and ectodermal (neuroblasts and glia) origin [8,18,21,22]. Both proteins not only are mitogenic and chemotactic for vascular endothelial cells in culture but also induce blood vessel growth *in vivo* [23–25]. This largely common set of target cells is consistent with the observation, based on radioreceptor competition binding and ligand–receptor cross-linking, that both FGFs share at least one common receptor [26,27]. The presence or absence of other receptors that bind preferentially to one of these FGFs has not yet been rigorously established.

These two prototypic FGFs also bind tightly to heparins, glycosaminoglycans released by mast cell degranulation, and heparans, proteoglycans found in all basement membranes and on the plasma membranes of some cells. The tight binding to these highly sulphated and structurally related oligosaccharides might be a

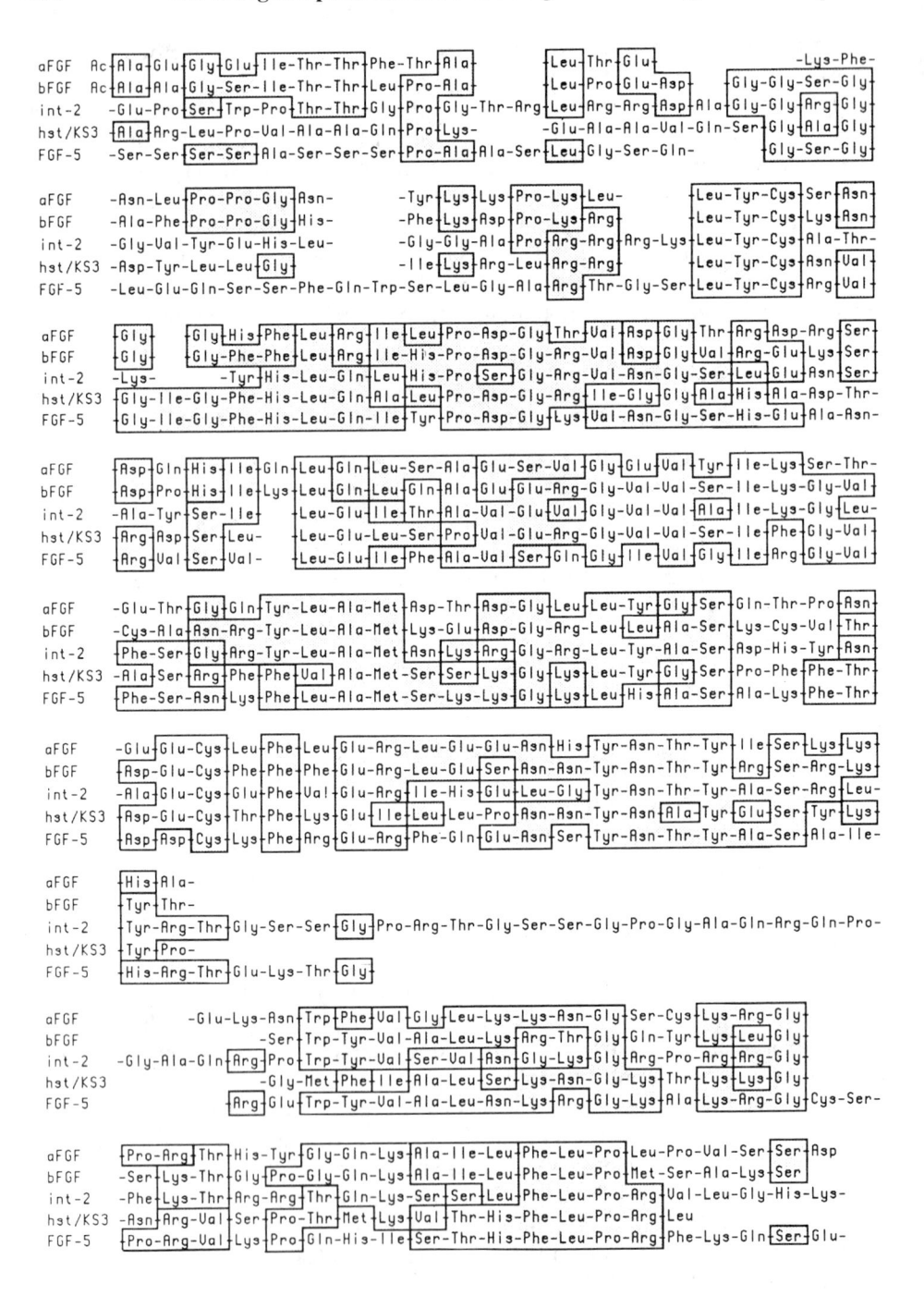

Fig. 1 — Amino acid sequence homologies within the FGF family. Human aFGF, human bFGF, mouse int-2, human hst/KS3 (also called KFGF) and human FGF-5 are aligned with identical residues enclosed in boxes. The long unique amino and carboxy terminal extensions that are encoded by the *hst*, *int*-2 and *FGF*-5 genes are not shown.

physiologically relevant means of selectively partitioning FGFs onto basement membranes, structures which appear to contribute to the generation and stability of normal tissue morphology. Heparins and, perhaps, heparans also protect aFGF and bFGF from inactivation *in vitro* by elevated temperatures, extreme pHs and degradation by some proteases [28–30]. Furthermore, heparins and heparans augment the mitogenic activity of aFGF, but not bFGF, in culture perhaps by stabilizing the active conformation of the relatively labile aFGF.

No clear indication exists that the genes for either aFGF or bFGF are oncogenic in naturally occurring tumours. Although bFGF has been found in solid tumours [18,21,31–34] and both aFGF and bFGF have been reported to exist in transformed cells in culture [35–40] they are also present in normal tissues and cells [18,21,41]. Their expression seems to be elevated in actively dividing cells. The levels of bFGF mRNA are low in sparse and serum-deprived cultures but are rapidly elevated on either replating the cells at subconfluence or adding serum [42]. The presence of these FGFs in tumours might, therefore, reflect expression as a consequence of active cell division that is primarily caused by other oncogenic stimuli. However, when a solid tumour reaches a critical size and becomes necrotic any leaderless FGFs released as a consequence of tumour cell lysis might contribute to both autocrine tumour stimulation and paracrine support of local hyperplasia including tumour angiogenesis.

Expression of elevated levels of either aFGF [43] or bFGF [44,45] from transfected genes results in an altered cellular morphology and growth to higher density in culture. In the case of bFGF these effects could be inhibited by an anti-bFGF antibody [44] demonstrating that mitogenesis was mediated by release of the growth factor into the medium. Since cellular turnover is observed in culture, bFGF might have been passively released by dying or dead cells. The aFGF-expressing cells were only able to form small non-progressive hyperplastic foci in nude mice [43].

However, if the leaderless FGFs were fused to an artificial hydrophobic leader sequence then their high and persistent expression coupled with secretion might support tumour growth. This hypothesis has been confirmed by construction of a fusion protein composed of an immunoglobulin leader sequence and bFGF (IgbFGF) [46]. Transfection of NIH 3T3 cells with a plasmid containing this fusion gene resulted in the growth of cells that, following selection for altered morphology, grew as large aggregates that adhered poorly to the culture dish. Although the growth of these cells in soft agar, an indicator of transformation, was poor they produced rapidly growing tumours in all 50 syngeneic mice tested. In contrast, no tumours were induced by the untransformed NIH 3T3 cells and only one tumour was found in 27 mice implanted with leaderless bFGF-expressing cells. Steady state levels of secreted bFGF in the media were below the limits of detection. The mitogen might have bound to its receptor and heparin proteoglycans either in a common secretory compartment or following secretion. Alternatively, the protein might have been rapidly degraded in solution.

A similar result has been found [47] following expression in NIH 3T3 cells of a fusion product between the secretory leader sequence of growth hormone and bFGF (GHbFGF). In contrast to the IgbFGF construct in which steady state levels of soluble extracellular bFGF were not detected, 80% of the total bFGF immunocrossreactive material was found in conditioned media of the GHbFGF-expressing cells compared with only 3% in the leaderless bFGF controls. However, the immuno-cross-reactive forms detected in the conditioned medium of GHbFGF-expressing cells were mainly high molecular weight forms (35–68 kDa) that were not detected either in extracts from the same cells or in the conditioned media of cells expressing bFGF without the leader sequence. These larger forms probably had either less or no activity since the specific mitogenic activity of total secreted immunocross-reactive protein was less than that of fully active lower mass bFGF. The GHbFGF-expressing cell lines also formed large colonies in soft agar and induced tumours in nude mice whereas the leaderless controls were inactive in both assays. The discrepancy between the IgbFGF- and GHbFGF-expressing cells in both the levels of bFGF in the media and the ability of the cells to grow in soft agar might reflect different efficiencies of expression, processing and secretion. Moreover, these results are consistent with the hypothesis that not only enhanced expression but also efficient secretion of the mitogens are needed to observe FGF-supported tumour growth.

To date, no spontaneously rearranged aFGF or bFGF genes that contain fused leader sequences have been found. Very recently, however, the FGF family has been recognized to contain at least three additional members [3]. These FGF-encoding genes, *int*-2, *hst*/*KS*3, and *FGF*-5, were all identified by either oncogene or growth factor expression cloning screens. As shown in Fig. 1, extensive amino acid sequence homology, ranging from approximately 35% to 55% in the common overlapping regions, exists among the five members of this homologous family of proteins. Although substantial variations exist in the lengths of amino and carboxyl terminal extensions, illustrated schematically in Fig. 2, all three transforming FGF genes encode proteins containing hydrophobic leader sequences as shown in Fig. 3.

3. *int*-2

One of the most common means by which retroviruses induce cell transformation is through 'insertional mutagenesis', a process by which these viruses integrate either within or adjacent to cellular proto-oncogenes causing diminished control of their expression levels [48,49]. Although *int*-2 mRNA is not detectably expressed in normal mouse mammary glands, insertion of the mouse mammary tumour virus (MMTV) into cellular DNA adjacent to the *int*-2 proto-oncogene results in its low level expression (<10 copies of int-2 mRNA/cell) in about 50% of MMTV-induced cancers. Viral integration at the *int*-2 locus occurs at multiple sites, all clustered at both ends of the *int*-2 gene spanning a total of at least 20 kb of cellular DNA [50,51]. Elevated *int*-2 expression is presumed to be mediated by orientation-independent enhancer sequences within the inserted viruses. Nevertheless, the orientation of both insertion site clusters is opposite, each directing transcription away from the central *int*-2 domain. The selection for the non-random orientational distribution of the MMTV sequences within the *int*-2 locus is not understood.

Comparison of the *int*-2 genomic [51] and cDNA [51,52] sequences reveals an open reading frame composed of three exons of 220, 104 and 411 nucleotides separated by two introns of 1.7 and 1.8 kb at sites similar to those identified in the aFGF and bFGF genes. Although a TATA box which is normally present near the RNA polymerase II binding site preceding the coding sequence has not been identified in the *int*-2 gene, GC-rich clusters are present which could serve the same

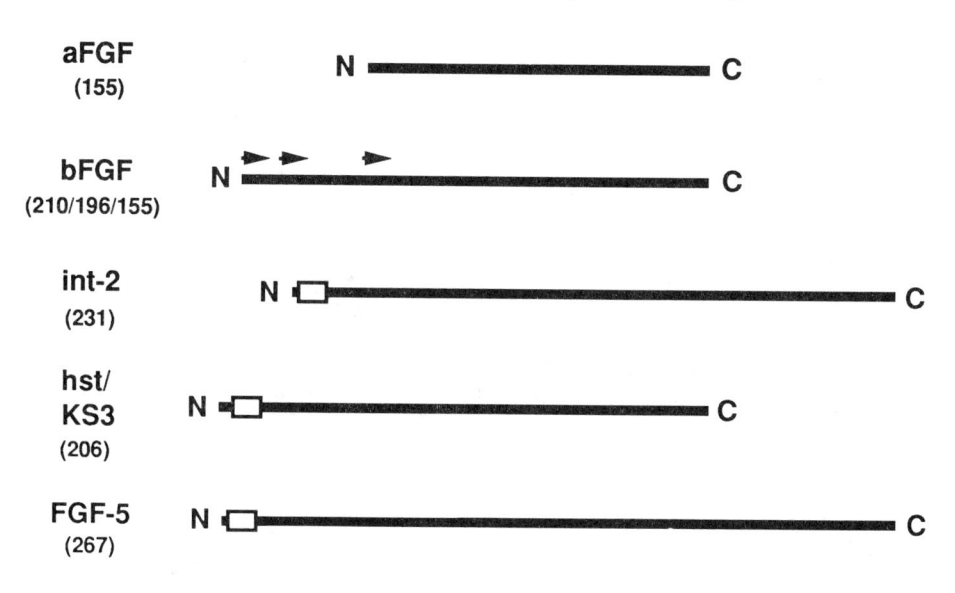

Fig. 2 — Schematic alignments of FGFs. The proteins have been aligned based on the amino acid homologies showed in Fig. 1. The homologous region is contained in the common overlapping central portion of the alignment, although an insertion of 20 residues has to be removed from int-2 to maximize the alignment. The length, in amino acid residues, is shown in parentheses for each protein. Leader sequences are represented as open boxes. Horizontal arrows above bFGF show the amino terminal positions of the 210, 196 and 155 amino acid residue forms of the protein that result from alternative translation initiation sites.

function. The full-length transcript encodes a basic protein of 245 amino acids, with an estimated molecular mass of 27 kDa, sharing a high degree of homology with aFGF and bFGF (Fig. 1). The deduced amino acid sequence contains a single potential Asn glycosylation site [53]. In contrast to aFGF and bFGF, the int-2 protein contains a hydrophobic amino terminus that could act as a secretory signal sequence (Fig. 3). The increased efficiency with which the int-2 protein product could be secreted compared with either aFGF or bFGF might contribute to a greater ability of this oncogenic protein to induce autocrine stimulation.

Transcripts of the *int*-2 gene have been found in 9 of 40 primary breast carcinomas. The single-copy gene was amplified from 2- to 20-fold in all 9 positive samples [54]. In three of the positive samples lymph node metastases were available and also observed to contain *int*-2 gene amplifications. Amplification of *int*-2 sequences is found not only in breast tumours but also in squamous carcinomas of the head and neck [55]. A correlation was found between *int*-2 gene amplification in breast carcinomas and the loss of an EcoR1 restriction site, located in the first intron of the gene, implying that the amplified genes are frequently altered in some as-yet undefined way [54]. Alterations of c-*myc*, *HER2/neu* and *int*-2 have been detected in 22%, 19% and 23% respectively of primary breast tumours [56]. No correlation was found between alterations in c-*myc* and *int*-2 although amplifications of *HER2/neu* and *int*-2 were inexplicably negatively correlated [54].

The normal function of *int*-2 is unknown. It is expressed at detectable levels in

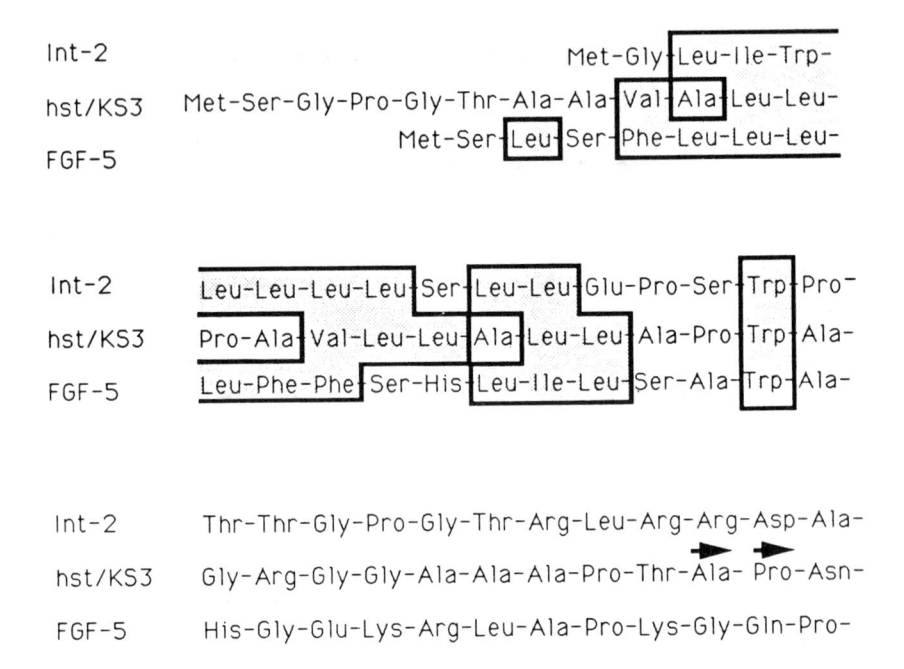

```
Int-2                                          Met-Gly-Leu-Ile-Trp-
hst/KS3     Met-Ser-Gly-Pro-Gly-Thr-Ala-Ala-Val-Ala-Leu-Leu-
FGF-5                         Met-Ser-Leu-Ser-Phe-Leu-Leu-Leu-

Int-2       Leu-Leu-Leu-Leu-Ser-Leu-Leu-Glu-Pro-Ser-Trp-Pro-
hst/KS3     Pro-Ala-Val-Leu-Leu-Ala-Leu-Leu-Ala-Pro-Trp-Ala-
FGF-5       Leu-Phe-Phe-Ser-His-Leu-Ile-Leu-Ser-Ala-Trp-Ala-

Int-2       Thr-Thr-Gly-Pro-Gly-Thr-Arg-Leu-Arg-Arg-Asp-Ala-
hst/KS3     Gly-Arg-Gly-Gly-Ala-Ala-Ala-Pro-Thr-Ala- Pro-Asn-
FGF-5       His-Gly-Glu-Lys-Arg-Leu-Ala-Pro-Lys-Gly-Gln-Pro-
```

Fig. 3 — FGF secretory leader sequences. The amino terminal sequences containing either the potential or known secretory leaders of the mouse int-2, human hst/KS3 and human FGF-5 are shown. Large hydrophobic residues in or near the central hydrophobic regions of the leaders are enclosed in stippled boxes. The two arrows over Ala_{31} and Pro_{32} of hst/KS3 denote the amino terminal residues identified in the secreted protein by amino terminal amino acid sequence analysis [67].

early embryonic tissue [52,57,58]. A qualitative correlation has been found between int-2 mRNA expression and the migration both of mesodermal cells from the primitive streak and of parietal endodermal cells [59]. These observations suggest that the int-2 protein product not only acts as either an autrocrine or a paracrine mitogenic factor but also might support chemokinesis and chemotaxis. Chemotactic activities have also been described for other growth factors including aFGF and bFGF [60–62]. From this perspective, the presence of int-2 mRNA in metastases raises the intriguing question about its contribution to the autocrine control of tumour cell migratory activity.

4. *hst/KS3*

An oncogene, denoted either *hst* or *KS3*, was originally identified by NIH 3T3 transfection assays. The *hst* oncogene has been found in DNAs from a primary human stomach tumour (hst), a non-cancerous portion of stomach mucosa from the same patient, a lymph node metastasis of a stomach cancer from a different patient, one colon tumour, three hepatomas and three gastric cancers [63]. The same oncogene was independently found in DNA from a Kaposi's sarcoma (KS), a tumour generally considered to be an angiosarcoma that appears with high frequency in immunosuppressed individuals including AIDS patients [64].

The *hst/KS3* cDNA predicts a full-length 206 amino acid residue protein with a mass of 22 kDa [64,65]. The amino terminal 70 residues appear to be unique to this FGF whereas the carboxyl terminal 136 amino acid residue sequence is homologous to other members of the family (Fig. 1). The existence of longer forms of the protein cannot be eliminated since no in-frame stop codons have been identified 5' of the initiator AUG. Comparison of the genomic and cDNA sequences reveals that the coding region is split over three exons separated by two introns of 0.61 and 0.53 kb located at sites equivalent to those at which introns are present in other members of the FGF family [66]. A consensus TATA box is present 40–50 nucleotides upstream of the beginning of the cDNA coding sequence, 278 nucleotides 5' of the initiation codon. The gene contains three putative Sp1 transcription factor binding sites and 8 homologous copies of the classical enhancer core sequence. The 3' end of the gene is devoid of either a standard polyadenylation signal or any of its known variants.

The *hst/KS3* gene encodes a protein with a standard hydrophobic leader sequence (Fig. 3). Translation of the message *in vitro* results in the expected full-length 22 kDa protein. However, as a result of processing of the leader sequence following secretion into the endoplasmic reticulum, the amino terminus of the mature 19 kDa polypeptide begins at either Ala$_{31}$ or Pro$_{32}$ [67]. The actual mass as monitored by SDS electrophoresis in polyacrylamide gels is still 22 kDa. The compensating mass is attributable to glycosylation of the polypeptide chain as shown by recovery of the expected 19 kDa secretory form from cells grown in the presence of tunicamycin, an inhibitor of *N*-glycosylation. Since treatment with tunicamycin resulted in a lower steady state level of the int-2 protein in conditioned media, either glycosylation is required for efficient secretion or the absence of the oligosaccharide chains facilitates proteolytic degradation of the secreted protein.

As expected for a member of the FGF family, the hst/KS3 protein binds tightly to heparin–Sepharose eluting with 1.0–1.2 M NaCl, the salt concentration required to elute aFGF [68]. Subsequent reversed-phase HPLC results in pure protein as assessed by electrophoresis in SDS polyacrylamide gels. The secreted hst/KS3 protein is mitogenic for NIH 3T3 cells with half-maximal stimulation achieved at 200 and 80 pg/ml in the absence and presence of heparin respectively. As is also seen with other characterized members of the FGF family, the hst/KS3 protein is a potent vascular endothelial cell mitogen in culture with an ED$_{50}$ of 30 pg/ml [68].

The effect of heparin on the specific mitogenic activity of the hst/KS3 protein varies depending on the assay conditions and target cells. The activity of COS1 cell-expressed protein is not increased by heparin when assayed in serum-free media on NIH 3T3 cells [64], whereas an almost total dependence on heparin is observed if the mitogen is assayed in 1% serum on bovine capillary endothelial cells [67]. Although the mechanism by which heparin potentiates the activity of this mitogen is not certain, its half-life after secretion from COS1 cells is increased by heparin [67]. The stability of the active mitogen could either be decreased by endogenous surface-bound and extracellular proteases or increased by cell surface and extracellular heparan proteoglycans. The potentiation of the mitogenic activity by exogenous heparin might most readily be seen on cells that express levels of heparan proteoglycans that are inadequate to provide protection from the proteases that they also express.

Normal expression of the *hst/KS3* gene is observed in the 11–14 day old mouse foetus but is not detectable either at 17 days of gestation or in new born mice [69]. Therefore, as is the case with other growth factors in general and FGFs in particular, the hst/KS3 protein appears to be involved in supporting early embryonic development. Variable amounts of cross-hybridizing transcripts of the gene are also expressed in transfected embryonic cells including a human immature teratoma cell line (NCC-IT) and 5 of 9 surgical specimens of human germ cell tumours [69].

Both the human *hst/KS3* and *int*-2 genes have been mapped to very similar locations in chromosome 11 band q13 [70–73]. Southern blot analysis of overlapping restriction fragments from several human cancer cell lines has shown that these two genes are less than 45 kb apart. Further Southern blot analysis of overlapping cosmid clones obtained from a human genomic library made from DNA of peripheral leukocytes of a healthy individual has shown that *hst/KS3* is about 35 kb downstream of *int*-2 in the same transcriptional orientation [71]. The proximity of these two genes is compatible with the likelihood that they arose during evolution by duplication of a common ancestral gene and that their expression during development might still be controlled by a common regulatory element. The proximity of these two FGFs is consistent with their 3- to 6-fold co-amplification in several tumours including a melanoma [73], a stomach tumour [72], and A-431 cells of vulvar carcinoma origin that express very high levels of the EGF receptor [71]. The significance of the amplification of these two genes in specific tumours is unknown but supports the possibility that amplification, also seen in the case of the *int*-2 gene, may be a mechanism of activation of proto-oncogenes of the FGF family.

5. *FGF-5*

The *FGF-5* gene was originally identified in DNA from human tumour cell lines using NIH 3T3 cell transfection assays. Transformed cells were selected for growth factor expression by their proliferation in a defined medium lacking FGFs and platelet-derived growth factors [74]. During transfection of the 3T3 cells the *FGF-5* gene was activated by recombination with a co-transfected plasmid carrying a murine retrovirus promoter–enhancer element. This fortuitous cloning artefact resulted in the positioning of the retroviral transcriptional enhancer next to the native promoter of the *FGF-5* gene. A cDNA library was constructed from cells expressing the mitogenic activity [75]. DNA sequence analysis of the cloned *FGF-5* gene revealed two AUG-initiated open reading frames, ORF-1 and ORF-2, encoding a 38-residue polypeptide and a 267-residue FGF homologue (Fig. 1). The two reading frames overlapped with the ORF-1 UGA termination codon situated one nucleotide downstream from the ORF-2 initiator AUG. Whether or not ORF-1 has a regulatory effect on the expression of the *FGF-5* gene is unknown. The longer FGF-like ORF-2 open reading frame also contains the pattern of three exons and two introns typical of the other members of the FGF gene family. The total size for the introns in the case of *FGF-5* is 19 kb.

As is the case for both the *int*-2 and *hst/KS3* oncogene protein products, the *FGF-5* gene encodes a protein with a hydrophobic amino terminal secretory leader sequence (Fig. 3). The hydrophobic leader sequence appears to be functional. Conditioned media from NIH 3T3 cells, transformed either by the rearranged *FGF*-5 gene or by the native *FGF*-5 gene linked to molony leukaemia virus long terminal repeat (LTR) enhancer element, were mitogenic for BALB/c 3T3 fibroblasts and, in the presence of heparin and 1.5% calf serum, for foetal bovine heart endothelial cells [75].

The mitogenically active FGF-5 protein binds to heparin–Sepharose and is released in the 1.0–1.5 M concentration range of NaCl required to elute most other known FGFs. Although the protein sequence contains a consensus *N*-glycosylation site, no direct evidence is available concerning either the presence or function of glycosylation [75].

Expression of cross-hybridizable mRNA to an *FGF*-5 probe has been observed in 3 of 13 human tumour lines of solid tumour origin including a hepatoma, bladder carcinoma and endometrial carcinoma. The 1.6 and 4.4 kb mRNA hybridizing bands found in the originally rearranged clone are seen in all three positive cell lines [75]. The *FGF*-5 gene has been mapped in the q21 band of chromosome 4, the same chromosome on which bFGF is located [76].

6. CONCLUSIONS

Knowledge of the FGF family of homologous growth factors has expanded rapidly in the last few years. Although five members are currently recognized, there is no reason to think that others will not be discovered. All FGFs appear to be potent mitogens and probably stimulate many, if not all, of the same cells, perhaps through identical, or similar, plasma membrane bound receptors. Nevertheless, hypotheses regarding the physiological functions of not only these FGFs but also most other non-haematopoietic growth factors are mainly extrapolations from activities observed in homogeneous cell cultures *in vitro*. A full appreciation of the potential roles of FGFs in pathologies including cancer will depend, in large part, on a better understanding of their normal roles *in vivo*.

Curiously, all FGFs characterized to date bind tightly to heparin–Sepharose. Although heparin is released by mast cells *in vivo*, the structurally related haparan proteoglycan, found in basement membranes and on some cell surfaces, is probably the more abundant sulphated oligosaccharide to which FGFs might bind. The embryonic generation and subsequent maintenance of morphological structures are thought to be guided and supported by basement membrane associated proliferation. Heparan binding of embryonically expressed FGFs could act to partition them onto basement membranes and, thereby, to support these activities.

Although the FGFs differ in the lengths of their amino and carboxyl terminal extensions, they all share homologous core structures presumably containing not only the heparin binding region but also the receptor binding sites. The functions, if any, of the terminal polypeptide extensions are unknown. All five identified FGF genes also have the same intron–exon structure with the size of some introns reaching 17 kb. If the unusually large aFGF and bFGF introns are conserved by selection, then they presumably have an advantageous, yet unrecognized, function.

The three new members of the FGF family were all identified in screens designed to recognize either oncogenes or growth factors. All three FGF oncogenes are assumed to encode mitogens that normally support controlled cell division. Their escape from appropriate regulation probably results, at least in part, from the previously recognized mechanisms of insertional mutagenesis and gene amplification. Since they all appear to have functional secretory leader sequences, their persistent expression in FGF responsive cells is probably sufficient to support pathological autocrine growth stimulation.

The requirement for a leader sequence to support efficient mitogenic autostimulation is further demonstrated by the tumorigenicity of bFGF fused to a leader sequence compared with normal leaderless bFGF. The absence of a leader might explain why potent mitogens such as aFGF and bFGF have not been well correlated with transformation and have not been identified in typical oncogene screens. One would predict, however, that if either aFGF or bFGF spontaneously fused to an in-frame leader sequence in an FGF-responsive cell then its persistent expression would transform the cell. Given the fact that such a naturally occurring construct has not yet been found in tumour DNA, such gene rearrangements are probably very infrequent events *in vivo*.

ACKNOWLEDGEMENT

We thank Jerry DiSalvo for preparing Fig. 1.

REFERENCES

[1] Doolittle, R. F., Hunkapiller, M. W., Hood, L. E., Devare, S., G., Robbins, K. C., Aaronson, S. A. & Antoniades, H. N. (1983). Simian sarcoma virus oncogene, v-*sis*, is derived from the gene (or genes) encoding a platelet-derived growth factor. *Science* **221** 275–277

[2] Waterfield, M. D., Scrace, G. T., Whittle, N., Stroobant, P., Johnsson, A., Wasteson, A., Westermark, B., Heldin, C.-H., Huang, J. S. & Deuel, T. F. (1983). Platelet-derived growth factor is structurally related to the putative transforming protein p28^{sis} of simian sarcoma virus. *Nature (London)* **304** 35–39

[3] Thomas, K. A. (1988). Transforming potential of fibroblast growth factor genes. *Trends Biochem. Sci.* **13** 327–328

[4] Trowell, O. A., Chir, B. & Willmer, E. N. (1939). Studies on growth of tissues *in vitro*: effects of some tissue extracts on growth of periosteal fibroblasts. *J. Exp. Biol.* **16** 60–70

[5] Hoffman, R. S. (1940). Growth-activating effect of extracts of adult and embryonic tissues of rat on fibroblast colonies *in vitro*. *Growth* **4** 361–376

[6] Thomas, K. A., Rios-Candelore, M. & Fitzpatrick, S. (1984). Purification and characterization of acidic fibroblast growth factor from bovine brain. *Proc. Natl. Acad. Sci. USA* **81** 357–361

[7] Bohlen, P., Baird, A., Esch, F., Ling, N. & Gospodarowicz, D. (1984). Isolation and partial molecular characterization of pituitary fibroblast growth factor. *Proc. Natl. Acad. Sci. USA* **81** 5364–5368

[8] Thomas, K. A. & Gimenez-Gallego, G. (1986). Fibroblast growth factors:

broad spectrum mitogens with potent angiogenic activity. *Trends. Biochem. Sci.* **11** 81–84

[9] Folkman, J. & Klagsbrun, M. (1987). Angiogenic factors. *Science* **235** 442–447

[10] Gimenez-Gallego, G., Conn, G., Hatcher, V. B. & Thomas, K. A. (1986). The complete amino acid sequence of human brain-derived acidic fibroblast growth factor. *Biochem. Biophys. Res. Commun.* **138** 611–617

[11] Jaye, M., Howk, R., Burgess, W., Ricca, G. A., Chiu, I.-M., Ravera, M. W., O'Brien, S. J., Modi, W. S., Maciag, T. & Drohan, W. N,. (1986). Human endothelial cell growth factor: cloning, nucleotide sequence, and chromosome localization. *Science* **233** 541–545

[12] Abraham, J. A., Mergia, A., Whang, J. L., Tumolo, A., Friedman, J., Hjerrild, K. A., Gospodarowicz, D. & Fiddes, J. C. (1986). Nucleotide sequence of a bovine clone encoding the angiogenic protein, basic fibroblast growth factor. *Science* **233** 545–548

[13] Kurokawa, T., Sasada, R., Iwane, M. & Igarashi, K. (1987). Cloning and expression of cDNA encoding human basic fibroblast growth factor. *FEBS Lett.* **213** 189–194

[14] Alterio, J., Halley, C., Brou, C., Soussi, T., Courtois, Y. & Laurent, M. (1988). Characterization of a bovine acidic FGF cDNA clone and its expression in brain and retina. *FEBS lett.* **242** 41–46

[15] Sommer, A., Brewer, M. T., Thompson, R. C., Moscatelli, D., Presta, M. & Rifkin, D. B. (1987). A form of human basic fibroblast growth factor with an extended amino terminus. *Biochem. Biophys. Res. Commun.* **144** 543–550

[16] Shimasaki, S., Emoto, N., Koba, A., Mercado, M., Shibata, F., Cooksey, K., Baird, A. & Ling, N. (1988). Complementary DNA cloning and sequencing of rat ovarian basic fibroblast growth factor and tissue distribution study of its mRNA. *Biochem. Biophys. Res. Commun.* **157** 256–263

[17] Abraham, J. A., Whang, J. L., Tumolo, A., Mergia, A., Friedman, J., Gospodarowicz, D. & Fiddes, J. (1986). Human basic fibroblast growth factor: nucleotide sequence and genomic organization. *EMBO J.* **5** 2523–2528

[18] Gospodarowicz, D., Ferrara, N., Schweigerer, L. & Neufeld, G. (1987). Structural characterization and biological functions of fibroblast growth factor. *Endocr. Rev.* **8** 95–114

[19] Moscatelli, D., Joseph-Silverstein, J., Manejias, R. & Rifkin, D. B. (1987). M_r 25,000 heparin-binding protein from guinea pig brain is a high molecular weight form of basic fibroblast growth factor. *Proc. Natl. Acad. Sci. USA* **84** 5778–5782

[20] Prats, H., Kaghad, M., Prats, A. C., Klagsbrun, M., Lelias, J. M., Liauzun, P., Chalon, P., Tauber, J. P., Amalric, F., Smith, J. A. & Caput, D. (1989). High molecular mass forms of basic fibroblast growth factor are initiated by alternative CUG codons. *Proc. Natl. Acad. Sci. USA* **86** 1836–1840

[21] Gospodarowicz, D., Neufeld, G. & Schweigerer, L. (1987). Fibroblast growth factor: structural and biological properties. *J. Cell. Physiol. Suppl.* **5** 15–26

[22] Thomas, K. A. (1987). Fibroblast growth factors. *FASEB J.* **1** 434–440

[23] Thomas, K. A., Rios-Candelore, M., Gimenez-Gallego, G., DiSalvo, J., Bennett, C., Rodkey, J. & Fitzpatrick, S. (1985). Pure brain derived fibroblast growth factor is a potent angiogenic vascular endothelial cell mitogen with sequence homology to interleukin-1. *Proc. Natl. Acad. Sci. USA* **82** 6409–6413

[24] Esch, F., Baird, A., Ling, N., Ueno, N., Hill, F., Denoroy, L., Klepper, R., Gospodarowicz, D., Bohlen, P. & Guillemin, R. (1985). Primary structure of bovine pituitary basic fibroblast growth factor (FGF) and comparison with the amino-terminal sequence of bovine brain acidic FGF. *Proc. Natl. Acad. Sci. USA* **82** 6507–6511

[25] Lobb, R. R., Alderman, E. M. & Fett, J. W. (1985). Induction of angiogenesis by bovine brain derived class I heparin binding growth factor. *Biochemistry* **24** 4969–4973

[26] Neufeld, G. & Gospodarowicz, D. (1986). Basic and acidic fibroblast growth factors interact with the same cell surface receptors. *J. Biol. Chem.* **261** 5631–5637

[27] Olwin, B. B. & Hauschka, S. D. (1986). Identification of the fibroblast growth factor receptor of Swiss 3T3 cells and mouse skeletal muscle myoblasts. *Biochemistry* **25** 3487–3492

[28] Gospodarowicz, D. & Cheng, J. (1986). Heparin protects basic and acidic FGF from inactivation. *J. Cell Physiol.* **128** 475–484

[29] Saksela, O., Moscatelli, D., Sommer, A. & Rifkin, D. B. (1988). Endothelial cell-derived heparan sulfate binds basic fibroblast growth factor and protects it from proteolytic degradation. *J. Cell Biol.* **107** 743–751

[30] Sommer, A. & Rifkin, D. B. (1989). Interaction of heparin with human basic fibroblast growth factor: protection of the angiogenic protein from proteolytic degradation by a glycosaminoglycan. *J. Cell. Physiol.* **138** 215–220

[31] Shing, Y., Folkman, J., Sullivan, R., Butterfield, C., Murray, J. & Klagsbrun, M. (1984). Heparin affinity: purification of a tumor-derived capillary endothelial cell growth factor. *Science* **223** 1296–1299

[32] Mydlo, J. H., Heston, W. D. W. & Fair, W. R. (1988). Characterization of a heparin-binding growth factor from adenocarcinoma of the kidney. *J. Urol.* **140** 1575–1579

[33] Nishi, N., Matuo, Y., Kunitomi, K., Takenaka, I., Usami, M., Kotake, T. & Wada, F. (1988). Comparative analysis of growth factors in normal and pathologic human prostates. *Prostate* **13** 39–48

[34] Mydlo, J. H., Michaeli, J., Heston, W. D. W. & Fair, W. R. (1988). Expression of basic fibroblast growth factor mRNA in benign prostatic hyperplasia and prostatic carcinoma. *Prostate* **13** 241–247

[35] Klagsbrun, M., Sasse, J., Sullivan, R. & Smith, J. A. (1986). Human tumor cells synthesize an endothelial cell growth factor that is structurally related to basic fibroblast growth factor. *Proc. Natl. Acad. Sci. USA* **83** 2448–2452

[36] Libermann, T. A., Friesel, R., Jaye, M., Lyall, R. M., Westermark, B., Drohan, W., Schmidt, A., Maciag, T. & Schelessinger, J. (1987). An angiogenic growth factor is expressed in human glioma cells. *EMBO J.* **6** 1627–1632

[37] Presta, M., Moscatelli, D., Joseph-Silverstein, J. & Rifkin, D. B. (1986). Purification from a human hepatoma cell line of a basic fibroblast growth factor-like molecule that stimulates capillary endothelial cell plasminogen activator production, DNA synthesis, and migration. *Mol. Cell. Biol.* **6** 4060–4064

[38] Witte, D. P., Stambrook, P. J., Feliciano, E., Jones, C. L. A. & Libermann, T. A. (1988). Growth factor production by a human megakaryocytic tumor cell line. *J. Cell. Physiol.* **137** 86–94

[39] Rizzino, A., Kuszynski, C., Ruff, E. & Tiesman, J. (1988). Production and utilization of growth factors related to fibroblast growth factor by embryonal carcinoma cells and their differentiated cells. *Dev. Biol.* **129** 61–71

[40] Schweigerer, L., Neufeld, G., Mergia, A., Abraham, J. A., Fiddes, J. C. & Gospodarowicz, D. (1987). Basic fibroblast growth factor in human rhabdomyosarcoma cells: implications for the proliferation and neovascularization of myoblast-derived tumors, *Proc. Natl. Acad. Sci. USA* **84** 842–846

[41] Moscatelli, D., Presta, M., Joseph-Silverstein, J. & Rifkin, D. B. (1986). Both normal and tumor cells produce basic fibroblast growth factor. *J. Cell. Physiol.* **129** 273–276

[42] Murphy, P. R., Sato, R., Sato, Y. & Friesen, H. G. (1988). Fibroblast growth factor messenger ribonucleic acid expression in a human astrocytoma cell line: regulation by serum and cell density. *Mol. Endocrinol.* **2** 591–598

[43] Jaye, M., Lyall, R. M., Mudd, R., Schlessinger, J. & Sarver, N. (1988). Expression of acidic fibroblast growth factor cDNA confers growth advantage and tumorigenesis to Swiss 3T3 cells. *EMBO J.* **7** 963–969

[44] Sasada, R., Kurokawa, T., Iwane, M. & Igarashi, K. (1988). Transformation of mouse BALB/c 3T3 cells with human basic fibroblast growth factor cDNA. *Mol. Cell. Biol.* **8** 588–594

[45] Neufeld, G., Mitchell, R., Ponte, P. & Gospodarowicz, D. (1988). Expression of human basic fibroblast growth factor cDNA in baby hamster kidney-derived cells results in autonomous cell growth. *J. Cell Biol.* **106** 1385–1394

[46] Rogelj, S., Weinberg, R. A., Fannin, P. & Klagsbrun, M. (1988). Basic fibroblast growth factor fused to a signal peptide transforms cells. *Nature (London)* **331** 173–175

[47] Blam, S. B., Mitchell, R., Tischer, E., Rubin, J. S., Silva, M., Silver, S., Fiddes, J. C., Abraham, J. A. & Aaronson, S. A. (1988).Addition of growth hormone secretion signal to basic fibroblast growth factor results in cell transformation and secretion of aberrant forms of the protein. *Oncogene* **3** 129–136

[48] Bishop, J. M. (1983). Cellular oncogenes and retroviruses. *Annu. Rev. Biochem.* **52** 301–354

[49] Varmus, H. E. (1984). The molecular genetics of cellular oncogenes. *Annu. Rev. Genet.* **18** 533–612

[50] Dickson, C., Smith, R., Brookes, S. & Peters, G. (1984). Tumorigenesis by mouse mammary tumor virus: proviral activation of a cellular gene in the common integration region *int*-2. *Cell* **37** 529–536

[51] Moore, R., Casey, G., Brookes, S., Dixon, M., Peters, G. & Dickson, C. (1986). Sequence, topography and protein coding potential of mouse *int*-2; a putative oncogene activated by mouse mammary tumour virus. *EMBO J.* **5** 919–924

[52] Mansour, S. L. & Martin, G. R. (1988). Four classes of mRNA are expressed from the mouse *int*-2 gene, a member of the FGF gene family. *EMBO J.* **7** 2035–2041

[53] Dickson, C. & Peters, G. (1987). Potential oncogene product related to growth factors. *Nature (London)* **326** 833

[54] Varley, J. M., Walker, R. A., Casey, G. & Brammar, W. J. (1988). A common

alteration to the *int-2* proto-oncogene in DNA from primary breast carcinomas. *Oncogene* **3** 87–91

[55] Zhou, D. J., Casey, G. & Cline, M. J. (1988). Amplification of human *int-2* in breast cancers and squamous carcinomas. *Oncogene* **2** 279–282

[56] Varley, J. M., Swallow, J. E., Brammar, W. J., Whittaker, J. L. & Walker, R. A. (1987). Alterations to either c-*erbB-2 (neu)* or c-*myc* proto-oncogenes in breast carcinomas correlate with poor short-term prognosis. *Oncogene* **1** 423–430

[57] Jakobovits, A., Shackleford, G. M., Varmus, H. E. & Martin, G. R. (1986). Two proto-oncogenes implicated in mammary carcinogenesis, *int-1* and *int-2*, are independently regulated during mouse development. *Proc. Natl. Acad. Sci. USA* **83** 7806–7810

[58] Smith, R., Peters, G. & Dickson, C. (1988). Multiple RNAs expressed from the *int-2* gene in mouse embryonal carcinoma cell lines encode a protein with homology to fibroblast growth factors. *EMBO J.* **7** 1013–1022

[59] Wilkinson, D. G., Peters, G., Dickson, C. & McMahon, P. (1988). Expression of the FGF-related proto-oncogene *int-2* during gastrulation and neurulation in the mouse. *EMBO J.* **7** 691–695

[60] Terranova, V. P., DiFlorio, R., Lyall, R. M., Hic, S., Friesel, R. & Maciag, T. (1985). Human endothelial cells are chemotactic to endothelial cell growth factor and heparin. *J. Cell Biol.* **101** 2330–2334

[61] Senior, R. M., Huang, S. S., Griffin, G. L. & Huang, J. S. (1986). Brain-derived growth factor is a chemoattractant for fibroblasts and astroglial cells. *Biochem. Biophys. Res. Commun.* **141** 67–72

[62] Herbert, J. M., Cottineau, M., Driot, F., Pereillo, J. M. & Maffrand, J. P. (1988). Activity of pentosan polysulphate and derived compounds on vascular endothelial cell proliferation and migration induced by acidic and basic FGF *in vitro*. *Biochem. Pharmacol.* **37** 4281–4288

[63] Sakamoto, H., Mori, M., Taira, M., Yoshida, T., Matsukawa, S., Shimizu, K., Sekiguchi, M., Terada, M. & Sugimura, T. (1986). Transforming gene from human stomach cancers and a noncancerous portion of the stomach mucosa. *Proc. Natl. Acad. Sci. USA* **83** 3997–4001

[64] Delli Bovi, P., Curatola, A. M., Kern, F. G., Greco, A., Ittman, M. & Basilico, C. (1987). An oncogene isolated by transfection of Kaposi's sarcoma DNA encodes a growth factor that is a member of the FGF family. *Cell* **50** 729–737

[65] Taira, M., Yoshida, T., Miyagawa, K., Sakamoto, H., Terada, M. & Sugimura, T. (1987). cDNA sequence of human transforming gene *hst* and identification of the coding sequence required for transforming activity. *Proc. Natl. Acad. Sci. USA* **84** 2980–2984

[66] Yoshida, T., Miyagawa, K., Odagiri, H., Sakamoto, H., Little, P. F. R., Terada, M. & Sugimura, T. (1987). Genomic sequence of *hst*, and transforming gene encoding a protein homologous to fibroblast growth factors and the *int-2*-encoded protein. *Proc. Natl. Acad. Sci. USA* **84** 7305–7309

[67] Delli-Bovi, P., Curatola, A. M., Newman, K. M., Sato, Y., Moscatelli, D., Hewick, R. M., Rifkin, D. B. & Basilico, C. (1988). Processing secretion, and biological properties of a novel growth factor of the fibroblast growth factor family with oncogenic potential. *Mol. Cell. Biol.* **8** 2933–2941

[68] Miyagawa, K., Sakamoto, H., Yoshida, T., Yamashita, Y., Mitsuy, Y., Furusawa, M., Maeda, S., Takaku, F., Sugimura, T. & Terada, M. (1988). hst-1 transforming protein: expression in silkworm cells and characterization as a novel heparin-binding growth factor. *Oncogene* **3** 383–389

[69] Yoshida, T., Tsutsumi, M., Sakamoto, H., Miyagawa, K., Teshima, S., Sugimura, T. & Terada, M. (1988). Expression of the *hst*-1 oncogene in human germ cell tumors. *Biochem. Biophys. Res. Commun.* **155** 1324–1329

[70] Casey, G., Smith, R., McGillivray, D., Peters, G. & Dickson, C. (1986). Characterization and chromosome assignment of the human homolog of *int*-2, a potential proto-oncogene. *Mol. Cell. Biol.* **6** 502–510

[71] Wada, A., Sakamoto, H., Katoh, O., Yoshida, T., Yokota, J., Little, P. F. R., Sugimura, T. & Terada, M. (1988). Two homologous oncogenes, *hst*-1 and *int*-2, are closely located in human genome. *Biochem. Biophys. Res. Commun.* **157** 828–835

[72] Yoshida, M. C., Wada, M., Satoh, H., Yoshida, T., Sakamoto, H., Miyagawa, K., Yokota, J., Koda, T., Kakinuma, M., Sugimura, T. & Terada, M. (1988). Human HST1 (HSTF1) gene maps to chromosome band 11q13 and coamplifies with the INT2 gene in human cancer. *Proc. Natl. Acad. Sci. USA* **85** 4861–4864

[73] Adelaide, J., Mattei, M.-G., Marics, I., Raybaud, F., Planche, J., De Lapeyriere, O. & Birnbaum, D. (1988). Chromosomal localization of the *hst* oncogene and its co-amplification with the *int*-2 oncogene in a human melanoma. *Oncogene* **2** 413–416

[74] Zhan, X., Culpepper, A., Reddy, M., Loveless, J. & Goldfarb, M. (1987). Human oncogenes detected by a defined medium culture assay. *Oncogene* **1** 369–376

[75] Zhan, X., Bates, B., Hu, X. & Goldfarb, M. (1988). The human *FGF*-5 oncogene encodes a novel protein related to fibroblast growth factors. *Mol. Cell. Biol.* **8** 3487–3495

[76] Nguyen, C., Roux, D., Mattei, M.-G., De Lapeyriere, O., Goldfarb, M., Birnbaum, D. & Jordan, B. R. (1988). The FGF-related oncogenes *hst* and *int*-2 and the BCL.1 locus are contained within one megabase in band Q13 of chromosome 11 while the *FGF*-5 oncogene maps to 4Q21. *Oncogene* **3** 703–708

8

Transcription factors in normal and malignant cells

Peter Herrlich, Helmut Ponta, Bernd Stein, Carsten Jonat, Stephan Gebel, Harald König, Richard O. Williams, Vladimir Ivanow and **Hans J. Rahmsdorf**
Kernforschungszentrum Karlsruhe, Institut für Genetik und Toxikologie, and Universität Karlsruhe, Institut für Genetik, P.O. Box 3640, D-7500 Karlsruhe 1, FRG

1. INTRODUCTION

Normal cells and cancer cells differ in obvious features of cellular phenotype, e.g. in various degrees of escape from normal growth control mechanisms and of invasiveness [1]. These massive changes in phenotype are based on an altered program of expressed genes. Although cancer cells constantly alter their properties, i.e. they progress, portions of the altered program are stable over many cell generations or even established permanently, probably because of selective pressure. Genes affected in the altered program code for growth factor receptors and growth factors [2–4], for components of or enzymes acting at the extracellular matrix [5–12] and for various intracellular and membrane-bound proteins [13–15]. It is reasonable to assume that the alterations of the genetic program are related to the mutations and gene rearrangements that make up the cancer cell. Advances in two separate areas of research, the process of carcinogenesis and the mechanisms of gene regulation, permit suggestions about the molecular links between carcinogenic mutation and gene expression, as well as suggestions for their experimental verification. The targets for carcinogenic mutations seem to be cellular oncogenes which code for protein components that function in intra- and intercellular communication and transfer various signals from outside and from the plasma membrane to the nucleus [16–19]. The effector ends of these communication pathways are transcription factors whose activities become modulated. Interestingly genes coding for transcription factors can act as cellular oncogenes themselves [20,21]. As viral derivatives they are constituents of acutely transforming viruses: v-*fos* [22] and v-*jun*[23]. Although it is still not certain how many mutational steps are required to convert a normal cell into a cancer cell, both types of mutations recognized so far appear to influence signal transfer [24]: 'gain of function mutations' leading to 'dominant oncogenes' change the structure or amount of an oncogene product such that normal external signalling becomes redundant, and transcription factor activity may then remain elevated by sustained signalling. 'Loss of function mutations' in so-called tumour suppressor genes ('recessive oncogenes') although less well explored may directly affect transcription factors. The candidates for these two types of mutations and their molecular roles will be covered in other chapters. We will discuss here how the activity and synthesis of two transcription factors our own laboratory is mainly concerned with, i.e. AP-1 and NFκB, are modulated transiently by carcinogenic and tumour-promoting agents. We will then describe recent experiments with an invasive lymphoma that support the proposed link between transforming principle and transcription factor activity.

2. CIS- AND TRANS-ACTING ELEMENTS DETERMINE THE EXPRESSION OF GENES

The rate of transcription of any one gene depends on specific sequences within the gene, in the promoter region and/or the flanking regions of the gene. In most protein-coding genes these sequences or cis-acting elements are assembled in the 5' flanking region. Elements that are fairly position-dependent without loss of function are called promoter-proximal and include the TATA and CAAT boxes found in many genes [25,26]. Distance- and orientation-independent (or relatively independent) elements are called promoter distal or enhancers and silencers depending on the positive or negative effect on transcriptional rate [27–30]. Trans-acting proteins (transcription factors) bind to all these elements [31,32]. Mutations in cis-acting elements that abolish binding also obliterate transcriptional function [33,34]. It is thought that transcription factors form the initiation complex which permits RNA polymerase to start. The nature of these protein–protein interactions is not understood and is under intense study [32, 35–38]. However, it is clear (e.g. from studies in which genes coding for transcription factors are overexpressed) that the concentration of active transcription factor determines the rate of transcription. The activity of a gene thus depends on the cis-acting elements it possesses and the availability of trans-acting factors for these cis-acting elements in a given cell at a given time.

We will deal with transcription factors whose abundance and/or activity are not constant but rather modulated in response to extracellular stimuli. Both transcription factors, AP-1 and NFκB, operate at many genes carrying the corresponding recognition sequence. We will use two target genes to exemplify transcription factor action: the human gene for collagenase which is regulated by AP-1 [34], and the human immunodeficiency virus whose promoter is regulated by NFκB [39]. Both genes are relatively simple in their regulation in that their transcription depends critically on one enhancer (one-domain enhancer) element which binds naturally one transcription factor. Deletion or point mutation of this enhancer obliterates promoter activity. Experiments describing these two systems can be found elsewhere [34,39–43].

The collagenase enhancer is located between positions -72 and -65 (Fig. 1) and it binds the factor AP-1. Although in the collagenase gene the enhancer is located in promoter proximity, it can be moved and shows the positional flexibility characteris-

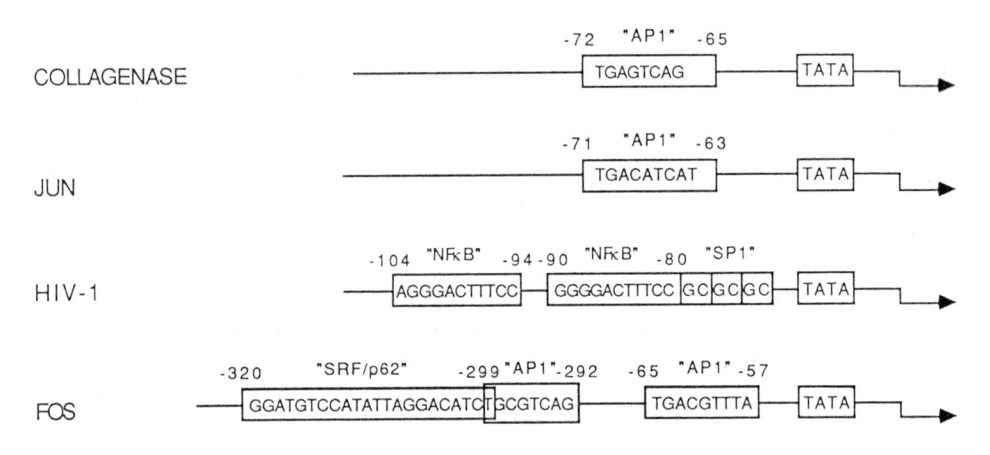

```
                              -72   "AP1"  -65
COLLAGENASE  ─────────────────────┤ TGAGTCAG ├──────┤ TATA ├──▶

                              -71   "AP1"  -63
JUN          ─────────────────────┤ TGACATCAT ├──────┤ TATA ├──▶

                -104  "NFκB"  -94 -90  "NFκB"  -80  "SP1"
HIV-1        ─────┤ AGGGACTTTCC ├──┤ GGGGACTTCC ├┤GCGCGC├──┤ TATA ├──▶

            -320    "SRF/p62"        -299 "AP1" -292    -65  "AP1" -57
FOS          ──┤ GGATGTCCATATTAGGACATCT│GCGTCAG ├─────────┤ TGACGTTTA ├──┤ TATA ├──▶
```

Fig. 1 — Essential cis-acting elements and transcription factors of four promoters covered in the chapter. "AP-1" denotes members of the AP-1 family, and the transcription factors SP-1, NFκB and serum response factor SRF that is associated with protein p62, are indicated similarly. The AP-1-like elements bind purified AP-1 (Jun–Fos) *in vitro*, not necessarily *in vivo*. See text and [82].

tic of an enhancer. Point mutation of the enhancer sequence leaving all other promoter sequences between positions −517 and +63 unchanged severely reduces both the basal and the induced transcriptional activity of the promoter (unpublished). The collagenase promoter responds to a large variety of conditions, some of which will be discussed later. These include the induced transcription of the gene after treatment of cells with serum growth factors, phorbol ester tumour promoters and DNA damaging agents. These agents affect transcription through the collagenase enhancer in conjunction with AP-1.

Similarly, the human immunodeficiency virus (HIV-1) promoter carries one major enhancer sequence (Fig. 1). Its sequence motif resembles the kB motif in the gene for the immunoglobulin light chain and is repeated in tandem between positions −105 and −78. Point mutations in both repeats severely reduce basal and inducible transcription. A transcription factor originally detected in B lymphocytes, NFκB, binds to the HIV enhancer. HIV transcription in appropriate cells is induced by mitogens, growth factors, phorbol esters and DNA damaging agents. These inductions are mediated through the HIV enhancer.

3. STRUCTURE OF AP-1

Small amounts of AP-1 are present in many differentiated cells and are detected by sensitive gel retardation assays which measure the *in vitro* binding of AP-1 to a synthetic radioactive oligonucleotide comprising the recognition sequence (Fig. 2). Using the cognate sequence bound to a matrix, AP-1 can be purified. Purified preparations always contain several proteins as observed by multiple bands in denaturing gels [34,44,45]. As it turns out, AP-1 is in fact a family of related proteins that have similar binding characteristics and sequences [46–52]. *In vitro* competition in binding assays using related sequences and titration experiments with bacterially

Ch. 8]   **Transcription factors in normal and malignant cells**   153

Detection of AP-1 subunits by crosslinking to DNA

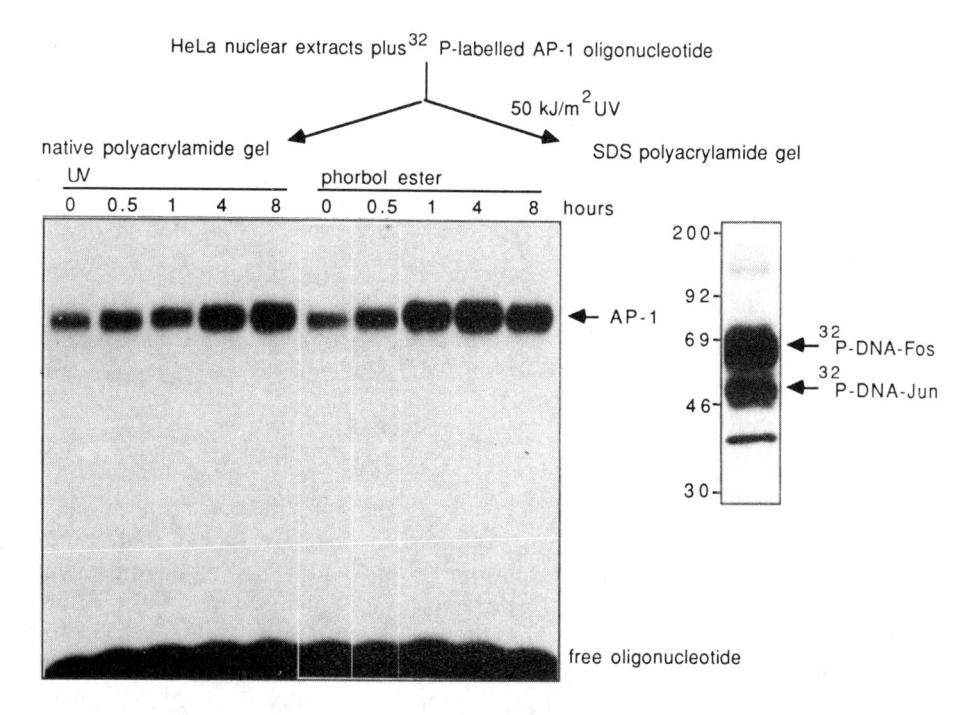

Fig. 2 — The transcription factor AP-1 consists of two subunits: Fos and Jun. The AP-1–DNA complex forms a single retarded band in a gel retardation experiment. The complex can be cross-linked to the DNA, precipitated by antibodies directed against Fos and resolved by denaturing polyacrylamide gel electrophoresis. Both subunits cross-link to DNA. Their apparent molecular weight is increased by the oligonucleotide single strand attached. Both phorbol ester and UV treatment of cells enhances the activity level of AP-1, part of which is due to new synthesis.

expressed AP-1 components suggest that the collagenase enhancer has preference for one specific transcription factor of the AP-1 family (unpublished). After characterization of the genes coding for this specific transcription factor it has been possible to show that *in vivo* this specific factor is also required for collagenase transcription [53,54].

It is particularly interesting to note that the genes *fos* and *jun* coding for the transcription factor AP-1 had been recognized previously as nuclear oncogenes: variants of these genes are the transforming principles of rodent and avian retroviruses respectively [23,55]. Antibodies to these proteins detect antigenic material exclusively in the cell nucleus. The cellular proteins differ in size and can easily be distinguished by SDS gel electrophoresis. Using nuclear or whole cell extracts from HeLa cells, AP-1 can be bound *in vitro* to the collagenase enhancer sequence and can be cross-linked to the DNA by UV irradiation. Cross-linked proteins are detected by the radioactive label in the DNA probe. On SDS gel electrophoresis of the material two radioactive bands of about 1:1 stoichiometry are detected (Fig. 2). By antibody techniques the two bands are recognized as Fos and Jun. Such experiments have shown that the collagenase enhancer binds preferentially a heterodimeric complex of the proteins coded for by c-*fos* and c-*jun*. Both components bind in similar UV cross-linkable proximity to DNA, and presumably to opposite strands of the palindromic sequence since no 'two-protein complexes' were detected.

Clones with the *fos* and *jun* genes have been used to produce the proteins in bacteria [20,56] and these bacterially expressed proteins have been used in our laboratory for *in vitro* DNA binding studies. It is from these *in vitro* titration and DNA competition experiments that we feel that Fos protein does not itself bind and that the heterodimer of Fos and Jun binds better than Jun alone [57]. This is in agreement with data of other laboratories [47,52,56,58]. By antisense RNA techniques *in vivo* either Jun or Fos can be eliminated from cells. Elimination of either one severely reduces collagenase transcription [53,54], suggesting that *in vivo* (in cultured HeLa and 3T3 cells) the Fos–Jun heterodimer is the active transcription factor.

4. REGULATION OF TRANSCRIPTION FACTOR ACTIVITY

The activity of a transcription factor defines the rate of transcription of the dependent gene. How is this activity changed in response to external stimuli? Collagenase is induced by serum, phorbol esters and DNA damaging agents. Induction by these agents requires the presence of the enhancer and of its transcription factor AP-1 (Fos–Jun) [34, 42]. Thus the level of activity of AP-1 must be a target of the signal transduction pathways stimulated by these agents.

Theoretically, changes in either AP-1 protein concentration or AP-1 function (DNA binding and/or transactivation) could lead to the same result: altered expression of collagenase. Both of these possibilities are in fact used in the mammalian cells we looked at. AP-1 activity is modulated by both post-translational modification and by increased transcription of *fos* and *jun*. The fact that the enhanced transcription of *fos* and *jun* also occurs in the absence of ongoing protein synthesis suggests that other transcription factors that operate at the *fos* and *jun* promoters must be regulated at the post-translational level.

5. POST-TRANSLATIONAL UP- AND DOWN-MODULATION OF AP-1 ACTIVITY

Growing cells in culture, serum-starved cells as well as various tissues in the intact animal maintain low levels of both Fos [59–62] and Jun [20,63,64]. Collagenase minimal promoter–reporter gene constructs (which contain just the collagenase enhancer in front of the herpes simplex thymidine kinase promoter and the CAT gene) can be induced in such cultured cells by phorbol esters or UV irradiation in the presence of cycloheximide or anisomycin [42], conditions that do not permit new synthesis of AP-1. Since AP-1 is required for transcription, it must be the preformed factor which is altered in its activity under the influence of signal transduction pathways elicited by the inducing agents. Although both Fos and Jun proteins are modified post-translationally by phosphorylation and glycosylation [34,55,65–67], a correlation between a change in activity and any specific modification event has as yet not been found. Site-directed mutagenesis of *fos* and *jun* will be required to remove modification sites individually before such a correlation can be detected and proven.

AP-1 activity is not only up-regulated by post-translational modification; it is also down modulated. Glucocorticoid hormone has been known for a long time to counteract both proliferative responses and tumour promotion [68]. It inhibits strongly the synthesis of collagenase in HeLa cells, both basal level expression and induced expression (Jonat et al., in preparation). By mutational analysis of the collagenase promoter it became clear that the only sequence necessary and sufficient for glucocorticoid–dependent repression is the enhancer sequence. This sequence is different from the enhancer element recognized by the glucocorticoid receptor [69–73]. The hormone receptor is, however, necessary for repression: in receptor-negative CV-1 cells glucocorticoids can no longer repress collagenase synthesis but they do repress in CV-1 cells that had been transfected with a suitable receptor cDNA clone. Repression occurs also in the presence of anisomycin. Thus new protein synthesis is not required. Neither Fos nor Jun protein synthesis is reduced by glucocorticoid hormone. Also, collagenase repression is rapid: the abundance of collagenase RNA is reduced severely within 6 h, and phorbol ester dependent induction is blocked totally when glucocorticoid is given at the same time as phorbol ester. From antisense experiments we know that it takes a much longer time than 6 h to deplete cells of pre-existing Fos and Jun. This argues for a post-translational down- modulation of AP-1 activity. One could suggest that AP-1 was simply diverted to some other glucocorticoid receptor dependent process requiring AP-1 or one of its two components. We favour, however, a direct repression mechanism because repression is established at a hormone concentration one order of magnitude below that needed for hormone-dependent activation of transcription. Thus we suggest, as the most straightforward interpretation, that the hormone receptor interferes directly with AP-1 action.

Another example of AP-1 down-modulation was found in cells expressing the adenovirus E1A protein [74]. A set of experiments similar to that described for glucocorticoid hormone suggests that AP-1 is the target of negative regulation by E1A (results in collaboration with R. Offringa). It is as yet not certain whether E1A or an E1A-induced protein acts as the repressor. Trans-activating and trans-repressing domains of E1A differ, however, which suggests a direct specific role of E1A protein in repression [75]. It is interesting that the repressing domain of E1A is also required for transformation. E1A with both of its actions, positive and negative, could be defined as an indirectly acting transcription factor since it does not seem to touch DNA itself. There is similarity to still another viral transforming oncogene: SV40 T antigen, which finds, however, specific DNA binding sites in the host genome [76] presumably serving in transactivating or transrepressing genes directly, and which inhibits the function of the transcription factor AP-2, perhaps through protein–protein interaction [77].

6. NEW AP-1 SYNTHESIS

Post-translational modulation is essential for the fast response to extracellular stimuli. New synthesis would be an advantage for a more sustained response of AP-1 dependent genes such as collagenase. To this end the transcription factors acting on the promoters of *fos* and *jun* would have to be activated in an immediate response.

fos has in fact been the first gene whose rapid and massive expression was detected after treatment of cells with a large number of agents including growth factors, phorbol esters and DNA damaging agents; for recent references see [78]. Also, *jun* transcription is induced rapidly after treatment of cells with growth factors, phorbol esters and DNA damaging agents [54,79–82]. Some of the transcription factors acting at the *fos* and *jun* promoters which form the primary end of the signal transduction pathways elicited by serum growth factors, phorbol esters or DNA damage are known. In *fos*, the protein(s) p67 or serum response factor [83] and p62 [84] that bind to the dyad symmetry element at −300 (Fig. 1), mediate the induction after either one of the above stimuli. In *jun*, a cis-acting sequence at position −70 resembling an AP-1 binding site (Fig. 1) is essential for the response to phorbol esters. *In vitro*, purified AP-1 can bind to this sequence, and overexpression of *jun* in embryonal carcinoma cells by transfection of an expression vector carrying v-*jun* causes *jun* expression from a transfected promoter construct [85]. It is, however, not clear which member of the AP-1 family is the natural activator of *jun*.

The activated transcription factors working at the *fos* and *jun* genes can of course not operate persistently: both *fos* and *jun* transcription are subjected to immediate down-regulation. *fos* RNA reaches a maximum of abundance at 40 min following induction in most cells [86,87]. *jun* RNA is maximal at 60–90 min [54]. Following these maxima both genes are no longer transcribed and the RNAs are degraded rapidly [88]. Also, both Fos and Jun proteins are labile [65,66] (and unpublished data). This rapid down regulation appears to safeguard and limit a response that would otherwise lead to continuous proliferation or transformation [89]. Physiologically, down regulation re-establishes responsiveness of cells to other stimuli. Several other mechanisms of extinguishing or counteracting received signals have been found (receptor internalization, down regulation of protein kinase C activity [90–92]) stressing the importance of tightly regulated transcription factor activities. Interestingly, the transcriptional turn-off of both *fos* and *jun* requires protein synthesis and shows autoregulatory features. Both Fos and Jun participate in the transcriptional turn-off of *fos* [82]. Fos appears to repress *jun* transcription [54].

The transient nature of induced AP-1 synthesis is not observed in certain exceptional conditions. In each of these only one of the two components has been examined and it is not known whether the other subunit of AP-1 follows the same pattern. In amniotic tissue highly elevated levels of *fos* RNA and Fos protein [93,94] suggest that the down-regulating mechanisms are absent or overruled. In macrophage cell lines elevated concentrations of cAMP lead to sustained high levels of *fos* RNA [95], and treatment of human fibroblasts with tumour necrosis factor α induces elevated levels of *jun* RNA for extended periods of time [96]. This type of permanent elevation is in fact what we expect to occur under transforming conditions. Apparently it is not sufficient for full transformation in all cells.

7. TRANSIENTLY ELEVATED EXPRESSION OF CYTOPLASMIC ONCOGENE PROTEINS AFFECTS THE ACTIVITY OF TRANSCRIPTION FACTORS IN THE NUCLEUS

In cells in culture and perhaps also in certain primary cells, the mutational activation of a single oncogene (see chapter 5) or the elevated expression of an oncogene will bring about part or even all of the transformed phenotype. For instance, NIH 3T3 cells stably transfected with an MMTV LTR *mos* or *ras* construct are rapidly pushed into transformation by treatment with glucocorticoid hormone which causes elevated levels of Mos or Ras proteins [97]. This type of experiment is uniquely suitable to demonstrate rapid onset of transcription in response to elevated *mos* or *ras* expression. Glucocorticoid hormone, which itself does not induce the *fos* or collagenase promoters, increases, via *mos* or *ras* expression, *fos* and collagenase transcription within a period of minutes to 1–2 h [53]. (In NIH 3T3 cells, the down-regulation of AP-1 activity is low for unknown reasons and thus does not interfere with the experiment described here.) Also, transient co-transfection experiments with cytoplasmic oncogenes such as *mos*, *ras*, *src* or *trk* cause increased transcription from the *fos* promoter, leading itself to the activation of collagenase transcription [53,54]. Since the same type of oncogene overexpression may obliterate the need for growth factor stimulation, we have argued that the cytoplasmic oncogenes participate in normal signal transfer and, under these experimental conditions, lead to an elevated level of transcription factor activity. Although not proven to our knowledge, we assume that permanent overexpression of an oncogene or mutational activation as occurs in cancer will also affect transcription factor activity, but presumably at a lower level, corresponding to a new equilibrium between oncogene-dependent stimulation and down-modulation. Interestingly, overexpression of a c-*jun* cDNA in primary rat cells does not transform but it does in combination with an activated *ras* [89]. v-*jun* in its viral gag-fusion form induces fibrosarcomas in chickens [23]. The role of the other subunit of AP-1, i.e. Fos, has not been addressed. A second-site revertant of a v-*fos* transformant has been isolated which lost transformability by *ras*, *abl* and *mos* [98] suggesting an oncogene transformation pathway. It will be interesting to examine whether the reversion occurred in c-*jun*. This would put AP-1 in a central position in transformation. Fos may, however, complex with other transcription factors in addition to *jun* [47,63].

8. POST-TRANSLATIONAL MODULATION OF NFκB

In contrast to AP-1, the protein acting on the HIV enhancer, NFκB, is stored in the cytoplasm and only translocated to the nucleus on stimulation of cells with mitogens, growth factors, phorbol esters or DNA damage [99,100]. NFκB is presumably also a member of a family of related proteins (PRDII-BF1 [101]; H2TF1 [102]; KBF1 [103]; EBP1 [104]) that recognize similar sequence elements and act as intracellular transducers of a variety of external influences in different cell types [33]. Similar sequence motifs which are recognized by the NFκB protein family have been found not only in growth factor and phorbol ester inducible genes but also in genes activated by interferons or by virus infection, and several obviously different proteins have been found to recognize these sequence motifs [33]. NFκB appears to be distinguished from other members of the family by predominantly post-translational regulation. Cytoplasmic NFκB is inactive and activity can be generated by treating the cytoplasmic extract with deoxycholate. This releases active transcription factor from a complex with a cytoplasmic inhibitor IkB. The release from IkB is one of the essential steps that is induced by stimulation of signal pathways and that causes nuclear translocation of the factor [99,100]. In addition to release from IkB, NFκB appears to require a second post-translational activation step, association with Fos protein [63, and unpublished]. Several lines of evidence suggest this association: HIV-1 induction is reduced by prior expression in cells of antisense *fos* RNA; the complex formed *in vitro* with an HIV-1 enhancer oligonucleotide is destroyed by antibodies to Fos proteins; a large excess of the collagenase enhancer oligonucleotide competes for complex formation at the HIV enhancer presumably by reducing the limiting concentration of Fos; Fos antibodies precipitate from UV cross-linked ^{32}P-DNA–protein complexes formed at the HIV enhancer, a labelled protein of the migration behaviour of NFκB.

9. CONSTITUTIVE ACTIVATION AND ELEVATED NEW SYNTHESIS OF NFκB IN A PARASITE-DRIVEN LYMPHOMA

The intracellular tick-borne bovine parasite *Theileria parva* causes a proliferative T cell disease killing cattle in East Africa [105,106]. Infected T cells grow as invasive lymphomas and are easily cultivated *in vitro*. Various parameters of their behaviour and gene expression in these tumour cells have been characterized. For instance, the tumour cells express interleukin-2 (IL-2) receptor and an interleukin-2-like growth factor [107]. Drugs that affect preferentially ATP generation in the parasite kill *Theileria parva* and return proliferating cells to an apparently normal state. Within 5–7 days the proliferating T cells stop growth and resume the morphology of resting peripheral T lymphocytes. Concomitantly, the expression of IL-2 receptor and growth factor production cease [108]. Thus *Theileria parva* imitates a mitogenic or growth factor dependent proliferative response of host cells in a completely reversible manner.

The IL-2 receptor gene (as other genes, e.g. MHC class I and class II genes) is regulated by a κB motif and by the level of active NFκB (as well as other factors). Using the κB motif of the HIV-1 promoter as a probe to monitor the NFκB activity in gel retardation assays, we have shown that *Theileria parva* infected cells maintain elevated levels of NFκB in the nucleus, indicating continuous post-translational modification (Fig. 3). However, also the total amount of cytoplasmic deoxycholate-activatable NFκB is increased in comparison with concanavalin A stimulated lymphocytes (Fig. 3). *Theileria parva* apparently increases the stimulation level of a pathway that induces both new synthesis and post-translational activation of NFκB. This elevated level drops within 24 h of treatment with the antiparasitic drug and disappears within 4–5 days.

10. CONCLUDING REMARKS

Using two examples, i.e. NIH 3T3 cells overexpressing *ras* or *mos*, and *Theileria parva* infected invasive T lymphocytes, we have tried to demonstrate that the state of transformation parallels the activation and new synthesis of transcription factors. The sustained activation of a transcription factor such as AP-1 in a cancer cell, although not documented directly, may be the cause of elevated collagenase expression in many cancer cells. It appears that many oncogene products are connected to each other via a net of signal-transferring components [109]. According to this hypothesis, elevated expression of any one component, a mutation that keeps the oncogene product in the 'on' configuration, or 'signal firing' by an intracellular

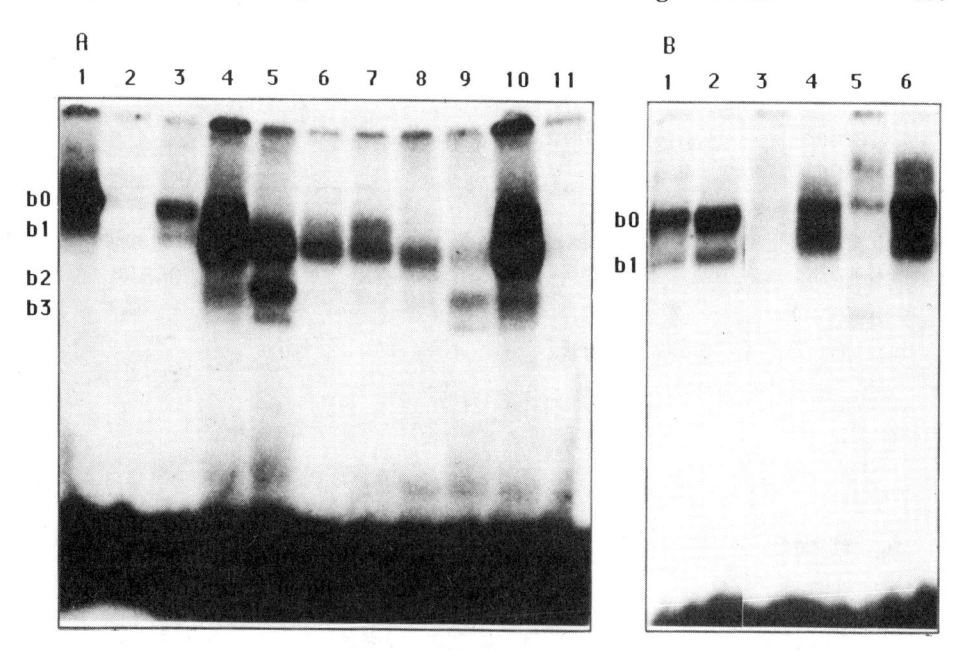

Ch. 8]   **Transcription factors in normal and malignant cells**   159

Fig. 3 — High levels of expression of NfκB in *Theileria* infected cells. Band shift assays using the NFκB binding sequence (Fig. 1) and whole cell, cytoplasmic or nuclear extracts from different cells. Extracts were prepared from 2×10^8 cells as described in [113] at the indicated times of culture following passage. The binding reactions were carried out by incubating end-labelled DNA (2000 cpm) with 2 μg of nuclear or cytoplasmic protein and 2 μg of poly[dI-dC] in a buffer containing 10 mM HEPES (pH 7.9), 60 mM KCl, 4% Ficoll, 1 mM DTT and 1 mM EDTA [114]. After 30 min at room temperature, the reaction mixtures were loaded onto a 4% polyacrylamide gel in 0.25×TBE buffer (89 mM Tris, 89 mM boric acid, 2 mM EDTA) and electrophoresed at 10 V/cm and room temperature for 2 h. For competition experiments a 200-fold molar excess of cold oligonucleotide was added to the reaction mixture prior to the addition of labelled DNA. HeLa cells were treated with the phorbol ester TPA at 50 ng/ml for 18 h before the extracts were prepared. To eliminate the parasite from infected cells, the cultures were treated with the theilericidal drug BW 720c, a hydronaphthoquinone derivative with potent broad spectrum antiprotozoal activity [108]. BW 720c was a gift from Coopers Animal Health Ltd., UK, and was used at a concentration of 50 ng/ml. The compound is not toxic for mammalian cells even at concentrations much higher than used here. The isolation of bovine lymphocytes and the generation of concanavalin A (ConA) stimulated lymphoblasts have been described previously [115]. Several specific DNA–protein complexes at the HIV-1 enhancer sequence were detected (b0–b3) [116]. The composition of the bands is yet not clear. (A) Lane 1: total cell extract from HeLa cells. Lane 2: as lane 1 in the presence of excess of unlabelled NFκB binding oligonucleotide. Lane 3: nuclear extract from ConA-stimulated bovine lymphocytes. Lanes 4 and 5: nuclear extracts of *Theileria* infected bovine T lymphocytes prepared on days 1 and 3 respectively following culture passage. Lanes 6–9: nuclear extracts of *Theileria* infected T lymphocytes treated with BW 720 c for 18 h (lane 6), 24 h (lane 7), 48 h (lane 8) and 72 h (lane 9). Lane 10: identical to lane 4. Lane 11: as lane 10 but in the presence of excess of unlabelled NFκB binding oligonucleotide. (B) Cytoplasmic extracts from ConA stimulated lymphocytes (lanes 1 and 2) and from *Theileria* infected cells (lanes 3–6). In lanes 2, 4 and 6 the extracts were treated during the binding reaction with the dissociating agents sodium deoxycholate (0.6%) and NP-40 (1.2%) as described previously [99] to release NFκB from its cytoplasmic inhibitor protein.

parasite, cause similar effects at the nuclear level (Fig. 4). Transcription factors are envisaged as the receiving ends of the communication network. We note that evidence for elevated transcription factor activity in cancer is yet missing, and the

A. normal signal transfer

	plasma membrane	cytoplasm	nucleus	
a.	gfr	oncogenes	transcription factors	effector genes
b.	GF GFR	ONCOGENES	TRANSCRIPTION FACTORS	EFFECTOR GENES

B. cancer

plasma membrane | cytoplasm | nucleus

a. a cytoplasmic oncogene as transforming principle

	gfr	ONCOGENES	TRANSCRIPTION FACTORS	EFFECTOR GENES

b. transcription factors as oncogenes

	gfr	oncogenes	TRANSCRIPTION FACTORS	EFFECTOR GENES

c. Theileria infected T cells

		PARASITE →	TRANSCRIPTION FACTORS	EFFECTOR GENES
	GF GFR	ONCOGENES	TRANSCRIPTION FACTORS	

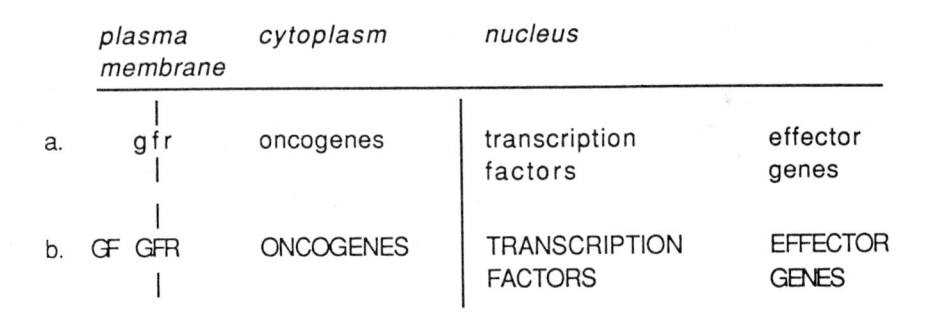

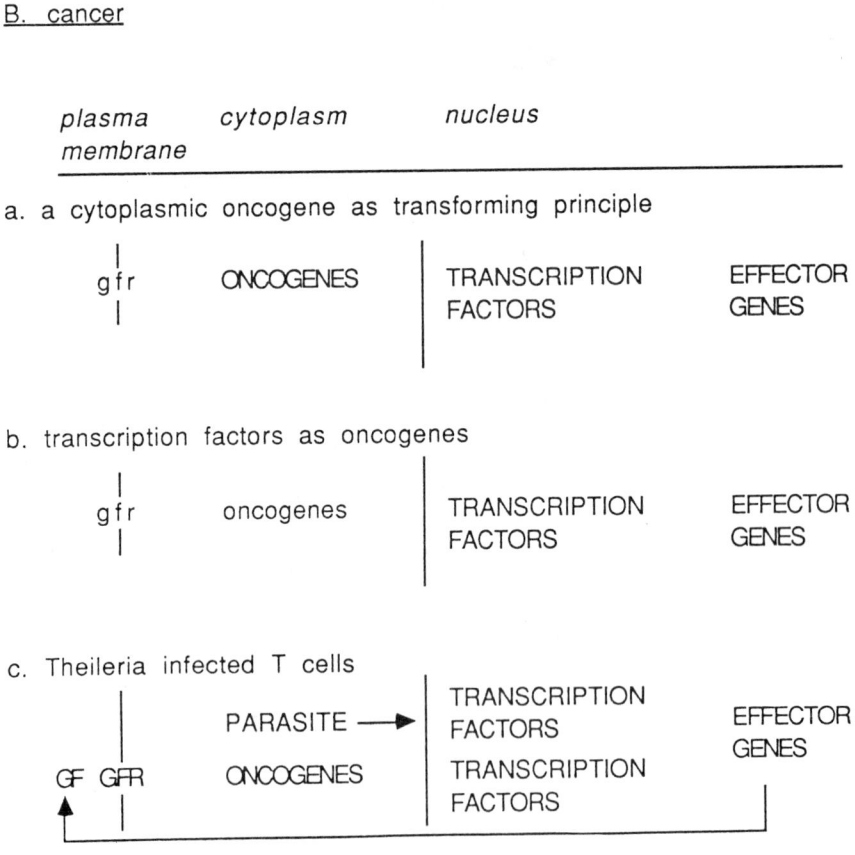

Fig. 4 — Components of signal transfer and flow of information in normal and cancerous cells (gfr, growth factor receptor; gf, growth factor). Capital letters indicate the 'on' state of signal transduction components and of genes.

NFκB activity of the parasite-driven lymphoma is indeed the first case to our knowledge. As a result of stimulated transcription factors, genes (effector genes) would be transcribed that carry the appropriate cis-acting elements and that then participate in the numerous changes of phenotype that distinguish a cancer cell from a normal cell. For instance, growth factors will be produced and secreted, possibly leading to autocrine stimulation, which in turn would elicit signal transfers through the components of the oncogene net. Consequently, genes coding for transcription factors should themselves exhibit transforming abilities. AP-1 is the first example of a transcription factor coded entirely by oncogenes. ErbA is another example coding for the thyroid hormone receptor [110,111]. Various other candidates are expected to follow, e.g. *rel* [112], *myb* [54]. In these cases, 'signals' would be 'generated' within the nucleus (Fig. 4). With respect to the phenotype, many of the different carcinogenic mutations or changes may be indistinguishable, all implicating transcription factor activations.

REFERENCES

[1] Franks, L. M. & Teich, N. (1986) *Introduction to the Cellular and Molecular Biology of Cancer*, Oxford University Press, Oxford

[2] Thomas, K. A. (1988) Transforming potential of fibroblast growth factor genes. *TIBS* **13** 327–328

[3] Heldin, C. & Westermark, B. (1984) Growth factors: mechanism of action and relation to oncogenes. *Cell* **37** 9–20

[4] Waterfield, M. D. (1986) The role of growth factors in cancer. In: L. M. Franks & N. Teich (eds) *Introduction to the Cellular and Molecular Biology of Cancer*, Oxford University Press, Oxford, pp. 251–276.

[5] Fischer, A. (1925) Beitrag zur Biologie der Gewebezellen. Eine vergleichend-biologische Studie der normalen und malignen Gewebezellen *in vitro*. *Arch. Entwicklungsmech. Org. (Wilhelm Roux)* **104** 210

[6] Chiffon, E. E. & Grossi, C. E. (1955) Fibrinolytic activity of human tumors as measured by the fibrin plate method. *Cancer* **8** 1146

[7] Nagy, B., Ban, J. & Brdar, B. (1977) Fibrinolysis associated with human neoplasia: production of plasminogen activator by human tumors. *Int. J. Cancer* **19** 614

[8] Reich, E., Rifkin, D. B. & Shaw, E. (1975) Proteases and biological control. *Cold Spring Harbor Conf. Cell Proliferation* **2** 333–335

[9] Yamada, K. M. & Olden, K., (1978) Fibronectins — adhesive glycoproteins of cell surface and blood. *Nature (London)* **275** 179–184

[10] Liotta, L. A. (1986) Tumor invasion and metastases — role of the extracellular matrix: Rhoads Memorial Award lecture. *Cancer Res.* **46** 1–7

[11] Cajot, J., Kruithof, E. K. O., Schleuning, W., Sordat, B. & Bachmann, F. (1986) Plasminogen activators, plasminogen activator inhibitors and procoagulant analyzed in twenty human tumor cell lines. *Int. J. Cancer* **38** 719–727

[12] Mignatti, P., Robbins, E. & Rifkin, D. B. (1986) Tumor invasion through the human amniotic membrane: requirement for a proteinase cascade. *Cell* **47** 487–498

[13] Sugioka, Y., Fujii-Kuriyama, Y., Kitagawa, T. & Muramatsu, M. (1985)

Changes in polypeptide pattern of rat liver cells during chemical hepatocarcinogenesis. *Cancer Res.* **45** 365–378

[14] Leavitt, J. & Kakunaga, T. (1980) Expression of a variant form of actin and additional polypeptide changes following chemical-induced *in vitro* neoplastic transformation of human fibroblasts. *J. Biol. Chem.* **255** 1650–1661

[15] Chadwick, C. M. (1986) Antigens of normal and neoplastic cells. In: C. M. Chadwick (ed.) *Receptors in Tumour Biology*, Cambridge University Press, Cambridge, pp. 169–188

[16] Willecke, K. & Schäfer, R. (1984) Human oncogenes. *Hum. Genet.* **66** 132–142

[17] Barbacid, M. (1986) Mutagens, oncogenes and cancer. *TIG* **2** 188–192

[18] Bister, K. & Jansen, H. W. (1986) Oncogenes in retroviruses and cells: biochemistry and molecular genetics. *Adv. Cancer Res.* **47** 99–188

[19] Bishop, J. M. (1987) The molecular genetics of cancer. *Science* **235** 305–311

[20] Angel, P., Allegretto, E. A., Okino, S. T., Hattori, K., Boyle, W. J., Hunter, T. & Karin, M. (1988) Oncogene jun encodes a sequence-specific transactivator similar to AP-1. *Nature (London)* **332** 166–171

[21] Bohmann, D., Bos, T. J., Admon, A., Nishimura, T., Vogt, P. K. & Tjian, R. (1987) Human proto-oncogene c-jun encodes a DNA binding protein with structural and functional properties of transcription factor AP-1. *Science* **238** 1386–1392

[22] Verma, I. M. (1986) Proto-oncogene fos: a multifaceted gene. *TIG* **2** 93–96

[23] Maki, Y., Bos, T. J., Davis, C., Starbuck, M. & Vogt, P. K. (1987). Avian sarcoma virus 17 carries the jun oncogene. *Proc. Natl. Acad. Sci. USA* **84** 2848–2852

[24] Knudson, A. G., Jr. (1986) Genetics of human cancer. *Ann. Rev. Genet.* **20** 231–251

[25] Dynan, W. S. & Tjian, R. (1985) Control of eukaryotic messenger RNA synthesis by sequence-specific DNA-binding proteins. *Nature (London)* **316** 774–778

[26] McKnight, S. & Tjian, R. (1986) Transcriptional selectivity of viral genes in mammalian cells. *Cell* **46** 795–805

[27] Banerji, J., Rusconi, S. & Schaffner, W. (1981) Expression of a β-globin gene is enhanced by remote SV40 DNA sequences. *Cell* **27** 299–308

[28] Benoist, C. & Chambon, P. (1981) *In vivo* sequence requirements of the SV40 early promoter region. *Nature (London)* **290** 304–310

[29] Schlokat, U. & Gruss, P. (1986) Enhancers as control elements for tissue-specific transcription. In: P. Kahn and T. Graf (eds) *Oncogenes and Growth Control*, Springer, Berlin, pp. 226–234

[30] Serfling, E., Jasin, M. & Schaffner, W. (1985) Enhancers and eukaryotic gene transcription. *TIG* **1** 224–230

[31] Guarente, L. (1988) UASs and enhancers: common mechanism of transcriptional activation in yeast and mammals. *Cell* **52** 303–305

[32] Ptashne, M. (1988) How eukaryotic transcriptional activators work. *Nature (London)* **335** 683–689

[33] Lenardo, M. J., Fan, C.-M., Maniatis, T. & Baltimore, D. (1989) The

involvement of NF-κB in β-interferon gene regulation reveals its role as widely inducible mediator of signal transduction. *Cell* **57** 287-294

[34] Angel, P., Imagawa, M., Chiu, R., Stein, B., Imbra, R. J., Rahmsdorf, H. J., Jonat, C., Herrlich, P. & Karin, M. (1987) Phorbol ester-indible genes contain a common cis element recognized by a TPA-modulated trans-acting factor. *Cell* **49** 729-739

[35] Ma, J. & Ptashne, M. (1987) A new class of yeast transcriptional activators. *Cell* **51** 113-119

[36] Hope, I. A., Mahadevan, S. & Struhl, K. (1988) Structural and functional characterization of the short acidic transcriptional activation region of yeast GCN4 protein. *Nature (London)* **333** 635-640

[37] Horikoshi, M., Carey, M. F., Kakidani, H. & Roeder, R. G. (1988) Mechanism of action of a yeast activator: direct effect of GAL4 derivatives on mammalian TFIID-promoter interactions. *Cell* **54** 665-669

[38] Bushman, F. D. & Ptashne, M. (1988) Turning λ Cro into a transcriptional activator. *Cell* **54** 191-197

[39] Nabel, G. & Baltimore, D. (1987) An inducible transcription factor activates expression of human immunodeficiency virus in T cells. *Nature (London)* **326** 711-713

[40] Angel, P., Baumann, I., Stein, B., Delius, H., Rahmsdorf, H. J. & Herrlich, P. (1987) 12-O-tetradecanoyl-phorbol-13-acetate induction of the human collagenase gene is mediated by an inducible enhancer element located in the 5'-flanking region. *Mol. Cell. Biol.* **7** 2256-2266

[41] Stein, B., Rahmsdorf, H. J., Schönthal, A., Büscher, M., Ponta, H. & Herrlich, P. (1988) The UV induced signal transduction pathway to specific genes. In: E. Triedberg, P. Hanawalt (eds) UCLA Symposium on Molecular and Cellular Biology, Vol. 83. *Mechanisms and Consequences of DNA Damage Processing*, Alan R. Liss, New York, pp. 557-570

[42] Stein, B., Rahmsdorf, H. J., Steffen, A., Litfin, M. & Herrlich, P. (1989) UV-induced DNA damage is an intermediate step in the UV-induced expression of human immunodeficiency virus type 1, collagenase, c-fos and metallothionein. *Mol. Cell. Biol.* **9** 5169-5181

[43] Dinter, H., Chiu, R., Imagawa, M., Karin, M. & Jones, K. A. (1987) *In vitro* activation of the HIV-1 enhancer in extracts from cells treated with a phorbol ester tumor promoter. *EMBO J.* **6** 4067-4071

[44] Hai, T., Liu, F., Allegretto, E. A., Karin, M. & Green, M. R. (1988) A family of immunologically related transcription factors that includes multiple forms of ATF and AP-1. *Genes Dev.* **2** 1216-1226

[45] Lee, W., Mitchell, P. & Tjian, R. (1987) Purified transcription factor AP-1 interacts with TPA-inducible enhancer elements. *Cell* **49** 741-752

[46] Ryder, K., Lau, L. F. & Nathans, D. (1988) A gene activated by growth factors is related to the oncogene v-jun. *Proc. Natl. Acad. Sci. USA* **85** 1487-1491

[47] Nakabeppu, Y., Ryder, K. & Nathans, D. (1988) DNA binding activities of three murine jun proteins: stimulation by fos. *Cell* **55** 907-915

[48] Cohen, D. R. & Curran, T. (1988). fra-1: a serum-inducible, cellular immediate-early gene that encodes a fos-related antigen. *Mol. Cell. Biol.* **8** 2063-2069

[49] Rauscher III, F. J., Cohen, D. R., Curran, T., Bos, T. J., Vogt, P. K., Bohmann, D., Tjian, R. & Franza, B. R., Jr. (1988) Fos-associated protein p39 is the product of the jun proto-oncogene. *Science* **240** 1010–1016

[50] Chiu, R., Boyle, W. J., Meek, J., Smeal, T., Hunter, T. & Karin, M. (1988) The c-fos protein interacts with c-Jun/AP-1 to stimulate transcription of AP-1 responsive genes. *Cell* **54**, 541–552

[51] Sassone-Corsi, P., Lamph, W. W., Kamps, M. & Verma, I. M. (1988) fos-associated cellular p39 is related to nuclear transcription factor AP-1. *Cell* **54** 553–560

[52] Zerial, M., Toschi, L., Ryseck, R., Schuermann, M., Müller, R. & Bravo, R. (1989) The product of a novel growth factor activated gene, fos B, interacts with Jun proteins enhancing their DNA binding activity. *EMBO J.* **8** 805–813

[53] Schönthal, A., Herrlich, P., Rahmsdorf, H. J. & Ponta, H. (1988) Requirement for fos gene expression in the transcriptional activation of collagenase by other oncogenes and phorbol esters. *Cell* **54** 325–334

[54] Schönthal, A., Gebel, S., Stein, B., Ponta, H., Rahmsdorf, H. J. & Herrlich, P. (1988) Nuclear oncoproteins determine the genetic program in response to external stimuli. *Cold Spring Harbor Symp. Quant. Biol.* **53** 779–787

[55] Curran, T., Miller, A. D., Zokas, L. & Verma, I. M. (1984) Viral and cellular fos proteins: a comparative analysis. *Cell* **36** 259–268

[56] Schuermann, M., Neuberg, M., Hunter, J. B., Jenuwein, T., Ryseck, R., Bravo, R. & Müller, R. (1989) The leucine repeat motif in fos protein mediates complex formation with Jun/AP-1 and is required for transformation. *Cell* **56** 507–516

[57] Gebel, S., Stein, B., König, H., Rahmsdorf, H. J., Ponta, H., Risse, G., Neuberg, M., Müller, R. & Herrlich, P. (1989) Two nuclear oncogene products cooperate in the formation of the transcription factor AP-1. In: H. Lother, R. Dernick, W. Ostertag (eds.) *Vectors as Tools for the Study of Normal and Abnormal Growth and Differentiation*, Springer, Berlin, 385–397

[58] Sassone-Corsi, P., Ransone, L. J., Lamph, W. W. & Verma, I. M. (1988) Direct interaction between fos and jun nuclear oncoproteins: role of the 'leucine zipper' domain. *Nature (London)* **336** 692–695

[59] Bravo, R., Burckhardt, J., Curran, T. & Müller, R. (1986) Expression of c-fos in NIH3T3 cells is very low but inducible throughout the cell cycle. *EMBO J.* **5** 695–700

[60] Dony, C. & Gruss, P. (1987) Proto-oncogene c-fos expression in growth regions of fetal bone and mesodermal web tissue. *Nature (London)* **328** 711–714

[61] de Togni, P., Niman, H., Raymond, V., Sawchenko, P. & Verma, I. M. (1988) Detection of fos protein during osteogenesis by monoclonal antibodies. *Mol. Cell. Biol.* **8** 2251–2256

[62] Müller, R. (1986) Cellular and viral fos genes: structure, regulation of expression and biological properties of their encoded products. *Biochem. Biophys. Acta* **823** 207–225

[63] Herrlich, P., Ponta, H., Stein, B., Gebel, S., König, H., Schönthal, A., Büscher, M. & Rahmsdorf, H. J. (1989) The role of fos in gene regulation. In:

C. E. Sekeris (ed.) *Molecular Mechanisms and Consequences of Activation of Hormone and Growth Factor Receptors,* in press

[64] Kaina, B., Stein, B., Schönthal, A., Rahmsdorf, H. J., Ponta, H. & Herrlich, P. (1989) An update of the mammalian UV response: gene regulation and induction of a protective function. In: M. W. Lambert *et al.* (eds) *DNA Repair Mechanisms and their Biological Implications in Mammalian Cells,* in press

[65] Müller, R., Bravo, R., Müller, D., Kurz, C. & Renz, M. (1987) Different types of modification in c-fos and its associated protein p39: modulation of DNA binding by phosphorylation. *Oncogene Res.* **2** 19–32

[66] Lee, W. M. F., Lin, C. & Curran, T. (1988) Activation of the transforming potential of the human fos proto-oncogene requires message stabilization and results in increased amounts of partially modified fos protein. *Mol. Cell. Biol.* **8** 5521–5527

[67] Jackson, S. P. & Tjian, R. (1988) O-glycosylation of eukaryotic transcription factors: implications for mechanisms of transcriptional regulation. *Cell* **55** 125–133

[68] Viaje, A., Slaga, T. J., Wigler, M. & Weinstein, I. B. (1977) Effects of antiinflammatory agents on mouse skin tumor promotion, epidermal DNA synthesis, phorbol ester-induced cellular proliferation, and production of plasminogen activator. *Cancer Res.* **37** 1530–1536

[69] Hynes, N., van Ooyen, A. J. J., Kennedy, N., Herrlich, P., Ponta, H. & Groner, B. (1983) Subfragments of the large terminal repeat cause glucocorticoid-responsive expression of mouse mammary tumor virus and of an adjacent gene. *Proc. Natl. Acad. Sci. USA* **80** 3637–3641

[70] Majors, J. & Varmus, H. E. (1983) A small region of the mouse mammary tumor virus long terminal repeat confers glucocorticoid hormone regulation on a linked heterologous gene. *Proc. Natl. Acad. Sci. USA* **80** 5866–5870

[71] Ponta, H., Kennedy, N., Skroch, P., Hynes, N. E. & Groner, B. (1985) Hormonal response region in the mouse mammary tumor virus long terminal repeat can be dissociated from the proviral promoter and has enhancer properties. *Proc. Natl. Acad. Sci. USA* **82** 1020–1024

[72] Klock, G., Strähle, U. & Schütz, G. (1987) Oestrogen and glucocorticoid responsive elements are closely related but distinct. *Nature (London)* **329** 734–736

[73] Chandler, V. L., Maler, B. A. & Yamamoto, K. R. (1983) DNA sequences bound specifically by glucocorticoid receptor *in vitro* render a heterologous promoter hormone responsive *in vivo*. *Cell* **33** 489–499

[74] Offringa, R., Smits, A. M. M., Houweling, A., Bos, L. J. L. & van der Eb, A. J. (1988) Similar effects of adenovirus E1A and glucocorticoid hormones on the expression of the metalloprotease stromelysin. *Nucleic Acids Res.* **16** 10973–10984

[75] van der Eb, A. J., Offringa, R., Timmers, H. T. M., van Dam, H., Meijer, I., van den Heuvel, S. J. L., Kast, W. M., Melief, C. J. M., Herrlich, P., Zantema, A. & Bos, J. L. (1989) Inhibition of cellular gene expression in adenovirus-transformed cells. In: *UCLA Symposium, Taos*

[76] Gruss, C., Wetzel, E., Baack, M., Mock, U. & Knippers, R. (1988) High-

affinity SV40 T-antigen binding sites in the human genome. *Virology* **167** 349–360

[77] Mitchell, P. J., Wang, C. & Tjian, R. (1987) Positive and negative regulation of transcription *in vitro:* enhancer-binding protein AP-2 is inhibited by SV40 T antigen. *Cell* **50** 847–861

[78] Büscher, M., Rahmsdorf, H. J., Litfin, M., Karin, M. & Herrlich, P. (1988) Activation of the c-fos gene by UV and phorbol ester: different signal transduction pathways converge to the same enhancer element. *Oncogene* **3** 301–311

[79] Quantin, B. & Breathnach, R. (1988) Epidermal growth factor stimulates transcription of the c-jun proto-oncogene in rat fibroblasts. *Nature (London)* **334** 538–539

[80] Ryseck, R. P., Hirai, S. I., Yaniv, M. & Bravo, R. (1988) Transcriptional activation of c-jun during the G_0/G_1 transition in mouse fibroblasts. *Nature (London)* **334** 535–538

[81] Lamph, W. W., Wamsley, P., Sassone-Corsi, P. & Verma, I. M. (1988) Induction of proto-oncogene Jun/AP-1 by serum and TPA. *Nature (London)* **334** 629–631

[82] König, H., Ponta, H., Rahmsdorf, U., Büscher, M., Schönthal, A., Rahmsdorf, H. J. & Herrlich, P. (1989) Autoregulation of fos: the dyad symmetry element as the major target of repression. *EMBO J.* **8** 2559–2566

[83] Treisman, R. (1987) Identification and purification of a polypeptide that binds to the c-fos serum response element. *EMBO J.* **6** 2711–2717

[84] Shaw, P. E., Schröter, H. & Nordheim, A. (1989) The ability of a ternary complex to form over the serum response element correlates with serum inducibility of the human c-fos promoter. *Cell* **56** 563–572

[85] Angel, P., Hattori, K., Smeal, T. & Karin, M. (1988) The jun proto-oncogene is positively autoregulated by its product, Jun/AP-1. *Cell* **55** 875–885

[86] Kruijer, W., Cooper, J. A., Hunter, T. & Verma, I. M. (1984) Platelet-derived growth factor induces rapid but transient expression of the c-fos gene and protein. *Nature (London)* **312** 711–716

[87] Greenberg, M. E. & Ziff, E. B. (1984). Stimulation of 3T3 cells induces transcription of the c-fos proto-oncogene. *Nature (London)* **311** 433–438

[88] Rahmsdorf, H. J., Schönthal, A., Angel, P., Litfin, M., Rüther, U. & Herrlich, P. (1987) Posttranscriptional regulation of c-fos mRNA expression. *Nucleic Acids Res.* **15** 1643–1659

[89] Schütte, J., Minna, J. D. & Birrer, M. J. (1989) Deregulated expression of human c-jun transforms primary rat embryo cells in cooperation with an activated c-Ha-ras gene and transforms Rat-1a cells as a single gene. *Proc. Natl. Acad. Sci. USA* **86** 2257–2261

[90] Schlessinger, J., Shechter, Y., Willingham, M. C. & Pastan, I. (1978) Direct visualization of binding, aggregation, and internalization of insulin and epidermal growth factor on living fibroblastic cells. *Proc. Natl. Acad. Sci. USA* **75** 2659–2663

[91] Rodriguez-Pena, A. & Rozengurt, E. (1984) Disappearance of Ca^{2+}-sensitive, phospholipid-dependent protein kinase activity in phorbol ester-treated 3T3 cells. *BBRC* **120** 1053–1059

[92] Pasti, G., Lacal, J., Warren, B. S., Aaronson, S. A. & Blumberg, P. M. (1986) Loss of mouse fibroblast cell response to phorbol esters restored by microinjected protein kinase C. *Nature (London)* **324** 375–377

[93] Müller, R., Slamon, D. J., Tremblay, J. M., Cline, M. J. & Verma, I. M. (1982) Differential expression of cellular oncogenes during pre- and postnatal development of the mouse. *Nature (London)* **299** 640–644

[94] Müller, R., Verma, I. M. & Adamson, E. D. (1983) Expression of c-onc genes: c-fos transcripts accumulate to high levels during development of mouse placenta, yolk sac and amnion. *EMBO J.* **2** 679–684

[95] Bravo, R., Neuberg, M., Burckhardt, J., Almendral, J., Wallich, R. & Müller, R. (1987) Involvement of common and cell type-specific pathways in c-fos gene control: stable induction by cAMP in macrophages. *Cell* **48** 251–260

[96] Brenner, D. A., O'Hara, M., Angel, P., Chojkier, M. & Karin, M. (1989) Prolonged activation of jun and collagenase genes by tumour necrosis factor-α. *Nature (London)* **337** 661–663

[97] Jaggi, R., Salmons, B., Muellener, D. & Groner, B. (1986) The v-mos and H-ras oncogene expression represses glucocorticoid hormone-dependent transcription from the mouse mammary tumor virus LTR. *EMBO J.* **5** 2609–2616

[98] Zarbl, H., Latreille, J. & Jolicoeur, P. (1987) Revertants of v-fos-transformed fibroblasts have mutations in cellular genes essential for transformation by other oncogenes. *Cell* **51** 357–369

[99] Baeuerle, P. A. & Baltimore, D. (1988) IκB: a specific inhibitor of the NF-κB transcription factor. *Science* **242** 540–546

[100] Baeuerle, P. A. & Baltimore, D. (1988) Activation of DNA-binding activity in an apparently cytoplasmic precursor of the NFκB transcription factor. *Cell* **53** 211–217

[101] Keller, A. D. & Maniatis, T. (1988) Identification of an inducible factor that binds to a positive regulatory element of the human β-interferon gene. *Proc. Natl. Acad. Sci. USA* **85** 3309–3313

[102] Singh, H., LeBowitz, J. H., Baldwin, A. S. Jr. & Sharp, P. A. (1988) Molecular cloning of an enhancer binding protein: isolation by screening of an expression library with a recognition site DNA. *Cell* **52** 415–423

[103] Yano, O., Kanellopoulos, J., Kieran, M., LeBail, O., Israel, A. & Kourilsky, P. (1987) Purification of KBF1, a common factor binding to both H-2 and β2-microglobulin enhancers. *EMBO J.* **6** 3317–3324

[104] Clark, L., Pollock, R. M. & Hay, R. T. (1988) Identification and purification of EBP1: a HeLa cell protein that binds to a region overlapping the 'core' of the SV40 enhancer. *Genes Dev.* **2** 991–1002

[105] Morrison, W. I., Lalor, P. A., Goddeeris, B. M. & Teale, A. J. (1985) Theileriosis: antigens and host-parasite interactions. In: T. W. Pearson (ed.) *Parasite Antigens, Toward New Strategies for Vaccines*, Dekker, New York, pp. 167–231

[106] Irvin, A. D. & Morrison, W. I. (1987) Immunopathology, immunology and immunoprophylaxis of *Theileria* infection. In: E. J. L. Soulsby (ed.) *Immune Responses in Parasite Infection: Immunology, Immunopathology and Immunoprophylaxis 3: Protozoa*, CRC Press, Boca Raton, FL, pp. 223–274

[107] Dobbelaere, D. A. E., Coquerelle, T. M., Roditi, I. J., Eichhorn, M. &

Williams, R. O. (1988) *Theileria parva* infection induces autocrine growth of bovine lymphocytes. *Proc. Natl. Acad. Sci. USA* **85** 4730—4734

[108] Hudson, A. T., Randall, A. W., Fry, M., Ginger C. D., Hill, B., Latter, V. S., McHardy, N. & Williams, R. B. (1985) Novel anti-malarial hydronaphthoquinones with potent broad spectrum anti-protozoal activity. *Parasitology* **90** 45–55

[109] Herrlich, P. & Ponta, H. (1989) "Nuclear" oncogenes convert extracellular stimuli into changes in the genetic program. *TIG* **5** 112–116

[110] Sap, J., Munoz, A., Damm, K., Goldberg, Y., Ghydael, J., Leutz, A., Beug, H. & Vennström, B. (1986) The c-erb-A protein is a high-affinity receptor for thyroid hormone. *Nature (London)* **324** 635–640

[111] Weinberger, C., Thompson, C. C., Ong, E. S., Lebo, R., Gruol, D. J. & Evans, R. M. (1986) The c-erb-A gene encodes a thyroid hormone receptor. *Nature (London)* **324** 641–646

[112] Gelinas, C. & Temin, H. M. (1988) The v-rel oncogene encodes a cell-specific transcriptional activator of certain promoters. *Oncogene* **3** 349–356

[113] Dignam, J. D., Lebovitz, R. M. & Roeder, R. G. (1983) Accurate transcription initiation by RNA polymerase II in a soluble extract from isolated mammalian nuclei. *Nucleic Acids Res.* **11** 1475–1489

[114] Barberis, A., Superti-Furga, G. & Busslinger, M. (1987) Mutually exclusive interaction of the CCAAT-binding factor and a displacement protein with overlapping sequences of histone gene promoter. *Cell* **50** 347–359

[115] Mastro, A. M. & Pepin, K. G. (1980) Suppression of lectin-stimulated DNA synthesis in bovine lymphocytes by the tumor promoter TPA. *Cancer Res.* **40** 3307–3312

[116] Ivanov, V., Stein, B., Baumann, I., Dobbelaere, D. A. E., Herrlich, P. & Williams, R. O. (1989). Infection with the intracellular protozoan parasite *Theileria parva* induces constitutively high levels of NF-κB in bovine T lymphocytes. *Mol. Cell. Biol.* **9** 4677–4686

9

Aberrant steroid-thyroid receptors as onocogenes

Mels Sluyser
Division of Tumour Biology, The Netherlands Cancer Institute, Plesmanlaan 121, 1066 CX Amsterdam, The Netherlands

1. INTRODUCTION

In 1985 Sluyser and Mester [1] proposed that the products of certain oncogenes may bear a structural resemblance to steroid hormone receptors. This idea was based on the hypothesis that the loss of hormonal dependency of certain tumours might be due to the appearance of mutated or truncated steroid-receptor-like proteins that act constitutively, i.e. enhance transcriptional activity even in the absence of hormone. These aberrant proteins would either be produced by mutations in steroid receptor genes or be encoded by oncogenes that have homology with steroid receptor genes.

Subsequently, investigations in several laboratories have shown that steroid receptors have structural homology with the avian erythroblastosis virus (AEV) oncogene v-*erb*-A (for reviews see refs. [2] and [3]). In fact, steroid receptors and *erb*-A have turned out to belong to a large superfamily of genes whose products are ligand-responsive transcription factors. As well as the steroid receptor genes, this superfamily contains various other genes including the vitamin D3 receptor genes, and thyroid hormone receptor genes [4].

It is not the purpose of this chapter to give a complete reviews of this superfamily, but rather to look at certain members that might be involved in malignant transformation and/or tumour progression.

2. GENERAL STRUCTURE OF STEROID–THYROID RECEPTORS

Steroid hormone receptor proteins contain different domains which have been designated A–F [3]. The assignment of these domains is based on degreees of homology between steroid receptors, i.e. regions A, C and E are highly conserved between these receptors, whereas this is less the case for regions B, D and F (Table

1). Subsequent studies revealed that these regions serve different functions. Domain C contains two zinc-binding 'fingers' that interact with DNA. Domain E is the region to which the hormone molecule binds. The D region separating C and E is called the 'hinge region' because it is thought to act as a hinge between the DNA-binding and hormone-binding domains. The A–B domain may serve to modulate the receptor–DNA interactions, and may also be involved in interactions with other molecules that regulate DNA transcription (Fig. 1).

Thyroid hormone receptors also have this general structure, and therefore the superfamily encoding this type of proteins is generally named the steroid and thyroid receptor gene superfamily. This superfamily also includes genes for retinoic acid receptors, vitamin D3 receptors, and dioxin receptors. More members of this superfamily are likely to exist [4]. The genomic organization of these genes probably follows the pattern found for the progesterone and oestrogen receptor genes. As shown in Fig. 2, these genes are split into eight exons. The A–B region is almost entirely encoded within a single exon, and each of the putative 'zinc fingers' of region C is encoded separately. Region E is assembled from five exons [80, 84]. Some members of the steroid–thyroid hormone receptor superfamily are shown in Fig. 3.

3. VIRAL *erb*-A AND THYROID HORMONE RECEPTORS

Viral *erb*-A was originally discovered when it was found that the ability of avian erythroblastosis virus (AEV) to transform erythroblasts is determined by two oncogenes named v-*erb*-A and v-*erb*-B [5–9]. Both *erb*-A and *erb*-B oncogenes are homologous to avian and mammalian chromosomal DNA sequences c-*erb*-A and c-*erb*-B [10]. Cellular *erb*-B is a truncated form of the epidermal growth factor (EGF) receptor [11]. Wild-type AEV-transformed erythroblasts are completely blocked in their maturation at the colony–forming unit stage. Studies on deletion mutants in *erb*-A and *erb*-B indicate that *erb*-B yields transformed erythroblast-like cells at different stages of maturation, whereas *erb*-A alone induced no transformation. Therefore *erb*-A potentiates the transforming activity of *erb*-B and appears to be responsible for the early blockage of cell differentiation within the erythroid lineage [12]. However, in transfection experiments, the v-*erb*-A oncogene product is sufficient to transform erythrocyte cells [127].

The human [13] and chicken [14] c-*erb*-A genes, the cellular counterparts of the viral oncogene v-*erb*-A, represent thyroid hormone (T3) receptors. The *erb*-A genes encode a cysteine-rich domain that shows high homology with the putative DNA-binding domain of steroid hormone receptors [13,14,16–21]. The amino acid sequence of this domain in v-*erb*-A is almost identical to that of the glucocorticoid receptors [22,23] and oestrogen receptors of various species [24,25]. A thyroid hormone receptor called *erb*-A-T has been described for the human testis which is closely related to the chicken *erb*-A [26].

Interestingly, several c-*erb*-A related genes exist in the human genome. Of the multiple human homologues to v-*erb*-A [15] the most closely related homologue is the hc-*erb*-A1 gene on human chromosome 17 [27] that probably encodes *erb*-A-T. Another homologue, the hc-*erb*-A-β gene, is located on human chromosome 3 [13].

The existence of multiple thyroid hormone receptors suggests that these recep-

Table 1 — Protein sequence homologies in regions C (DNA binding) and E (hormone binding) of steroid hormone receptors

Receptor	Region	ER	PR	GR	c-*erb*-A
ER	C	—	62	62	55
	E	—	28	29	15
PR	C	62	—	91	45
	E	28	—	55	18
GR	C	62	91	—	46
	E	29	55	—	20

The percentages of identical amino acid residues in regions C and E are shown in oestrogen, progesterone and glucocorticoid receptors and human c-*erb*-A. Comparison of v-*erb*-A and chicken c-*erb*-A with the steroid hormone receptors gives values which are identical to those above with human c-*erb*-A within ±2%. Data taken from Green *et al.* [3].

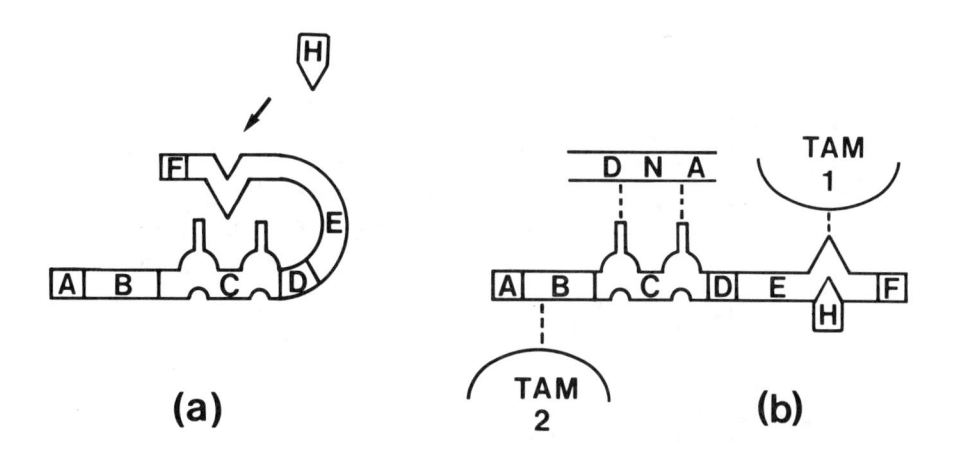

Fig. 1 — Schematic representation of steroid–thyroid receptor activation. The receptor contains different domains (A–F). The hormone (H) binds to region E, thereby inducing a conformational change in the receptor that unmasks the DNA-binding 'fingers' of region C and allows the E region to interact with component(s) of the transcription activation machinery (TAM). The A–B region may also interact with the TAM.

tors play different roles from one tissue to another. This is shown clearly by studies in the rat where, of three thyroid hormone receptors thus far detected, one (called r-*erb*Aβ-2) is expressed only in the pituitary gland [112] whereas other homologues of the v-*erb*-A oncogene are expressed in other tissues. Of interest is the fact that alternative splicing generates messages encoding rat c-*erb*-A proteins that do not bind thyroid hormone [114] (Table 2).

The thyroid hormone receptor, in the absence of its ligand, blocks the activity of a

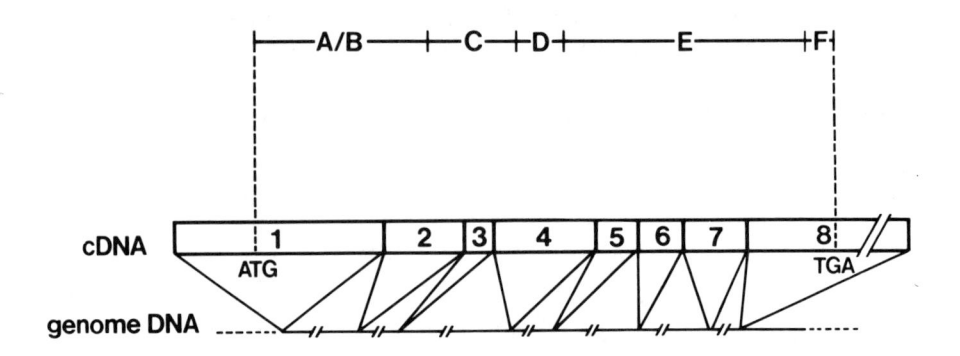

Fig. 2 — Genomic organization of the human ER gene. The eight exons are shown as solid boxes numbered 1–8. The positions of the translation initiation (ATG) and termination (TGA) codons are indicated. The corresponding division of the hER protein in regions A–F is shown above. Data from Ponglikitmongkol et al. [80].

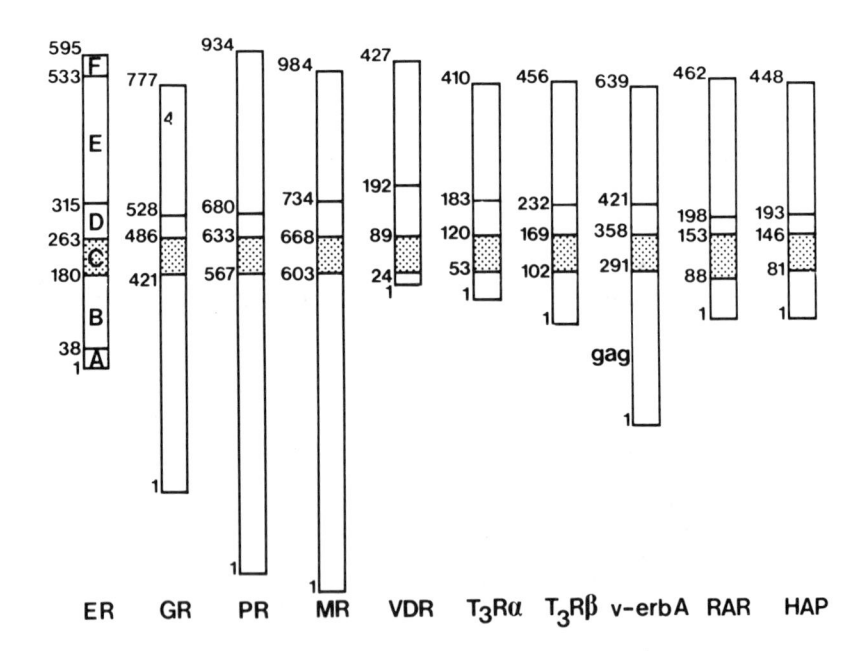

Fig. 3 — Schematic amino acid comparison of members of the steroid–thyroid hormone receptor superfamily. Primary amino acid sequences have been aligned on basis of regions of maximum amino acid similarity. Amino acid numbers are those for the human receptors with the exception of v-erb-A. Abbreviations: ER, oestrogen receptor; GR, glucocorticoid receptor; PR, progesterone receptor; MR, mineralocorticoid receptor; VDR, vitamin D_3 receptor; T_3R, thyroid hormone receptor; RAR, retinoic acid receptor; HAP, hepatoma-associated protein (epithelial-type retinoic acid receptor, RAR_E). The DNA-binding regions (C) are shaded [3, 4].

Table 2 — Cellular homologue to the v-*erb*-A oncogene in the rat

Homologue	T$_3$ binding	Expression
r-erbAα−1	+	Skeletal muscle, brown fat
r-erbAα−2[a]	−	Brain, hypothalamus
r-erbAβ−1	+	Kidney, liver
r-erbAβ−2	+	Pituitary gland

[a]r-erbAα−2 is a truncated form of the thyroid hormone receptor that lacks the thyroid hormone (T$_3$) binding domain. Data derived from ref. [112].

responsive promoter. This suppression is abolished when thyroid hormone is added to the system, which stimulates expression. The oncogenic analogue of the thyroid hormone receptor, v-*erb*-A, acts as a constitutive repressor in this system and, when co-expressed with the receptor, blocks activation [125].

The v-*erb*-A has a lower affinity than the normal receptor for binding to a thyroid-hormone-responsive element in the long terminal repeat of Moloney murine leukaemia virus that binds c-*erb*-A-α protein. Overexpressed v-*erb*-A protein may therefore negatively interfere with normal transcriptional control mechanisms [124], indicating that in fact v-*erb*-A is a transcriptional regulator run amok [111].

4. RETINOIC ACID RECEPTORS

The discovery that the DNA-binding domain of the steroid and thyroid hormone receptors is highly conserved has led to the use of the DNA sequences encoding these domains as hybridization probes to scan the genome for related, but novel, ligand receptors. In this way a cDNA was cloned encoding a 462 amino acid polypeptide which turned out to be a receptor for the morphogen retinoic acid [28,29].

Dejean *et al.* [30] have found an integration of hepatitis B virus next to a liver cell DNA sequence with a striking resemblance to that of v-*erb*-A and steroid receptors. They proposed that this gene (called *hap*) is usually silent or transcribed at a very low level in normal hepatocytes, but becomes inappropriately expressed as a consequence of the hepatitis B viral integration, thus contributing to the cell transformation.

Cloning and sequence analysis of the corresponding complementary DNA from a human cDNA library revealed that the cellular *hap* gene has an open reading frame of 448 amino acids, corresponding to a predicted polypeptide of relative molecular mass 51 kDa [31].

The *hap* gene product displays two regions similar to the DNA- and hormone-binding domains of steroid–thyroid hormone receptors, but differs too much from these to be a variant of one of them.

Subsequent studies by Benbrook *et al.* [32] revealed that the *hap* gene product is a retinoic acid receptor. This was demonstrated by using the 'finger-swap' approach in which the DNA-binding ('finger') domain of the unknown receptor is exchanged with the DNA-binding domain of a receptor for which specific hormone-responsive DNA elements (HRE) have been identified. The hybrid receptor is used to transfect susceptible tissue culture cells together with a reporter gene containing the relevant HRE. In the transient co-transfection experiment by Benbrook *et al.* [32] chloramphenicol acetyl transferase (CAT) reporter gene was linked to a promoter containing the oestrogen-responsive element [ERE]. It was then found that only retinoic acid induced CAT gene expression strongly at a physiological concentration (2.5×10^{-8} M).

This finding is at variance with the report by De Thè *et al.* [31] who stated that the *hap* protein did not specifically bind retinoic acid. The apparent discrepancy is resolved by the fact that specific binding of retinoic acids to their receptors cannot be carried out with protein *in vitro*, as done by De Thè *et al.*, but can only be demonstrated indirectly by using the finger-swap approach.

In the rat, this novel retinoic acid receptor gene is strongly expressed in various organs but not liver. The size of the transcript ranges from 2.4 to 9.5 kb in these organs. This, retinoic acid receptor gene shows a considerable specificity for epithelial-type tissue including skin and has therefore been named RAR_E[30]. The discovery of the RAR_E gene apparently contributing to a hepatocarcinoma is surprising as vitamin A and other retionoids are generally considered to have anti-tumour activity. The RAR_E gene might, however, contribute to tumour development when expressed erroneously in liver tissue, where it is normally silent. Retinoids are known to maintain the proliferative state of epithelial cells and this type of cell proliferation, when induced in other types of tissues by erroneous RAR_E expression in the presence of retinoic acid, may lead to tumour development [32].

5. GLUCOCORTICOID RECEPTOR MUTANTS

The glucocorticoid receptor has been cloned [21], and it is now known which amino acid residues are involved in attachment of the ligand [38–41]. The hormone-binding region of hGR resembles the hormone-binding domains of the mineralocorticoid and progesterone receptors, and to a lesser extent those of other receptors [4].

Two separate sequences, NL1 and NL2, within the glucocorticoid receptor act as signals for the hormone-dependent nuclear localization of the receptor [109]. The carboxyl terminus of hGR contains a 30 amino acid sequence (named 'tau 2') that functions as an activation domain. A similar and independent activity has also been identified in the amino terminus of the receptor. These two sequences in the molecule are both acidic but are structurally unrelated [33].

Removal of 29 amino acids from the carboxy terminus of the rat glucocorticoid receptor leads to only very little (1%) loss of hormone–inducible activity, but further deletions cause loss of hormone-inducible activity [35, 40]. While deletions up to 180 carboxy terminal residues failed to produce a biological response in either the presence of the absence of steroid, longer carboxy terminal deletions or truncation of the amino terminal part of the domain induced transcriptional activation even in the absence of hormone (Fig. 4). Truncation at both ends of the molecule resulted in a polypeptide of only 150 amino acids which was still effective in constitutive activation [34–39].

Regions in the carboxyl and amino termini of the glucocorticoid receptor that increase transcription but are not involved in DNA binding may be moved to other parts of the receptor or attached to heterologous binding domains, and still maintain

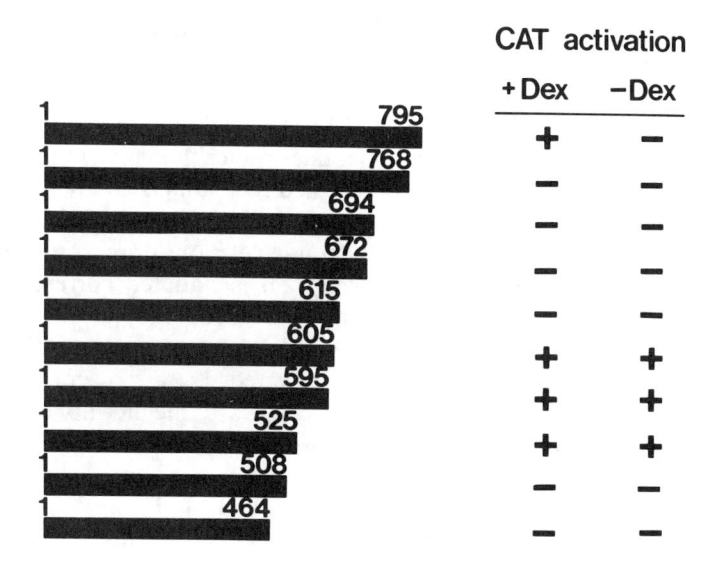

Fig. 4 — Activation of chloramphenicol acetyl transferase (CAT) by the wild-type rat glucocorticoid receptor and deletion mutants. The CAT gene was linked to a promoter cointaining the glucocorticoid-responsive element (GRE) and co-transfected with the glucocorticoid receptor gene in the presence or absence of dexamethasone. The receptor mutants comprising amino acid residues 1–605, 1–595 and 1–525 presented constitutive CAT activation. Data from Godowski et al. [35].

function by increasing transcription [33]. Such studies can also be done using systems in which glucocorticoid receptors exert negative effects. The results suggest that the negative effects on transcription that glucocorticoid receptors exert in some systems are generated via steric hindrance. The amino terminus is not critical for this repression but both the DNA- and the hormone-binding domains are required for efficient repression to occur [37].

These data on receptor structure and function are of interest because certain cells (e.g. mouse T cell lymphomas and human lymphoblastic leukaemia cells) are known to contain mutant glucocorticoid receptors with structural defects which causes an inability to mediate in glucocorticoid-induced cell lysis. As growth of the normal ('wild-type') cells is inhibited by adding glucocorticoids, this makes isolation of glucocorticoid-resistant cell variants quite easy. The majority of these resistant cells possess glucocorticoid receptors with structural and functional defects. Four major abnormalities have been identified: r⁻ (receptor deficient), nt⁻ (nuclear transfer deficient), nt^i (nuclear transfer increased), and act^l (activation labile). The r⁻ phenotype occurs most often and is characterized by low or undetectable hormone binding. Some r⁻ cells contain a polypeptide that cross-reacts with anti-receptor antibodies and has the same molecular weight as the wild-type receptor (94 kDa). This suggests a mutation in the hormone-binding domain [42–44]. The r⁻ phenotype may in some cells also arise from lack of expression of receptor with no gene product or specific mRNA detectable [45]. Whatever the cause of the r⁻ phenotype may be, this phenomenon is not due to gross DNA rearrangements or deletions and therefore probably is due to point mutations [46].

The glucocorticoid receptor of nt^i phenotype is a 40 kDa polypeptide that represents the amino-terminally truncated form of wild-type receptor. These nt^i receptors are synthesized as shorter polypeptides rather than being processed from larger molecules. The mRNA of these truncated receptors is about 1.5 kb shorter than the wild-type mRNA from which the 5' sequences are missing. However, Southern blot analysis did not reveal any genomic deletions or rearrangements [44, 46]. Therefore it seems likely that transcription is initiated correctly and that aberrant splicing is responsible for the truncated nt^i mRNA [39].

Receptors of the act^l phenotype are relatively labile in the sense that the hormone is dissociated easily when these receptors are activated. The act^l receptors, however, behave in a similar manner as wild-type receptors if the hormone is attached covalently by cross-linking labelling and so is unable to dissociate on activation of the complex [47].

Glucocorticoid receptors have also been studied at the protein level. Mild treatment with proteases generates fragments that structurally resemble the nt^i receptors of glucocorticoid-resistant cells, and that resemble these nt^i receptors in exhibiting increased non-specific affinity for DNA *in vitro*. However, this does not mean that these moelcules can activate transcription since other studies show that an intact amino terminus (A–B region) is required for the glucocorticoid receptor to be able to activate efficiently the HRE of the mouse mammary tumour long-terminal repeat (MMTV-LTR) [34,38]. Analysis of the normal rat liver glucocorticoid receptor protein indicated that the amino terminus is blocked [48]. Glucocorticoid receptors are phosphoproteins [49–51], and serine [49] or threonine [51] residues are phosphorylated depending on the conditions. Whether these phosphorylations play a role in receptor functioning, and if so in what way, is unclear at present (for a review see ref. [52]).

Steroid receptors that are isolated from cytosols are often found to be complexed with 90 kDa heat shock protein (hsp90) [53]. There is a debate as to whether these complexes play a role in steroid hormone action or are just artefacts of the isolation procedure. Whatever the answer may be, phosphorylation of the receptor does not appear to play a role in the association or dissociation of glucocorticoid receptor from hsp90 [54].

When rat thymus cells are depleted of ATP by anaerobiosis, the specific glucocorticoid binding capacity of these cells disappears, and rapidly reappears when ATP levels are restored. In cells deprived of ATP the glucocorticoid receptor is present in a 'null receptor' form that cannot bind hormone and that is bound in the nuclei of the ATP-depleted cells [122]. It is possible that the null receptor is present as the dominant form in r$^-$ mutant cells described above. The finding also suggests that the release of steroid receptors from nuclei may be an energy-requiring process, or that ATP-dependent phosphorylation is required for activity.

6. PROGESTERONE RECEPTOR A AND B FORMS

Progesterone receptors (PRs) in chicken and human tissues are represented by two molecular forms, A and B, with molecular weights of 79 kDa and 109 kDa respectively. Equimolar ratios of these A and B forms have been found in the cytosols of chicken oviduct tubular gland cells, and in human breast cancer T-47D cells [55–57]. Immunoanalysis and peptide mapping of the photoaffinity-labelled proteins indicate that the A and B forms are structurally related [56,58,59]. Furthermore, both forms bind to DNA [58]. This finding has been confirmed by DNAseI footprinting experiments using the hormone-responsive element of the chicken lysozyme gene [60].

The complete mRNA sequence of the chicken PR has been determined and has revealed DNA-binding and hormone-binding structures that resemble those of other steroid–thyroid hormone receptors [62–64,68,79]. Expression of the cloned cDNA produced a protein that resembles the natural B form (109 kDa). A protein corresponding in size to form A (79 kDa) was produced by expressing an amino-terminally truncated receptor starting at Met128, or by internal initiation during *in vitro* translation.

Evidence has been presented that the A and B proteins are derived by alternate initiation of translation from a single mRNA transcript [65]. The synthesis of two forms of a given protein by alternate initiation of translation from two in-phase AUG signals in a single mRNA transcript is rare but has been reported before [66]. Comparison of the sequences surrounding the AUG triplets in rabbit [67], human [68] and chicken [61] PRs indicates only partial conservation and does not provide an explanation for only a single form of PR (the B form) in rabbit, in contrast to the two protein forms (A and B) found in humans and chickens. It has therefore been proposed that the secondary structural context of these sequences on the mRNA may play a role in determining whether alternate initiation of translation will take place [65].

The A–B region of the chicken progesterone receptor plays a crucial role in the differential activation by progesterone of the ovalbumin gene and the MMTV–HRE. This is shown by the fact that, whereas progesterone receptor form B activates preferentially the MMTV-LTR promoter, its naturally occurring amino-terminally truncated form (form A) can activate the promoter of both genes [69].

The chicken progesterone receptor gene appears to be present as a single copy and to consist of a 38 kb transcription unit [84]. An alternative polyadenylation signal is present near the 5' end of the second intron, and this results in a truncated mRNA. The putative protein product of this variant (1.8 kb) mRNA would contain only the amino terminal region and half of the DNA-binding region. If actually present in cells, such truncated receptors might compete with normal receptors for available steroid-regulatory elements on target genes. Alternatively, such molecules may exist as dangerous cellular variants if they retain any biological activity, since the repressive hormone-binding regulatory domain is absent [84].

7. OESTROGEN RECEPTOR VARIANTS

The sequence of the human oestrogen receptor (hER) protein has been deduced from the cloned gene [17,18,85]. The hER gene has been mapped to human chromosome 6 [3,74], and the mouse ER gene to mouse chromosome 10 [78]. Comparison of the ERs of various species reveals complete (100%) homology between the DNA-binding domains of these species [17,25,76,77].

The ER of the cold-blooded organism *Xenopus laevis* exhibits high similarity in amino acid sequence with the human and avian ERs. In the putative DNA-binding region, its amino acid sequence differs at only 1 of 83 amino acids [81].

The DNA-binding region (region C) plays a role in specific recognition and binding to the oestrogen-responsive elements of target genes, whereas region E is indispensable for hormone binding. Receptors that lack the hormone-binding domain, however, still recognize specific responsive DNA elements, while the isolated hormone-binding domain binds oestradiol with wild-type affinity [24,75]. The length of the hinge region D, which joins the DNA- and hormone-binding domains, can be substantially altered without affecting ER function [75].

In order to elucidate the role of the hormone-binding domains of hER and hGR, these carboxy-terminally located sequences were joined to the DNA-binding domain of the yeast transcription factor GAL4 [116]. Stimulation of transcription by these chimaeric receptors from GAL4 responsive reporter genes was hormone dependent. The chimaeric receptor only bound tightly to nuclei when hormone or anti-hormone was present. However, the anti-hormone did not cause transcription activation, indicating that the hormone-binding region of receptors has a dual function by both causing hormone (or anti-hormone) dependent binding to nuclei, and hormone (but not anti-hormone) dependent activation of transcription.

Of interest also was the fact that GAL-ER(147–251) binds tightly to nuclei even in the absence of hormone, and that it competes for GAL4-activated transcription. Apparently, the ER sequence 147–251 does not efficiently mask the DNA-binding domain in GAL-ER(147–251), and hence this chimaeric receptor can by its own accord attach to the GAL4 binding sites in the absence of hormone [116].

The transcriptional activation function located in the hormone-binding domain of hER is encoded by separate codons. The results suggest that the protein surface responsible for the activation function is generated by the three-dimensional folding of the hormone-binding domain, and is probably created from dispersed elements [117]. This finding is of interest because the location of transcriptional activation domains within steroid receptors has been controversial. Whereas receptors that are deleted for the hormone-binding domain show constitutive activation, the magnitude of this activation ranges from only 5% to full wild-type activity depending on the receptor and the experimental conditions. This indicates that the *N*-terminal A–B region is important for activation, an idea that is also supported by the fact that *N*-terminal deletions of the hER [75,69] and chicken PR [61,118] can reduce the ability of the receptor to activate transcription, but only on certain promoters, suggesting that the A–B region may be interacting with promoter-specific factors (see Fig. 1).

It has been reported that the human ERcDNA clone pOR8 obtained from MCF-7 cells contains a glycine-400 (GGG) to valine-400 (GTG) mutation compared with the ER of a human genomic library [80]. In addition, two silent mutations (Ser-10, TCC→TCT; Thr-594, ACG→ACA) are present in the MCF-7 ER cDNA. Interestingly, a glycine is found at position 400 in the chicken ER [77], mouse ER [25], rat ER [76] and *Xenopus* ER [81] sequences, suggesting that it is the MCF-7 ER which differs from the wild-type sequence. The Gly-400→Val-400 mutation is in the hormone-binding domain of the cloned hER and destabilizes the ER structure, thereby decreasing its apparent affinity for oestradiol at 25°C, but not at 4°C. This point mutation may be a cloning artefact, because the valine codon GTG was not found in the sequence of three other hERcDNA clones derived from MCF-7 cell RNA. However, it is also possible that a minor population of MCF-7 cells had this mutation or that a minor fraction of hERmRNA could be mutated post-transcriptionally [123].

Restriction fragment length polymorphism (RFLP) has been reported in ER genes. With restriction enzyme *Pvu* II, absence of a 0.6 kb restriction fragment was found to be associated with ER-negative human breast cancer cells; the ER-negative cells had a restriction fragment of 1.6 kb. The RFLP is probably located within sequences encoding the DNA- or hormone-binding regions [83].

The finding that ER-positive and ER-negative breast tumours differ significantly in RFLP is of interest since it suggests that ER activation or lack of activation is associated with different alleles (designated P1 and P2). The molecular basis of this phenomenon is unknown, but it might affect the proper splicing of ERmRNA. There is the possibility of point mutations' being responsible for the effect. Therefore sequence analysis of ERmRNA from breast cancer patients may elucidate this matter [83]. *Pss* 1 restriction enzyme analysis has identified a single two-allele polymorphism at either 1.7 or 1.4 kb; the 1.4 kb allele occurs in 91% of North American caucasians, the 1.7 kb allele in 9% [82].

Multiple ERmRNA variants have been reported for *Xenopus laevis*. The *Xenopus* ER encodes 4 mRNAs with lengths of approximately 9, 6.5, 2.8 and 2.5 kb. It is possible that these derive from different polyadenylation signals [81]. Transcription initiation sites at 10 major sites have been reported close together for the mouse ER gene; however, in the mouse uterus only a single ERmRNA transcript is observed [25]. Of interest is the fact that, when the mouse ERcDNA is tested for functional activity by transfection, it not only shows hormonally dependent activity but also some constitutive activity. The latter may possibly be due to the presence of mutated or truncated receptors [25].

The transcripts required to encode a protein the size of the human ER protein (6.5 kDa) are about 2 kb. In human uterus the size of the ERmRNA is, however, found to be about 4.2 kb, and therefore about half of the mRNA appears to be untranslated [90]. This is also the case with the ERmRNA of MCF-7 human breast cancer cells which show a short 5' and a very long 3' untranslated region [17,18].

Investigations have been carried out to identify truncated ER proteins in tissues from different sources. Studies at the protein level have revealed ER variants in various tissues. Such studies generally involve covalent labelling of cytosolic ER with [^{3}H]tamoxifen aziridine in the presence of protease inhibitors, and subsequent identification of the labelled receptors by sodium dodecyl sulphate polyacrylamide gel electrophoresis (SDS-PAGE) [89].

Investigations of this type carried out with mouse uterus demonstrated a major ER component of 65 kDa with minor fragments of 54 kDa and 37 kDa. Perhaps the low molecular weight forms originate from the 65 kDa holoreceptor by proteolytic degradation, but the possibility that the 65, 54 and 37 kDa species are different gene products cannot be ruled out [87].

Similar experiments in our laboratory using [^{3}H]tamoxifen aziridine tagging of oestrogen-dependent and -independent GR mouse mammary tumours show essentially similar results. In these tumours we detected 65, 50 and 35 kDa receptors. Of interest was a shift towards the low molecular weight forms when these mouse tumours became hormonally independent during serial transplantations in syngeneic mice [88]. Similar ER fragments have been described in rat uterus and MCF-7 cultured cells.

A protease with α-chymotrypsin activity that increased the affinity of ER for binding to DNA has been described in mammary tumours of C3H mice [92]. However, there is also evidence for a specific hormonally regulated modification of ER. The response, as detected by SDS-PAGE, is the formation if a closely spaced ER doublet of 65 and 66.5 kDa in nuclei of mouse uterus. This ER doublet displays a bimodal temporal pattern following administration of oestradiol *in vivo*. Furthermore, the relative proportions of the two bands are affected by the occupying ligand, with potent oestrogens such as oestradiol producing equivalent amounts of the upper and lower bands. It does not seem likely that the nuclear ER doublet is an artefact of proteolysis during the isolation procedure [90]. A possible explanation for the doublet might be a phosphorylative mechanism, as it has been reported that ER from calf uteri is phosphorlated and dephosphorylated by a nuclear kinase and a cytosolic phosphatase [91]. The upper band of the nuclear ER in mouse uteri appears to be associated with the salt-resistant fraction, and might thus represent receptor that is tightly bound to chromatin [90].

It should be pointed out that even if some of the low molecular weight forms of ER observed in some tissues turn out to be proteolytic cleavage products of the holoreceptor, this proteolysis may have functional significance. The cleavage is at the border regions of the functional domains and may serve to separate these regions. Furthermore, it has been reported that the ER molecule itself has proteolytic activity that is responsible for its own transformation to the active state [93].

8. LOSS OF HORMONAL DEPENDENCE

The loss of hormonal dependence in breast cancer is due to the emergence of autonomous cell clones that progressively achieve dominance in the tumour mass [94,95]. This apparent heterogeneity of the tumours suggests that combined chemohormonal therapy would be a more effective treatment for breast cancer than either single modality alone. Animal studies indicate this to be the case; however, these studies also show that even the combined treatment does not cure but only extends the latency period of the tumour [96, 97]. This is also observed in human studies where the combined approach causes higher response rates and longer relapse-free intervals than either modality used singly. Even with chemoendocrine combinations, however, complete remissions remain relatively uncommon and cure of metastatic diseases remains impossible [98]. The reason for this disappointing result apparently is that the cytotoxic drugs used in studies so far are not sufficiently effective in destroying the highly malignant subclones that emerge during the tumour progression. It seems likely that breast cancer progression is a highly complex affair, resulting in the emergence of increasingly malignant cells by natural selection. The question therefore is: where do these cells come from and what is the mechanism by which they are able to achieve dominance?

Assuming that a breast tumour originates from a single transformed breast tissue cell, it seems likely that the subsequent outgrowth of the tumour and the progression steps involve different oncogenes. Several oncogenes have been implicated in human mammary cancer including *myc*, *ras*, *src*, *neu* and perhaps others [99]. Many of the chromosomes frequently altered in human breast cancer (i.e. numbers 1,3,6,7 and 11) contain sequences which encode human cellular oncogenes. Examples of this are c-*blim*, N-*ras* and c-*ski* on chromosome 1, c-*raf* on chromosome 3, c-*myb* on chromosome 6, c-*erbB* and c-*met* on chromosome 7, H-*ras* and c-*ets* on chromosome 11 [119]. It has been reported that c-*myb* expression shows a strong association with human breast cancers that have high oestrogen receptor levels [120]. The *myb* gene, like the ER gene, is on human chromosome 6 [121] and mouse chromosome 10 [78]. Whether these genes are within the same region on these chromosomes is not known. If this turns out to be the case, it raises the interesting possibility that the *myb* oncogene is involved in the mechanism by which levels of ER are enhanced in hormone-dependent breast cancers.

Lippman's group [100] reported that, when human breast cancer MCF-7 cells were transfected with v-Ha-*ras*, these cells no longer responded to exogenous oestrogens in culture and were fully tumorigenic without oestrogen supplementation when tested in nude mice. These transformed cells still had high levels of ER. The authors concluded that MCF-7 cells that acquire an activated *onc* gene can bypass the hormonal regulatory signals that trigger the neoplastic growth of the breast cancer cell line [100]. Interestingly, transfection of the v-Ha-*ras* oncogene into MCF-7 cells causes these cells to secrete increased levels of several growth factor activities constitutively. This suggests that growth of hormonally independent breast cancer cells might be stimulated by these growth factors [102]. Cultured ER-negative cell lines constitutively secrete relatively high concentrations of several growth factors such as transforming growth factor α (TGF-α), insulin-like growth factor I (IGF-I), platelet-derived growth factor (PDGF), an epithelial cell colony-stimulating factor, mammary-derived growth factor, and autocrine motility factor [99]. Secretion of several of these growth factor activities in ER-positive lines is regulated by oestrogen: oestrogen deprivation or anti-oestrogen treatment reduces the growth factor secretion, whereas oestrogen administration increases it [99, 101]. Thus ER-negative cells might have a growth advantage as a result of constitutive growth factor secretion. However, results by Osborne *et al.* [103] are at variance with this hypothesis. These authors inoculated ER-negative MDA-231 human breast cancer cells in castrated female athymic nude mice, and found that the resulting tumours did not support growth of MCF-7 cells inoculated in the opposite flank, which required oestrogen supplementation for growth [103]. The data by Osborne *et al.* [103] suggest that growth factors are not capable of replacing oestrogen for growth simulation of MCF-7 breast cancer cells and make it seem unlikely that the loss of oestrogen dependence in breast cancer can simply be explained by the constitutive production of growth factors that replace oestrogen in stimulation of cell growth. If growth factors are not responsible for sustaining hormone-independent growth, how is this growth sustained? We have proposed that proteins resembling mutated or truncated steroid receptors are involved [1]. Of interest is the fact that defective oestrogen receptors have been reported in human mammary cancers [104, 113].

9. CONCLUSIONS

In recent years we have seen a tremendous revolution in the field of steroid hormone action. The exciting discovery that there is a relationship between the structures of steroid and thyroid receptors with the viral oncogene v-*erb*-A makes it tempting to speculate that activation of some of these receptor genes is implicated in oncogenesis.

That v-*erb*-A can exert its specific effects on erythroblasts even though it cannot bind thyroid hormones is apparently due to changes having occurred in hormone-binding domain E of c-*erb*-A. It seems possible that these mutations prevent the E region from masking the DNA-binding region (domain C), thereby permitting constitutive binding of the receptor to DNA. In the progesterone receptor gene there is an alternative polyadenylation signal near the 5' end of the second intron, and this could result in a truncated mRNA. The putative protein product of this variant (1.8 kb) mRNA would contain only the *N*-terminal domain and half of the DNA-binding domain of the complete receptor, and would lack hormone-binding activity. If actually present in cells, such truncated receptors might compete with normal receptor forms for available steroid-regulatory elements on target genes. Alternatively, such molecules may exist as dangerous cellular variants with oncogenic potential [84]. Of interest also is the difference between the complete progesterone receptor (form B) and its truncated form (form A) in ability to activate specific genes [69]. This raises the question of whether the low molecular weight forms of steroid receptors that have been found in various tissues serve a physiological function. It is important to establish which of these forms are due to changes at the genomic or mRNA level and which are caused by secondary events, e.g. phosphorylation or proteolysis. Proteases have been reported that can degrade oestrogen and progesterone receptors selectively [70].

Steroid receptors are particularly susceptible to proteases at sites in the molecule linking the various domains, and it is possible that proteolysis serves a physiological function by causing these regions to be released in the cell [70–72]. There is also the possibility that changes in steroid receptor genes cause increased susceptibility of the encoded proteins to endogenous proteases. Of interest is the report that gestodene, a synthetic progestin, binds to the ER of malignant breast tumours, but not to the ER of normal breast, liver, pancreas or endometrium. This suggests a malignant change in the ER of breast neoplasms [115].

Interesting findings are that human breast cancer cell line MCF-7 contains a glycine-to-valine mutation at amino acid residue 400 [80], and that restriction fragment length polymorphism in the ER gene is associated with the presence or absence of a functional ER protein in human breast cancer cells [83].

Immunochemical staining of ER with polyclonal anti-ER antibodies has identified two types of defective ER among human breast cancer: those that are unable to bind to the nucleus in a hormone-filled state, and those that bind to the nucleus as naked ER [104]. Furthermore, a variant oestrogen receptor mRNA has been found to be associated with reduced levels of oestrogen binding in human mammary tumours. The message modification is in the B region of the ER [113]. Therefore a subclassification of tumours based on functional abnormalities of ER may predict refractoriness to hormone therapy. Abnormal methylation of the ER gene has been reported to reduce ERmRNA levels in human endometrial carcinomas [86]. It would be of interest to establish whether this also is the case in breast cancers.

The steroid–thyroid hormone receptor superfamily is involved in liver carcinogenesis (*hap* gene) and induction of avian leukaemia (v-*erb*-A). Certain members of this superfamily may also play a role in other specific types of cancer. Changes in the coding region of the DNA-binding domain of glucocorticoid receptors have been reported between mouse lymphoma cells and an androgen-dependent mouse mammary tumour [110]. Therefore, changes in steroid receptors at the genomic level may be associated with certain malignant states.

Deletions of chromosome 3 have been reported in small cell lung carcinoma (SCLC) and many non-small cell tumours of the lung [105]. In the case of SCLC a consensus deletion (3p21–25) has been defined which contains a putative suppressor gene [106]. The thyroid hormone receptor $T_3R\beta$ (c-*erb*-Aβ) locus maps to this region on chromosome 3, and the *erb*-Aβ hybridization to DNA extracted from these tumours is reduced, suggesting that either $T_3R\beta$ or another *erb*-A-related gene found at this locus may be the suppressor gene. It is also possible that another suppressor gene is located at this locus, perhaps the RARβ gene which maps to chromosome 3p24 [107]. It has been proposed that in SCLC both alleles of the c-*erb*-A gene are inactivated, one by a chromosomal deletion and the other by a more subtle mutation. Thyroid hormone action is inhibited by a non-hormone binding form of the c-*erb*-A protein that is generated by alternative mRNA splicing [73,114].

Herkskowitz [126] has proposed that some oncogenes may be examples of naturally occurring dominant negative mutations. Both dominant and recessive mutations are known to cause cellular transformation. Recessive mutations, such as that causing retinoblastoma, involve loss of wild-type functions. By contrast, v-*erb*-A may be an example of a dominant negative mutation, as this oncogene acts as a constitutive repressor and, when co-expressed with the thyroid hormone receptor, blocks activation of the thyroid-hormone-regulated gene [125].

In conclusion, therefore, steroid receptors have moved from the role of prognostic guides to therapy into a new role, that of potential pathogenic factors in oncogenesis. This superfamily of receptors under certain circumstances may thus be classed as 'nuclear oncogenes' [108]. In the next few years we can expect to be able to define the structures and functions of the abnormal transcripts of these and other DNA-binding 'finger' proteins, and to establish their roles on oncogenesis and tumour progression in specific types of cancer.

REFERENCES

[1] Sluyser, M. & Mester, J. (1985) Oncogenes homologous to steroid receptors? *Nature (London)* **315** 546

[2] Green, S. & Chambon, P. (1986) A superfamily of potentially oncogenic hormone receptors. *Nature (London)* **324** 615

[3] Green, S., Gronemeyer, H. & Chambon, P. (1987) Structure and function of

steroid hormone receptors. In: M. Sluyser (ed.) *Growth Factors and Oncogenes in Breast Cancer.* Ellis Horwood, Chichester, p. 7

[4] Evans, R. M. (1988) The steroid and thyroid hormone receptor superfamily. *Science* **240** 889

[5] Bister, K. & Duesbuerg, P. H. (1979) Structure and specific sequences of avian erythroblastosis virus RNA. Evidence for mutiple classes of transforming genes among avian tumor viruses. *Proc. Natl. Acad. Sci. USA* **176** 5023

[6] Lai, M. M. C., Hu, S. S. F. & Vogt, P. K. (1979) Avian erythroblastosis virus: transformation-specific sequences form a contiguous segment of 3.25 kb located in the middle of the 6-kb genome. *Virology* **97** 366

[7] Roussel, M., Saule, S., Lagrou, C., Rommens, C., Beug, H., Graf, T. & Stehelin, D. (1979) Three new types of viral oncogene of cellular origin specific for haematopoietic cell transformation. *Nature (London)* **281** 452

[8] Vennström, B., Fanshier, L., Moscovici, C. & Bishop, J. M. (1980) Molecular cloning of the avian erythroblastosis virus genome and recovery of the oncogenic virus by transfection of chicken cells. *J. Virol.* **36** 575

[9] Debuire, B., Henry, C., Benaissa, M., *et al.* (1984) Sequencing the *erb*A gene of avian erythroblastosis virus reveals a new type of oncogene. *Science* **224** 1456

[10] Sergeant, A., Saule, S., Leprince, D., Begue, A., Rommens, C. & Stehelin, D. (1982) Molecular cloning of the chicken DNA locus related to the oncogene *erb*-B of the avian erythroblastosis virus. *EMBO J.* **1** 237

[11] Downward, J., Yarden, Y., Mayes, E., *et al.* (1984) Close similarity of epidermal growth factor receptor and v-*erb*-B oncogene protein sequences. *Nature (London)* **307** 521

[12] Graf, T. & Beug, H. (1983) Role of the v-*erb*-A and v-*erb*-B oncogenes of avian erythroblastosis virus in erythroid cell transformation. *Cell* **34** 7

[13] Weinberger, C., Thompson, C. C., Ong, E. S., Lebo, R., Gruol, D. J. & Evans, R. M. (1986) The c-*erb*-A gene encodes a thyroid hormone receptor. *Nature (London)* **324** 641

[14] Sap, J., Munoz, A., Damm, K., Goldberg, Y., Ghysdael, J., Leutz, A., Beug, H. & Vennström, B. (1986) The c-*erb*-A protein is a high affinity receptor for thyroid hormone. *Nature (London)* **324** 635

[15] Jansson, M., Philipson, L. & Vennström, B. (1983) Isolation and characterization of multiple human genes homolgous to the oncogenes of avian erythroblastosis virus. *EMBO J.* **2** 561

[16] Weinberger, C., Hollenberg, S. M., Rosenfeld, M. G. & Evans, R. M. (1985) Domain structure of human glucocorticoid receptor and its relationship to the v-*erb*-A oncogene product. *Nature (London)* **318** 670

[17] Green, S., Walter, P., Kumar, V., Krust, A., Bornert, J. M., Argos, P. & Chambon, P. (1986) Human oestrogen receptor cDNA: sequence expression and homology to v-*erb*-A. *Nature (London)* **320** 134

[18] Greene, G. L., Gilna, P., Waterfield, M., Baker, A., Hort, Y. & Shine, J. (1986) Sequence and expression of human estrogen receptor complementary DNA. *Science* **231** 1150

[19] Conneely, O. M., Sullivan, W. P., Toft, D. O., Birnbaumer, M., Cook, R. G.,

Maxwell, B. L., Zarucki-Schulz, T., Green, G. L., Schrader, W. T. & O'Malley, B. W. (1986) Molecular cloning of the chicken progesterone receptor. *Science* **233** 767

[20] McDonnell, D. P., Mangelsdorf, D. J., Pike, J. W., Haussler, M. R. & O'Malley, B. W. (1987) Molecular cloning of complementary DNA encoding the avian receptor for vitamin D. *Science* **235** 1214

[21] Hollenberg, S. M., Weinberger, C., Ong, E. S., Cerelli, G., Oro, A., Lebo, R., Thompson, E. B., Rosenfeld, M. G. & Evans, R. M. (1986) Primary structure and expression of a functional human glucocorticoid receptor cDNA. *Nature (London)* **318** 635

[22] Miesfeld, R., Rusconi, S., Godowski, P. J., Maler, B. A., Okret, S., Wikstrom, A. C., Gustafsson, J. A. & Yamamoto, K. R. (1986) Genetic complementation of a glucocorticoid receptor deficiency by expression of cloned receptor cDNA. *Cell* **46** 389

[23] Danielsen, M., Northrop, J. P. & Ringold, G. M. (1986) The mouse glucocorticoid receptor: mapping of functional domains by cloning, sequencing and expression of wild-type and mutant receptor proteins. *EMBO J.* **5** 2513

[24] Kumar, V., Green, S., Staub, A. & Chambon, P. (1986) Localisation of the oestradiol-binding and putative DNA binding domains of the human oestrogen receptor. *EMBO J.* **5** 2231

[25] White, R., Lees, J. A., Needham, M., Ham, J. & Parker, M. (1987) Structural organization and expression of the mouse estrogen receptor. *Mol. Endocrinol.* **1** 735

[26] Benbrook, D. & Pfahl, M. (1987) A novel thyroid hormone receptor encoded by a cDNA clone from a human testis library. *Science* **238** 788

[27] Spurr, N. K., Solomon, E., Jansson, M., Sheer, D., Goodfellow, P. N., Bodmer, W. F. & Vennström, B. (1984) Chromosomal localisation of the human homologues of the oncogenes *erb* A and B. *EMBO J.* **3** 159

[28] Giguère, V., Ong, E. S., Segui, P. & Evans, R. M. (1987). Identification of a receptor for the morphogen retinoic acid. *Nature (London)* **330** 624

[29] Petkovich, M., Brand, N. J., Krust, A. & Chambon, P. (1987) A human retinoic acid receptor which belongs to the family of nuclear receptors. *Nature (London)* **330** 444

[30] Dejean, A., Bougueleret, L., Grzeschik, K. H. & Tiollais, P. (1986) Hepatitis B virus DNA integration in a sequence homologous to v-*erb*-A and steroid receptor genes in a hepatocellular carcinoma. *Nature (London)* **322** 70

[31] De Thè, H., Marchio, A., Tiollais, P. & Dejean, A. (1987) A novel steroid thyroid receptor gene inappropriately expressed in human heptocellular carcinoma. *Nature (London)* **330** 677

[32] Benbrook, D., Lernhardt, E. & Pfahl, M. (1988) A new retinoic acid receptor identified from a heptocellular carcinoma. *Nature (London)* **333** 669

[33] Hollenberg, S. M. & Evans, R. M. (1988) Multiple and cooperative *trans*-activation domains of the human glucocorticoid receptor. *Cell* **55** 899

[34] Hollenberg, S. M., Giguere, V., Segui, P. & Evans, R. M. (1987) Colocalization of DNA-binding and transcriptional activation functions in the human glucocorticoid receptor. *Cell* **49** 39

[35] Godowski, P. J., Rusconi, S., Miesfeld, R. & Yamamoto, K. R. (1987) Glucocorticoid receptor mutants that are constitutive activators of transcriptional enhancement. *Nature (London)* **325** 365

[36] Danielsen, M., Northrop, J. P., Jonklaas, J. & Ringold, G. M. (1987) Domains of the glucocorticoid receptor involved in specific and nonspecific deoxyribonucleic acid binding, hormone activation, and transcriptional enhancement. *Mol. Endocrinol.* **1** 816

[37] Oro, A. E., Hollenberg, S. M. & Evans, R. M. (1988) Transcriptional inhibition by a glucocorticoid receptor-β-galactosidase fusion protein. *Cell* **55** 1109

[38] Miesfeld, R., Godowski, P. J., Maler, B. A. & Yamamoto, K. R. (1987) Glucocorticoid receptor mutants that define a small region sufficient for enhancer activation. *Science* **236** 423

[39] Gehring, U. (1987) Steroid hormone receptors: biochemistry, genetics and molecular biology. *Trends Biochem. Sci.* **12** 399

[40] Rusconi, S. & Yamamoto, K. R. (1987) Functional dissection of the hormone and DNA binding activities of the glucocorticoid receptor. *EMBO J.* **6** 1309

[41] Simons, S. S., Pumphrey, J. G., Rudikoff, S. & Eisen, H. J. (1987) Identification of cysteine 656 as the amino acid of hepatoma tissue culture cell glucocorticoid receptors that is covalently labeled by dexamethasone 21-mesylate. *J. Biol. Chem.* **262** 9676

[42] Westphal, H. M., Mugele, K., Beato, M. & Gehring, U. (1984) Immunochemical characterization of wild-type and variant glucocorticoid receptors by monoclonal antibodies. *EMBO J.* **3** 1493

[43] Northrop, J. P., Gametchu, B., Harrison, R. W. & Ringold, G. M. (1985) Characterization of wild type and mutant glucocorticoid receptors from rat hepatoma and mouse lymphoma cells. *J. Biol. Chem.* **260** 6398.

[44] Miesfeld, R., Okret, S., Wikström, A. C., Wrange, O., Gustafsson, J. A. & Yamamoto, K. R. (1984) Characterization of a steroid hormone receptor gene and mRNA in wild-type and mutant cells. *Nature (London)* **312** 779

[45] Gehring, U. (1986) Genetics of glucocorticoid receptors. *Mol. Cell. Endocrinol.* **48** 89

[46] Northrop, J. P., Danielsen, M. & Ringold, G. M. (1986) Analysis of glucocorticoid unresponsive cell variants using a mouse glucocorticoid receptor complementary DNA clone. *J. Biol. Chem.* **261** 11064

[47] Gehring, U. & Hotz, A. (1983) Photoaffinity labeling and partial proteolysis of wild-type and variant glucocorticoid receptors. *Biochemistry* **22** 4013

[48] Carlsted-Duke, J., Strönstedt, P. E., Wrange, O., Bergman, T., Gustafsson, J. A. & Jörnvall, H. (1987) Domain structure of the glucocorticoid receptor protein. *Proc. Natl. Acad. Sci. USA* **84** 4437.

[49] Housley, P. R. & Pratt, W. B. (1983) Direct demonstration of glucocorticoid receptor phosphorylation by intact L-cells. *J. Biol. Chem.* **258** 4630

[50] Grandics, P., Miller, A., Schmidt, T. J. & Litwack, G. (1984) Phosphorylation *in vivo* of rat hepatic glucocorticoid receptor. *Biochem. Biophys. Res Commun.* **120** 59

[51] Miller-Diener, A., Schmidt, T. J. & Litwack, G. (1985) Protein kinase activity

associated with purified rat hepatic glucocorticoid receptor. *Proc. Natl. Acad. Sci. USA* **82** 4003

[52] Litwack, G. (1988) The glucocorticoid receptor at the protein level. *Cancer Res.* **48** 2636

[53] Shuk, S., Yonemoto, W., Brugge, J., Bauer, V. J., Riehl, R. M., Sullivan, W. P. & Toft, D. O. (1985) A 90,000-Dalton binding protein common to both steroid receptors and the Rous sarcoma virus transforming protein, PP60V-SRC. *J. Biol. Chem.* **260** 14292

[54] Orti, E., Mendel, D. B. & Munck, A. (1989) Phosphorylation of glucocorticoid receptor-associated and free forms of the 90-kDa heat shock protein before and after receptor activation. *J. Biol. Chem.* **264** 231

[55] Schrader, W. T. & O'Malley, B. W. (1972) Progesterone-binding components of chick oviduct IV. Characterization of purified subunits. *J. Biol. Chem.* **247** 51

[56] Gronemeyer, H., Harry, P. & Chambon, P. (1983) Evidence for two structurally related progesterone receptors in chick oviduct cytosol. *FEBS Lett.* **156** 287

[57] Gronemeyer, H. & Govindan, M. V. (1986) Affinity labeling of steroid hormone receptors. *Mol. Cell Endocrinol.* **46** 1

[58] Gronemeyer, H., Govindan, M. V. & Chambon, P. (1985) Immunological similarity between the chick oviduct progesterone receptor forms A and B. *J. Biol. Chem.* **260** 6916

[59] Sullivan, W. P., Beito, T. G., Proper, J., Kroo, C. J. & Toft, D. O. (1986) Preparation of monoclonal antibodies to the avian progesterone receptor. *Endocrinology* **119** 1549

[60] Ahe, D. v.d., Renoir, J. M., Buchou, T., Baulieu, E. E. & Beato, M. (1986) Receptors for glucocorticosteroid and progesterone recognize distinct features of a DNA regulatory element. *Proc. Natl. Acad. Sci. USA* **83** 2817

[61] Gronemeyer, H., Turcotte, B., Quirin-Stricker, C., Bocquel, M. T., Meyer, M. E., Krozowski, Z., Jeltsch, J. M., Lerouge, T., Garnier, J. M. & Chambon, P. (1987) The chicken progesterone receptor sequence. Expression and functional analysis. *EMBO J.* **6** 3985

[62] Horwitz, K. B., Wei, L. L., Sedlacek, S. M. & D'Arville, C. N. (1985) Progestin action and progesterone receptor structure in human breast cancer: a review. *Recent Prog. Horm. Res.* **41** 249

[63] Logeat, F., Pamphile, R., Loosfelt, H., Jolivet, A., Fournier, A. & Milgrom, E. (1985) One-step immunoaffinity purification of active progesterone receptor. Further evidence in favor of the existence of a single steroid binding unit. *Biochemistry* **24** 1029

[64] Loosfelt, H., Logeat, F., Vu Hai, M. T. & Milgrom, E. (1984) The rabbit progesterone receptor: evidence for a single steroid-binding subunit and characterization of receptor mRNA. *J. Biol. Chem.* **259** 14196

[65] Conneely, O. M., Maxwell, B. L., Toft, D. O., Schrader, W. T. & O'Malley, B. W. (1987) The A and B forms of the chicken progesterone receptor arise by alternate initiations of translation of a unique mNRA. *Biochem. Biophys. Res. Commun.* **149** 493

[66] Strubin, M., Long, E. O. & Mack, B. (1986) Two forms of the Ia antigen-associated invariant chain result from alternative initiations at two in-phase AUGs. *Cell* **47** 619

[67] Loosfelt, H., Atger, M., Misrahi, M., Guiochon-Mantel, A., Meriel, C., Logeat, F., Bernarous, R. & Milgrom, E. (1986) Cloning and sequence analysis of rabbit progesterone-receptor complementary DNA. *Proc. Natl. Acad. Sci. USA* **83** 9045

[68] Misrahi, M., Atger, M., d'Auriol, L., Loosfelt, H., Meriel, C., Fridlansky, F., Guiochon-Mantel, A., Galibert, F. & Milgrom, E. (1987) Complete amino acid sequence of the human progesterone receptor deduced from cloned cDNA. *Biochem. Biophys. Res. Commun.* **143** 740

[69] Tora, L., Gronemeyer, H., Turcotte, B., Gaub, M. P. & Chambon, P. (1988) The N-terminal region of the chicken progesterone receptor specifies target gene activation. *Nature (London)* **333** 185

[70] Maeda, K., Tsuzimura, F., Nomura, Y., Sato, B. & Matsumoto, K. (1984) Partial characterization of protease(s) in human breast cancer cytosols that can degrade estrogen and progesterone receptors selectively. *Cancer Res.* **44** 996

[71] Lukola, A. & Punnonen, R. (1983) Characterization of the effect of sodium molybdate and diisopropylfluorophosphate on the human myometrial estrogen and progesterone receptors. *J. Steroid Biochem.* **18** 231

[72] Feil, P. D., Clarke, C. L. & Satyaswaroop, P. (1988) Progesterone receptor structure and protease activity in primary human endomytrial carcinoma. *Cancer Res.* **48** 1143

[73] Koenig, R. J. N., Lazar, M. A., Hodin, R. A., Brent, G. A., Larsen, P. R., Chin, W. W. & Moore, D. D. (1989) Inhibition of thyroid hormone action by a non-hormone binding c-*erb*-A protein generated by alternative mRNA splicing. *Nature (London)* **337** 659

[74] Gosden, J. R., Middleton, P. G. & Rout, D. (1986) Localization of the human oestrogen receptor gene to chromosome 6q24→q27 by *in situ* hybridization. *Cytogenet. Cell Genet.* **43** 218

[75] Kumar, V., Green, S., Stack, G., Berry, M., Jin, J. R. & Chambon, P. (1987) Functional domains of the human estrogen receptor. *Cell* **51** 941

[76] Koike, S., Sakai, M. & Muramatsu, M. (1987) Molecular cloning and characterization of rat estrogen receptor cDNA. *Nucleic Acids Res.* **15** 2499

[77] Krust, A., Green, S., Argos, P., Kumar, V., Walter, P., Bornert, J. M. & Chambon, P. (1986) The chicken estrogen receptor sequence: homology with v-*erb*-A and the human estrogen and glucocorticoid receptors. *EMBO J.* **5** 891

[78] Sluyser, M., Rijkers, A. W. M., De Goeij, C. C. J., Parker, M. & Hilkens, J. (1988) Assignment of estradiol receptor gene to mouse chromosome 10. *J. Steroid Biochem.* **31** 757

[79] Carson, M. A., Tsai, M. J., Conneely, O. M., Maxwell, B. L., Clark, J. H., Dobson, A. D. W., Elbrecht, A., Toft, D. O., Schrader, W. T. & O'Malley, B. W. (1987) Structure–function properties of the chicken progesterone receptor A synthesized from complementary deoxyribonucleic acid. *Mol. Endocrinol.* **1** 791

[80] Ponglikitmongkol, M., Green, S. & Chambon, P. (1988) Genomic organization of the human oestrogen receptor gene. *EMBO J.* **7** 3385

[81] Weiler, I. J., Lew, D. & Shapiro, D. J. (1987) The *Xenopus laevis* estrogen receptor sequence; homology with human and avian receptors and identification of multiple estrogen receptor messenger RNA species. *Mol. Endocrinol.* **1** 355

[82] Coleman, R. T., Taylor, J. E., Shine, J. J. & Frossard, P. M. (1988) Human estrogen receptor (ESR) gene locus: *Pss*I dimorphism. *Nucleic Acids res.* **16** 7208

[83] Hill, S. M., Fuqua, S. A. W., Chamness, G. C., Greene, G. L. & McGuire, W. L. (1989) Estrogen receptor expression in human breast cancer associated with an estrogen receptor gene restriction fragment length polymorphism. *Cancer Res.* **49** 145

[84] Huckaby, C. S., Conneely, O. M., Beattie, W. G., Dobson, A. D. W., Tsai, M. J. & O'Malley, B. W. (1987) Structure of the chromosomal chicken progesterone receptor gene. *Proc. Natl. Acad. Sci. USA* **84** 8380

[85] Walter, P., Green, S., Greene, G., Krust, A., Bornert, J. M., Jeltsch, J. M., Staub, A., Jensen, E., Scrace, G., Waterfield, M. & Chambon, P. (1985) Cloning of the human estrogen receptor cDNA. *Proc. Natl. Acad. Sci. USA* **82** 7889

[86] Piva, R., Kumar, V. L., Hanau, S., Rimondi, A. P., Pansini, S., Mollica, G. & del Seuno, L. (1989) Abnormal methylation of estrogen receptor gene and reduced estrogen receptor RNA levels in human endometrial carcinomas. *J. Steroid Biochem.* **32** 1

[87] Horigome, T., Golding, T. S., Quarmby, V. E., Lubahn, D. B., McCarthy, K. & Korack, K. S. (1987) Purification and characterization of mouse estrogen receptor under conditions of varying hormonal status. *Endocrinology* **121** 2099

[88] Moncharmont, B., De Goeij, C. C. J., Ramp, G., Rijkers, A. W. M., van Duijvenbode, S. & Sluyser, M. (1989) In preparation.

[89] Monsma, F. J., Katzenellenbogen, B. S., Miller, M. A., Ziegler, Y. S. & Katzenellenbogen, J. A. (1984) Characterization of the estrogen receptor and its dynamics in MCF-7 human breast cancer cells using a covalently attaching anti-estrogen. *Endocrinology* **115** 143

[90] Golding, T. S. & Korack, K. S. (1988) Nuclear estrogen receptor molecular heterogeneity in the mouse uterus. *Proc. Natl. Acad. Sci. USA* **85** 69

[91] Auricchio, F., Migliaccio, A., Castoria, G., Rotundi, A., Di Domenico, M. & Pagano, M. (1986) Activation–inactivation of hormone binding sites of the oestradiol-17β receptor is a multiregulated process. *J. Steroid Biochem.* **24** 39

[92] Baskevitch, P. P. & Rochefort, H. (1985) A cytosol protease from the estrogen-resistant C3H mammary carcinoma increases the affinity of the oestrogen receptor for DNA *in vitro*. *Eur. J. Biochem.* **146** 671

[93] Puca, G. A., Abbondanza, C., Nigro, V., Armetta, I., Medici, N. & Molinari, A. M. (1986) Estradiol receptor has proteolytic activity that is responsible for its own transformation. *Proc. Natl. Acad. Sci. USA* **83** 5367

[94] Sluyser, M. & Van Nie, R. (1974) Estrogen receptor content and hormone responsive growth of mouse mammary tumours. *Cancer Res.* **34** 3253

[95] Sluyser, M., Evers, S. G. & De Goeij, C. C. J. (1976) Sex hormone receptors in mammary tumors of GR mice. *Nature (London)* **263** 386

[96] Sluyser, M., De Goeij, C. C. J. & Evers, S. G. (1981) Combined endocrine

therapy and chemotherapy of mouse mammary tumors. *Eur. J. Cancer* **17** 155
[97] Sluyser, M. (1983) Clinical relevance of experimental mammary tumours. In: B. A. Stoll (ed.) *Reviews on Endocrine Related Cancer*, Vol. 14, ICI, p. 23
[98] Paridaens, R. J. & Piccart, M. H. (1987) Chemo-hormonal treatment of breast cancer: the state of the art. In: M. Sluyser (ed.) *Growth Factors and Oncogenes in Breast Cancer*, Ellis Horwood, Chichester, p. 193
[99] Gelmann, E. P. & Lippman, M. E. (1987) Understanding the role of oncogenes in human breast cancer. In: M. Sluyser (ed.) *Growth Factors and Oncogenes in Breast Cancer*, Ellis Horwood, Chichester, p. 29
[100] Kasid, A., Lippman, M. E., Papageorge, A. G., Lowy, D. R. & Gelmann, E. (1985) Transfection of v-ras^H DNA into MCF-7 human breast cancer cells bypasses dependence on oestrogen for tumorigenicity. *Science* **228** 725
[101] Dickson, R. B., Bates, S. E., McManaway, M. E. & Lippman, M. E. (1986) Characterization of estrogen responsive transforming activity in human breast cancer cell lines. *Cancer Res.* **46** 1707
[102] Dickson,. R. B., Kasid, A., Huff, K. T., Bates, S. E., Knabbe, C., Bronzert, D., Gelmann, E. P. & Lippman, M. E. (1987) Activation of growth factor secretion in tumorigenic states of breast cancer induced by 17β-estradiol or v-Ha-*ras* oncogene. *Proc. Natl. Acad. Sci. USA* **84** 837
[103] Osborne, C. K., Ross, C. R., Coronado, E. B., Fuqua, S. A. W. & Kitten, L. J. (1988) Secreted growth factors from estrogen receptor-negative human breast cancer do not support growth of estrogen receptor–positive breast cancer in the nude mouse model. *Breast Cancer Res. Treat.* **11** 211
[104] Raam, S., Robert, N., Pappas, C. A. & Tamura, H. (1988) Defective estrogen receptors in human mammary cancers: their significance in defining hormone dependence. *J. Natl. Cancer Inst.* **80** 756
[105] Kok, K., Osinga, J., Carritt, B., Davis, M. B., van der Hout, A. H., van der Veen, A. Y., Landsvater, R. M., De Leij, L. F. M. H., Berendsen, H. H., Postmus, P. E., Poppema, S. & Buys, C. H. C. M. (1987) Deletion of a DNA sequence at the chromosomal region 3p21 in a major type of lung cancer. *Nature (London)* **330** 578
[106] Dobrovic, A., Houle, B., Belouchi, A. & Bradley, W. E. C. (1988) Erb A-related sequence coding for DNA-binding hormone receptor localized to chromosome 3p21–3p25 and deleted in small cell lung carcinoma. *Cancer Res.* **48** 682
[107] Brand, N., Petkovich, M., Krust, A., Chambon, P., de Thé, H., Marchio, A., Tiollais, P. & Dejean, A. (1988). Identification of a second human retinoic acid receptor. *Nature (London)* **332** 850
[108] Fuller, P. J. (1988) Steroid receptors as oncogenes? *Mol. Cell. Endocrinol.* **59** 161
[109] Picard, D. & Yamamoto, K. R. (1987) Two signals mediate hormone-dependent nuclear localization of the glucocorticoid receptor. *EMBO J.* **6** 3333
[110] Nohno, T., Kasai, Y. & Saito, T. (1989) Novel cDNA sequence possibly generated by alternative splicing of a mouse glucocorticoid receptor gene transcript from Shionogi carcinoma 115. *Nucleic Acids Res.* **17** 445

[111] Bishop, J. M. (1986) Oncogenes as hormone receptors. *Nature (London)* **321** 112
[112] Hodin, R. A., Lazar, M. A., Wintman, B. I., Darling, D. S., Koenig, R. J., Larsen, P. R., Moore, D. D. & Chin, W. W. (1989) Identification of a thyroid hormone receptor that is pituitary-specific. *Science* **244** 76
[113] Garcia, T., Lehrer, S., Bloomer, W. D. & Schachter, B. (1988) A variant estrogen receptor messenger ribonucleic acid is associated with reduced levels of estrogen binding in human mammary tumors. *Mol. Endocrinol.* **2** 785
[114] Mitsuhashi, T., Tennyson, G. E. & Nikodem, V. M. (1988) Alternative splicing generates messages encoding c-*erb*-A proteins that do not bind thyroid hormone. *Proc. Natl. Acad. Sci. USA* **85** 5804
[115] Colletta, A., Igbal, M. J. & Baum, M. (1987) Oestrogen receptor, does it undergo malignant change? Evidence from steroid binding studies. In: *Abstracts Joint Ann. Meet. of the Association of Head Neck Oncology of Great Britain*, London, April 25–30, 1987, p.97
[116] Webster, N. J. G., Green, S., Jin, J. R. & Chambon, P. (1988) The hormone-binding domains of the estrogen and glucocorticoid receptor contain an inducible transcription activation function. *Cell* **54** 199
[117] Webster, N. J. G., Green, S., Tasset, D., Ponglikitmongkol, M. & Chambon, P. (1989) The transcriptional activation function located in the hormone-binding domain of human oestrogen receptor is not encoded in a single exon. *EMBO J.* **8** 1441
[118] Tora, L., Gaub, M. P., Mader, S., Deierich, A., Bellard, M. & Chambon, P. (1988) Cell-specific activity of a GGTCA half-palindromic oestrogen-responsive element in the chicken ovalbumin promoter. *EMBO J.* **7** 3771
[119] Trent, J. M., Yang, J. M., Thompson, F. H., Leibovitz, A., Villar, H. & Dalton, W. S. (1987) Chromosome alterations in human breast cancer. In: M. Sluyser, M. (ed.) *Growth Factors and Oncogenes in Breast Cancer*, Ellis Horwood, Chichester, p. 142
[120] Guérin, M., Barrois, M. & Riou, G. (1989) C-*myb* proto-oncogene expression in breast cancer: strong association with estrogen receptors. *Proc. Am. Assoc. Cancer. Res.* **30** Abstr. 1744
[121] Lalley, P. A. & McKusick, V. A. (1985) Report of the Committee on Comparative Mapping, Eighth International Human Gene Mapping Workshop, Helsinki. *Cytogenet. Cell Genet.* **40** 536
[122] Mendel, D. B., Bodwell, J. E. & Munck, A. (1986) Glucocorticoid receptors lacking hormone-binding activity are bound in nuclei of ATP-depleted cells. *Nature (London)* **324** 478
[123] Tora, L., Mullick, A., Metzger, D., Ponglikitmongkol, M., Park, I. & Chambon, P. (1989) The cloned human oestrogen receptor contains a mutation which alters its hormone binding properties. *EMBO J.* **8** 1981
[124] Sap, J., Munoz, A., Schmitt, J., Stunnenberg, H. & Vennström, B. (1989) Repression of transcription mediated at a thyroid hormone responsive element by the v-*erb*-A oncogene product. *Nature (London)* **340** 242
[125] Damm, K., Thompson, C. C. & Evans, R. M. (1989) Protein encoded by v-*erb*-A functions as a thyroid-hormone receptor antagonist. *Nature (London)*

339 593
- [126] Herskowitz, I. (1987) Functional inactivation of genes by dominant negative mutations. *Nature (London)* **329** 219
- [127] Gandrillon, O., Jurdic, P., Pain, B., Desbois, C., Madjar, J. J., Moscovici, M. G., Moscovici, C. & Samarut, J. (1989) Expression of the v-*erb*-A product an altered nuclear hormone receptor, is sufficient to transform erythrocytic cells *in vitro*. *Cell* **58** 115

10

Analysis of c-*myc* amplification and expression in xenotransplanted human prostatic carcinoma cells

Manabu Fukumoto and **Osamu Midorikawa**
Department of Pathology, Faculty of Medicine, Kyoto University, Yoshida-Konoe-cho, Sakyo-ku, Kyoto 606, Japan

1. INTRODUCTION

Gene amplification is known as one of the major mechanisms of gene activation by overreplication of particular genes. Gene amplification generally increases the transcriptional level of amplified genes, and cells can overcome the surrounding selective pressure by overproduction of the proteins which are encoded by the amplified genes [1,2]. The frequency of gene amplification is very high and appears to be one of the widespread phenomena for the acquisition of resistance to cytotoxic substances *in vitro* and *in vivo* (see Chapter 14).

Gene amplification is often accompanied by karyotypic abnormalities such as double minutes (DMs) and homogeneously staining regions (HSRs) where amplified genes are known to reside [3]. These chromosomal abberations are rare in normal mammalian cells and have been found in a number of tumour cell lines and tumour tissues which were not exposed to cytotoxic drugs [4,5]. In some of these cases cells contain oncogene amplification, implying that gene amplification is involved in neoplastic transformation.

Various kinds of oncogene amplifications are reported such as those of the *myc* family, *ras* family, *erb*B, *yes* and *myb* in sporadic and in particular malignant tumour cell lines and tissues (for reviews, see [6,7]). In certain cases, oncogene amplification in fresh tumour tissues has reportedly been associated with clinical and pathological stages. The N-*myc* gene amplification correlates with advanced stages of neuroblastomas and poor prognosis [8]. The c-*erb*B$_2$/*neu* gene amplification in breast cancer is related to the number of lymph node metastases and forecasts both overall survival and relapse time in patients with breast cancer [9].

The Ki-*ras* proto-oncogene is well known to become an activated oncogene by point mutation of specific sites; however, amplification of the wild-type Ki-*ras* proto-oncogene appears important for the oncogenic process of ovarian carcinoma [10,11].

Today the notion that neoplastic transformation is the result of sequential multiple steps in a carcinogenic process involving genetic changes is widely accepted and gene amplification is believed to occur at a relatively late stage of carcinogenesis [12]. To date the selective pressure which triggers and maintains oncogene amplification is unclear and it is an open question whether gene amplification in cultured cells is related to carcinogenesis *in vivo* or the result of the process of establishing stable cell lines derived from tumours.

Karyotypic analysis is an effective way to find gene amplification. However, there are several speculations about the early structure and mechanisms of gene amplification [13], so that it is reasonable to assume gene amplifications of early stages and/or low level amplifications that are undetectable under cytogenetic studies. In addition, the cytogenetic procedure is rather laborious, and therefore the molecular genetic approach to screening gene amplification is necessary for an understanding of its significance.

Recently we have developed a modified in-gel DNA renaturation method to detect gene amplification in general. Using this method, we analysed gene amplification in human solid tumour cell lines and leukaemia cell lines and found co-amplification of *erb*B$_2$ and *erb*A in a lung adenocarcinoma cell line Calu3. Also, 10- to 12-fold c-*myc* amplification was found in prostatic carcinoma cells after several passages in nude mice, whereas the parental PC-3 cell line did not have gene amplification, suggesting that c-*myc* gives PC-3 tumour cells an advantage for *in vivo* growth [14].

The expression and amplification of *myc* oncogenes in a wide variety of tumours lead us to speculate on their intimate involvement in tumorigenesis and tumour progression in general. Therefore, in order to investigate the biological relevance of oncogene amplification in the carcinogenic process, we examined the reproducibilty of c-*myc* gene amplification after xenograft transplantation of PC-3 cells into nude mice and also monitored its activation by *in situ* hybridization.

2. DETECTION OF GENE AMPLIFICATION

The in-gel DNA renaturation technique was originally developed by Roninson to detect and clone amplified genes without information about the structure, function or location of the amplified genes in the genome [15]. A new version of the in-gel DNA renaturation method which is a combination of in-gel DNA renaturation and Southern hybridization has been developed to detect 7- to 8-fold amplified genes of unknown nature. The principle of modified in-gel DNA renaturation has been summarized elsewhere [14,16]. It is briefly as follows. High molecular DNA extracted from cells or tissues by phenol–chloroform was digested with *Hin*dIII and fractionated by agarose gel electrophoresis. After electrophoresis the DNA fragments in agarose gel were put through two cycles of denaturation, renaturation and digestion with single-strand-specific nuclease S1. At the denaturation step double-stranded DNA becomes single stranded in alkaline solution; then neutralization allows single-stranded DNA to reanneal. The DNA fragments from amplified units

Analysis of c-*myc* amplification

and repeated sequences of sufficient local concentration reassociate and remain in the gel, whereas unannealed single-copy fragments are digested and washed away completely from the gel after S1 nuclease digestion and washing.

After the second S1 nuclease digestion, small heteroduplexes of *Alu* sequences remain in the gel (Fig. 1(A)) and make a smeary background. To avoid this

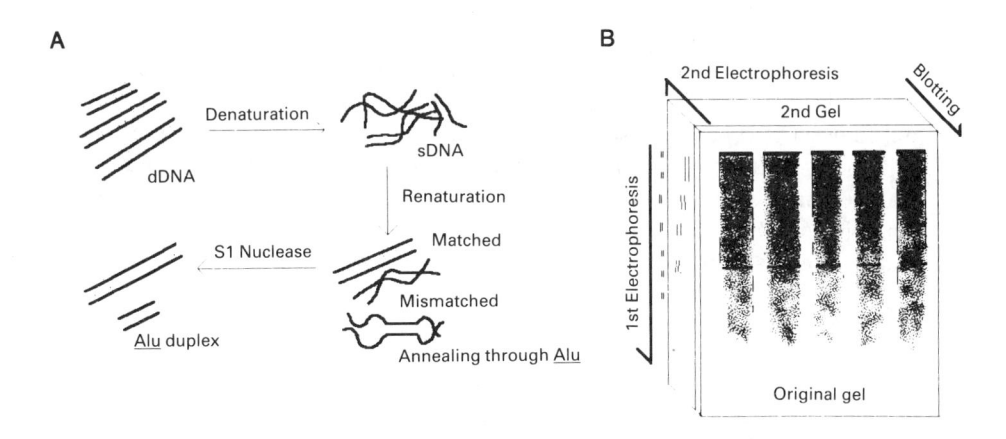

Fig. 1 — Schematic drawing of in-gel DNA renaturation. (A) After electrophoresis in 1% agarose gel, restriction fragments of genomic DNA are put through two rounds of denaturation, renaturation and S1 nuclease digestion in the gel. Fragments from repeated sequences and amplified sequences reanneal preferentially because of their higher local concentration. Association through *Alu* sequences occurs even between single copy fragments. This kind of *Alu* heteroduplex remains in the gel after S1 nuclease digestion because of its resistance to S1 nuclease. (B) To avoid smeary background from these heteroduplexes, 1.5% agarose is cast on the original gel to make the second gel layer. *Alu* heteroduplexes are eluted out from the gel sandwich by electrophoresis across the sandwich. The DNA fragments remaining in the gel are Southern transferred and *Alu*-containing amplified and repeated fragments are detected by hybridization to BLUR8 clone.

background, the following procedure was added (Fig. 1(B)). The gel was washed twice in 25 mM NaH_2PO_4 (pH 6.5; phosphate buffer) for 30 min per wash to terminate S1 nuclease digestion and to equilibrate the gel with the phosphate buffer. The gel was placed upside down in a casting tray, and a small amount of melted 1.5% agarose gel in phosphate buffer was poured along the edge of the gel to make the gel adhere to the substratum of the tray to prevent it from floating into the second gel layer. Once the agarose at the edges solidified, another quantity of 1.5% agarose in phosphate buffer at 60°C was cast to make an extra 6 mm gel layer. The double-layered gel 'sandwich' was trimmed to fit in the electroblotting chamber. The chamber was filled with 5 l of pre-cooled phosphate buffer in a cold room and the gel sandwich was placed into the chamber vertically (model TE42, Hoefer) or horizontally (model HRH, International Biotechnologies, Inc.). DNA was electrophoresed across the gel sandwich for 150 min at 1.1 A. The required duration of electrophoresis may be different depending on the type of electroblotting tank. The optimal duration of the second electrophoresis was assessed by using a standard set of DNA gel markers (1 kb ladder, Bethesda Research Laboratories). The optimal time is at the point when the 1.5 kb size marker is eluted out from the gel sandwich. The DNA fragments remaining in the gel were Southern transferred onto a nylon membrane (Biodyne A, Pall Biosupport) and hybridized to the oligolabelled [17] human *Alu* sequence BLUR8 (Amersham). The direction of the blotting is opposite to that of the second electrophoresis to minimize the smeary background from the second gel. A set of *Alu*-containing fragments from repetitive sequences and amplified units was detetected on the autoradiogram.

Compared with the original in-gel DNA renaturation method where a part of the restriction-digested DNA should be radiolabelled with T4 DNA polymerase, the technique described here has several advantages. This technique has about a three times higher sensitivity. The oligonucleotide labelling is a simpler procedure and does not require a large amount of radioisotope. Furthermore, the fact that the *Alu* sequence is human specific can avoid artifacts from contaminated microorganisms originating in the specimen.

3. C-*myc* AMPLIFICATION IN PC-3 CELLS

The human prostatic carcinoma cell line PC-3 was originally obtained from American Type Culture Collection and its subclones were a gift of D. H. Shevrin. The phylogeny of PC-3 and its derivatives are shown in Fig. 2(A).

The profile of amplified bands is conserved among all the subclones, indicating that the amplification occurred at the latest by the ninth generation of transplantation and that the structure of the amplified sequence was stable and major rearrangement did not take place after the amplification. A set of bands similar to that of subclones *in vitro* is also seen in the DNA preparation from the tumour tissue which was only passed through *in vivo*. This reveals that gene amplification was not an artifact of *in vitro* culture but occurred during passages *in vivo*. The amplified gene turned out to be a 10- to 12-fold amplification of the c-*myc* gene after regular Southern blot hybridization to several known oncogene probes. This suggests that c-*myc* amplification gives tumour cells an advantage for growth *in vivo*. The stage at which *myc* gene amplification occurs is a controversial subject. Wong *et al.* reported that gene amplification of c-*myc* and N-*myc* in small cell carcinoma of the lung occurs relatively early, that is, amplification was observed in a homogeneous fashion within individual patients before metastasis [18]. Brodeur *et al.* suggested that the N-*myc* copy number in human neuroblastoma is usually consistent within a tumour, not only at different tumour sites but also at different times *in vivo* [19]. On the contrary, Yokota *et al.* found heterogeneity of *myc* family oncogene amplification and rearrangement in human lung cancers, indicating that gene amplification occurs not at the time of transformation but during tumour progression [20]. In our case, PC-3 is an already established cell line *in vitro* from a human tumour and xenograft transplantation into nude mice must be a stronger selective pressure than *in situ* or *in vitro* growth. In this context, our result suggests that c-*myc* amplifcation takes place during tumour progression to alter biological behaviour but is not related to metastatic characteristics of tumour cells.

4. REPRODUCIBILITY OF C-*myc* AMPLIFICATION

In an attempt to determine the reproducibility of c-*myc* amplification in PC-3 cells, xenograft transplantation into nude mice was carried out. Cells (2×10^6) from the PC-3 cell line were subcutaneously inoculated in the back of male BALB/c nude mice; tumour tissues were resected when the size reached about 5 mm in diameter. The tissues were frozen in liquid nitrogen immediately and used for analysis. Simultaneously, tumours were minced and further transplanted into nude mice by trocar and cannula to maintain serial passages *in vivo*. The culture of tumour cells *in vitro* was also tried after trypsin treatment and mincing of the tumour tissues.

With the frozen tissues, DNA was extracted by phenol–chloroform and used for Southern hybridization after *Eco*RI digestion, and RNA was extracted by guanidine isothianate and CsCl density gradient for Northern hybridization analysis.

As c-*myc* probes, 1.5 kb *Hin*dIII–*Eco*Ri fragments of pDP-100 for exon 1 (New England Nuclear) and 1.6 kb *Cla*I–*Eco*RI fragments of pHSR-1 for exon 3 [21] were labelled with ^{32}P-dCTP by oligolabelling [17] for Southern and Northern hybridization and were labelled with biotin-7-dATP (Bethesda Research Laboratories) by nick translation [22] for *in situ* hybridization. Unincorporated precursors were removed using Quick Spin G-25 pre-pack columns (Boehringer Mannheim Biochemicals).

5. *IN SITU* HYBRIDIZATION

Although the sensitivity is less, non-radioisotopic detection was employed because this is a time-saving procedure with higher resolution and probes are stable and easier to handle compared with the radioisotopic detection method. The hybridization procedure was carried out basically according to Lawrence and Singer's method [23] with some modifications [24].

Glass slides were cleaned and treated with 0.05% poly-L-lysine (Sigma). The frozen tumour tissues were sectioned at 6 μm thickness. The tissue sections were air dried for 20 min on glass slides at room temperature and fixed in 4% paraformaldehyde in phosphate-buffered saline (PBS). After washing with PBS, slides were pretreated with 0.2 N HCl (10 min), 0.01% Triton X-100 (5 min) and 0.5 μg/ml proteinase K (10 min, 37°C). Proteinase K reaction was terminated by rinsing the slides with PBS; they were then post-fixed in 4% paraformaldehyde for 5 min and thereafter washed three times with 2×SSC. Prehybridization was carried out at 37°C for 2 h in a solution of 50% formamide, 5 mM EDTA, 4 mM vanadyl ribonucleoside complex, 2× Denhardt's, 100 μg/ml denatured salmon sperm DNA, 20 μg/ml yeast tRNA, 2×SSC (hybridization buffer). After this step, the following procedure was carried out in a humidified chamber. After blotting of the hybridization buffer, hybridization was performed at 44°C for 36 h in the hybridization buffer containing the heat-denatured probe at a concentration of 20 μg/ml. Thereafter, slides were washed serially in 50% formamide–2×SSC at 37°C, 2×SSC, and 1×SSC at room temperature.

The BlueGENE kit (Bethesda Research Laboratories) was used for the detection of biotinylated hybridized probes. After non-specific binding was blocked by 2.5% BSA–2×SSC for 1 h, and the sections were incubated with streptavidin–alkaline

A

```
PC-3, parental ────── lane 9
│ S.C., 9 passages
│                              │
rt. ax. ln                    S.C., 8 passages ───── lane 7 (431-P)
│ S.C., 2 passages            │ I.P.
med. ln                        Ascites
│ S.C.                         │ I.V.
med. ln                        b.m. metastasis
│ S.C.                         │ S.C.
                               b.m. metastasis
│                              │ S.C.
                               b.m. metastasis ─── lanes 1 & 8
                               │ S.C.
                               b.m. metastasis ─── lane 5
│
lt. ax. ln ──────────────────── lane 4 (C-505N)
│ I.P., 2 passages ─────────── lane 2
│ I.V., b.m. metastasis ────── lane 3
│ S.C., ln ─────────────────── lane 6 (C-673N)
```

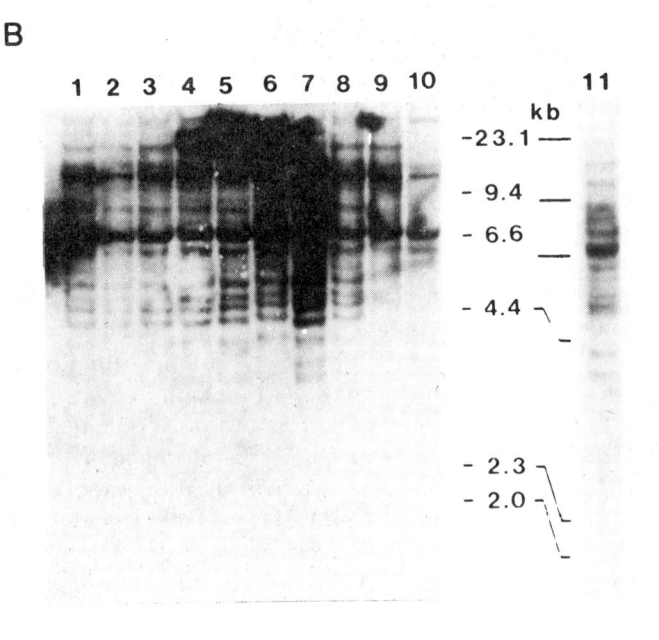

B

Lanes 1–11, kb markers: 23.1, 9.4, 6.6, 4.4, 2.3, 2.0

Fig. 2 — (A) Phylogeny of the parental PC-3 cell line and its derivative subclones and (B) their amplified DNA profile. Each lane number affixed to individual sublines in (A) corresponds to the lane number of (B). DNA extracted from pancreatic adenocarcinoma cell line Capan-1 is a negative control for gene amplification (lane 10). Lane 11 contains DNA from a tumour passed only through nude mice without culture *in vitro*. Each lane contains 15 μg of *Hind*III-digested DNA, run through a 9 mm wide comb. S.C., I.P. and I.V. represent subcutaneous, intraperitoneal and intravenous injection of cells respectively. The sites of metastasis are right axillary lymph nodes (rt. ax. ln.), mediastinal lymph nodes (med. ln.) and bone marrow (b.m.)

phosphatase conjugate, the procedure of Singer *et al.* [25] was used for visualization. After the colour development reaction was terminated by 10 mM Tris-HCl–1 mM EDTA (pH 7.5), the sections were cleared and mounted in Aquatex (Merck). Microscopic photographs were taken within a few days of staining before the colour became bleached. As a negative control for specimens RNase-treated sections were used, and for probes the labelled plasmid pUC18 was used.

6. CONCLUSION AND OUTLOOK

Reproducibility of c-*myc* gene amplification in PC-3 cell tumours could not be found up to the fourth generation of transplantation (PC-3-4) in nude mice (Fig. 3). *In situ*

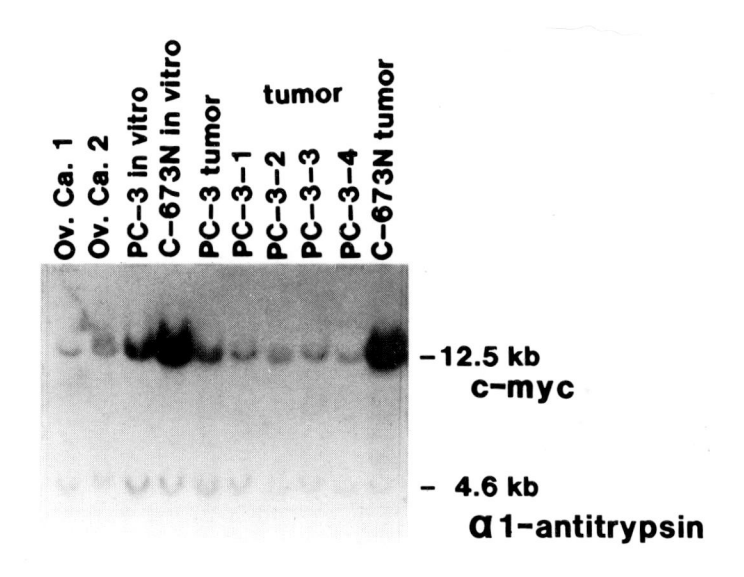

Fig. 3 — Southern hybridization to determine reproducibility of c-*myc* amplification. Each line contains 5 μg of *Eco*RI-digested DNA and the filter was hybridized with the exon 3 probe to estimate the c-*myc* copy number. An α_1-antitrypsin probe was used as a single-copy control to normalize the intensity of the c-*myc* bands. DNA extracted from ovarian carcinoma tissues was used for a single-copy control of the c-*myc* gene (Ov. Ca. 1, 2). Copy numbers of the c-*myc* gene in DNAs from the tumours induced by injection of PC-3 cells (PC-3 tumour) and its first (PC-3-1), second (PC-3-2), third (PC-3-3) and fourth (PC-3-4) xenograft transplantations in nude mice serially were compared with that from parental PC-3 cells *in vitro*. The notion that the intensity of the c-*myc* band is the same between DNAs from C-673N cells *in vitro* and tumour tissue means that the contamination of normal cells which reduces the estimate of the copy number of amplification is negligible in this analysis.

hybridization of c-*myc* is shown in Fig. 4. The c-*myc* transcripts are visible as darker dots. In the tumour tissue formed by inoculation of *in vitro* cultured PC-3 cells (PC-3 tumour), the hyperexpressive cells (positive cells) of non-coding exon 1 are seen dispersed in the marginal portion of the tumour (Fig. 4(a)), especially adjacent to capillaries (Fig. 4(b)), whereas exon 3 which is a part of the coding sequences is not

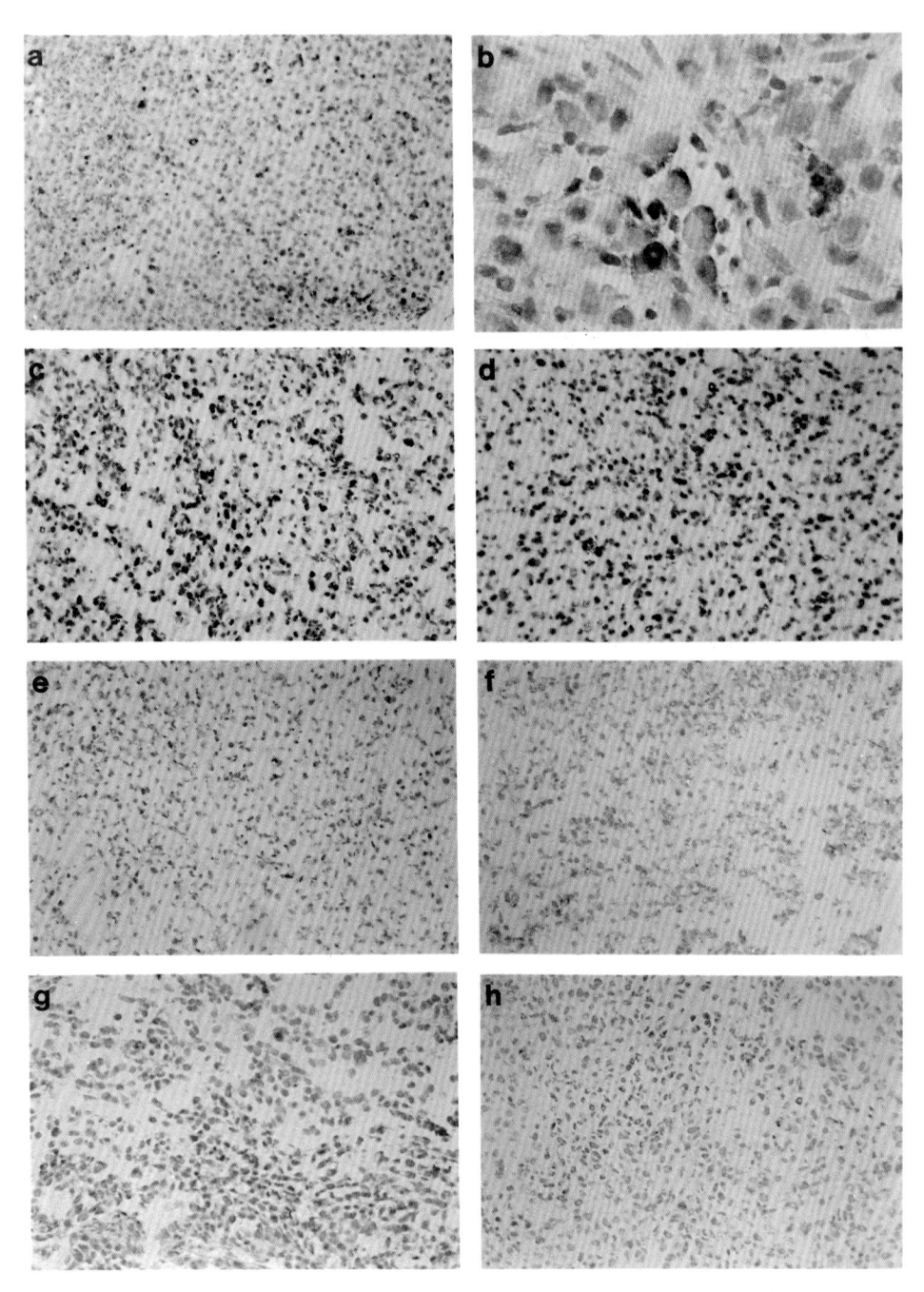

Fig. 4 — *In situ* hybridization to detect c-*myc* transcripts in tumour tissues. Strong hybridization to the c-*myc* probes is indicated by purple–red precipitates in the cytoplasm of cells and they are defined as positive cells. Exon 1 positive cells are seen in the marginal part of the tumour induced by injection of PC-3 cells (a); in particular, cells infiltrating to the juxta-capillary portion are positive (b). Almost all the cells in the tumour of the first xenograft transplantation (PC-3-1 tumour) are positively hybridized to both exon 1 (c) and exon 3 (d). All the cells in the tumour of the second xenograft (PC-3-2) are negative to both exon 1 (e) and exon 3 (f). The tumour of C-673N cells is also negative to both exon 1 (g) and exon 3 (h). The original magnifications were 100× for (a),(d)–(h) and 400× for (b).

seen in any tumour cells (data not shown). The quality of the probes and the efficiency of the hybridization are not completely comparable between exon 1 and exon 3. However, in the first transplanted tumour tissue (PC-3-1), almost all the tumour cells are both exon 1 (Fig. 4(c)) and exon 3 (Fig. 4(d)) positive. Thus the elevation of expression of exon 1 proceeded to that of exon 3. After the second transplant of tumour tissues (PC-3-2), both exon 1 and exon 3 positive cells have disappeared (Figs 4(e) and 4(f)). In the third (PC-3-3) and fourth (PC-3-4) transplanted tumours, neither exon 1 nor exon 3 positive cells were observed (data not shown). Interestingly, positive cells are not seen in the tumour made by inoculation of c-*myc* amplified subclone C-673N (C-673N tumour, Figs 4(g) for exon 1 and (4(h) for exon 3). Histological distribution and time-dependent fluctuation of the expression in the tumours suggests that the c-*myc* oncogene amplification is not a necessary process of carcinogenesis but a casual result of temporary demand for c-*myc* expression for the tumour progression *in vivo*. This result is not consistent with the finding of Schwab *et al.* that the number of DMs containing the c-*myc* gene of mouse SEWA cells increased after injection of cells into a mouse and was reduced when the cells were explanted back into culture; when the cells were re-introduced into the mouse, they again acquired multiple DMs [26]. Before injection into nude mice, c-*myc* amplification was not observed in the parental PC-3 cells, whereas in their case c-*myc* was significantly amplified prior to the first injection of cells, implying that SEWA cells originally had the predisposition of c-*myc* amplification. In either case, the c-*myc* oncogene expression may give cells advantages for the adaptation to the growth and invasion or establishment of tumours *in vivo*.

C-673N cells have c-*myc* gene amplification and more malignant characteristics such as higher tumorigenic activity and a higher growth rate of tumours in nude mice compared with parental PC-3 cells. In spite of amplification, the expression level of exon 3 of the c-*myc* gene in C-673N cells *in vitro* is slightly higher than that in parental cells but as low as that of other contol cell lines, and it is not significant compared with that in HL60 cells (Fig. 5). The profile of Northern blot analysis hybridized to exon 1 was comparable with that to exon 3 (data not shown). The size of the c-*myc* transcripts detected by exon 1 and exon 3 is the same at 2.4 kb. This means that the c-*myc* expression in PC-3 and its derivatives is regulated by the transcriptional level at the original promoter site which is located upstream of exon 1, not by elongation blockage in exon 1 which is known to regulate the c-*myc* expression in mononuclear cells [27]. Our results coincide with the observation of Sugimoto *et al.* that decreased expression of the amplified *mdr*1 gene in revertants of multidrug-resistant human myelogenous leukaemia K562 occurred without loss of amplified DNA [28]. This suggests that the expression of some kinds of amplified genes is still regulated by the demand for the genes.

Putting these results together, we conclude that the transient hyperexpression of the c-*myc* gene, especially exon 1, probably controls expression of other genes which are associated with tumorigenesis at early stages and has triggered their deregulation. The biological meaning of low level expression of the *c*-myc gene and its target genes in relation to tumorigenesis is at present under investigation.

Recently a cytogenetic study revealed that C-673N cells cultured *in vitro* for a long period do not have HSRs but contain only DMs (data not shown). This fact contradicts the general observation that gene amplification in the form of DMs is

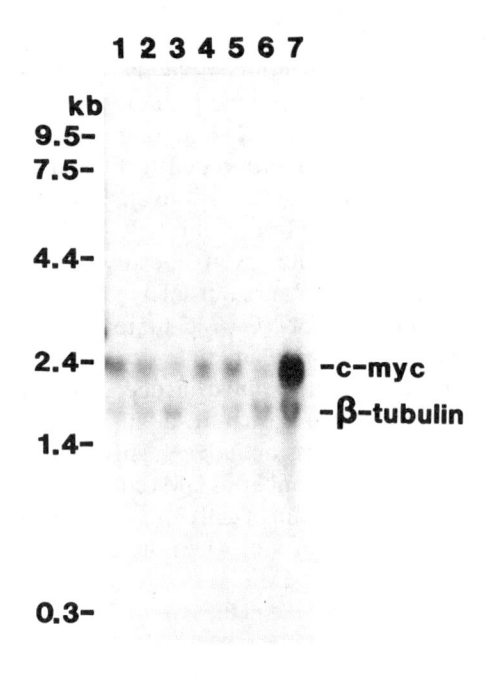

Fig. 5 — Northern blot analysis of c-*myc* expression in the PC-3 cell line and its subclones. 20 μg of total RNA extracted from myelogenous leukaemia cell lines KYO-1 (lane 1), KPB-M8 (lane 2), PC-3 parental cells (lane 3), C-673N cells (lane 4), and cell lines established from PC-3 tumour tissue (lane 5) and PC-3-2 tumour tissue (lane 6) were fractionated in 0.8% agarose gel containing 2.2 M formalydehyde. The filter was hybridized to the mixture of c-*myc* exon 3 probe. HL60 was used for the control of c-*myc* hyperexpression (lane 7). The intensity of the c-*myc* transcript was normalized for that of β-tubulin transcript.

very unstable compared with HSRs [1]. The selective pressure to retain c-*myc* amplification in C-673N in vitro as well as the preference for maintaining the amplified gene in the fashion of DMs also remains unclear.

Mariani-Constantini *et al.*, found that dense clusters of infiltrating lymphocytes present in breast carcinomas exhibited c-*myc* hybridization in addition to tumour cells by *in situ* hybridization using a radiolabelled method [29]. In PC-3 parental cells and their derivative cell tumours, the infiltrating inflammatory cells were always negative to either exon 1 or exon 3 probes of the c-*myc* gene, presumably because of the lower detectability of the method. However, this fact does not ruin the value of the non-radioisotopic method because c-*myc* expression of HL60 cells is visible by this method [24] and it is enough to detect the biologically significant level of gene expression.

It is widely believed that the normal function of cellular proto-oncogenes is in some ways concerned with the developmental process, but it has yet to be determined what this function is and how their expression is controlled. If we could understand the physiological role of these oncogenes, that would not only provide insights into the control of normal development but could also suggest ways in which aberration of that control might lead to malignancy. Oncogenes were originally

Analysis of c-*myc* amplification

discovered in acute transforming retroviruses, and cellular counterparts which do not have transforming activity were termed proto-oncogenes; these have the possibility of transforming activity by qualitative or quantitative changes. Several oncogenes have been found by the transfection of DNA from tumours into eukaryotic cells to produce foci *in vitro* or tumour-forming activity. The number of oncogenes discovered by searching for viral oncogenes and the DNA transfection system is increasing at a rate of a few a year. Only the in-gel DNA renaturation assay can provide a third way of finding oncogenes. Furthermore, this allows one to find and clone carcinogenesis-associated genes which cannot be made by the other two methods. Even though the amplified genes are remnants of a transient protein demand which gives cells advantages for selection, it would be valuable to investigate amplified genes to clarify their physiological and pathological function.

For detection of gene amplification by the in-gel DNA renaturation technique, the wider the comb used, the better the resolution obtained. Therefore, 13–15 μg of restriction-digested DNA is routinely loaded for electrophoresis to obtain a clear result. Using HL60 cells and PC-3 cells as a control system, we assessed the amount of DNA required to observe gene amplification (Fig. 6). If a 5 mm wide comb is used

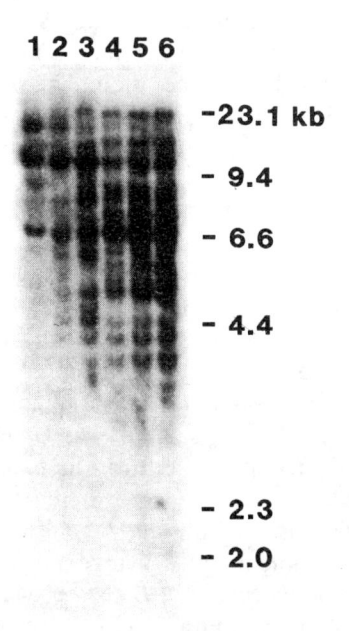

Fig. 6 — Small scale in-gel DNA renaturation. Using a 5 mm wide comb, 7.5 μg of DNA from PC-3 parental cells (lane 1) and from its derivatives C-673N (lane 2) and 431-P (lane 3), which have 10- to 12-fold c-*myc* amplification, were analysed. Lanes 4, 5 and 6 contain 4 μg, 6 μg and 8 μg respectively of DNA extracted from HL60 cells which have 30-fold c-*myc* amplification.

for the first electrophoresis, some of the amplified bands can be seen even at 4 μg of DNA extracted from HL60 cells which are known to have 30-fold c-*myc* amplification, and 7.5 μg of DNA are enough for detection of the 10- to 12-fold level, indicating that the applicability of this technique could be wider for the screening of gene amplification in general.

Southern hybridization to several oncogene probes with the DNA preparation where we found gene amplification is necessary to determine whether the amplified fragments are from reported oncogenes or not. However, the modified in-gel DNA renaturation method described here is a widely available technique for finding and cloning amplified genes of unknown nature, because a multidrug resistance gene in Chinese hamster cells [30] and a gene *gli* which was amplified in human glioma [31] have been cloned by the original in-gel DNA renaturation technique. The informative strategy of cloning amplified genes by this method is seen in the article by Roninson [32].

ACKNOWLEDGEMENTS

The establishment of the modified in-gel DNA renaturation technique was accomplished at Dr I. B. Roninson's laboratory and we thank him for his support and advice. We also thank Dr D. H. Shevrin for his permission to investigate PC-3 subclones in Japan, Dr Y. Hosokawa for the suggestion about *in situ* hybridization, Dr K. Kita for cell lines HL60, KPB-M8 and KYO-1, and Drs H. Tashiro, T. Ohbayashi and I. Kashu for their cooperation. This work was partly supported by grants from the Ichiro Kanehara Foundation and the Osaka Cancer Research Foundation to M.F.

REFERENCES

[1] Hamlin, J., Milbrant, J. D., Heintz, N. H. & Azizkhan, J. C. (1984) DNA sequence amplification in mammalian cells. *Int. Rev. Cytol.* **90** 31–82
[2] Stark, G. R. (1984) Gene amplification. *Ann. Rev. Biochem.* **53** 447–491
[3] Cowell, J. K (1982) Double minutes and homogeneously staining regions: gene amplification in mammalian cells. *Ann. Rev. Genet.* **16** 21–59
[4] Barker, P. E. (1982) Double minutes in human tumor cells. *Cancer Genet. Cytogenet.* **5** 81–94
[5] Gebhart, E., Brüderlein, S., Tulusan, A. H., Maillot, K. V. & Birkmann, J. (1984) Incidence of double minutes, cytogenetic equivalents of gene amplification, in human carcinoma cells. *Int. J. Cancer* **34** 369–373
[6] Taya, Y., Terada, M. & Sugimura, T. (1987) Role of oncogene amplification in tumor progression. In: G. Klein (ed.) *Advances in Viral Oncology*, Vol. 7, Raven Press, New York, pp. 141–153
[7] Masuda. H., Battifora, H., Yokota, J., Meltzer, S. & Cline, M. J. (1987) Specificity of proto-oncogene amplification in human malignant diseases. *Mol. Biol. Med.* **4** 213–227
[8] Seeger, R. C., Brodeur, G. M., Sather, H., Dalton, A., Siegel, S. E., Wong, K. Y. & Hammond, D. (1985) Association of multiple copies of the N-*myc* oncogene with rapid progression of neuroblastomas. *New Engl. J. Med.* **313** 1111–1116

[9] Slamon, D. J., Clark, G. M., Wong, S. G., Levin, W. J., Ulrich, A. & McGuire, W. L. (1987) Human breast cancer: correlation of relapse and survival with amplification of the HER-2/neu oncogene. *Science* **235** 177–182

[10] van't Veer, L. J., Hermans, R., van der Berg-Bakker, L. A. M., Cheng, N. C., Fleuren, G.-J., Bos, J. L., Cleton, F. J. & Schrier, P. I. (1988) Ras oncogene activation in human ovarian carcinoma. *Oncogene* **2** 157–165

[11] Fukumoto, M., Estensen, R. D., Sha, L., Oakley, G. J., Twiggs, L. B., Adcock, L. L., Carson, L. F. & Roninson, I. B. (1989) Association of Ki-*ras* with amplified DNA squences, detected in ovarian carcinomas by a modified in-gel renaturation assay. *Cancer Res.* **49** 1693–1697

[12] Bishop, J. M. (1987) The molecular genetics of cancer. *Science* **235** 305–311

[13] Schimke, R. T. (1984) Gene amplification in cultured animal cells. *Cell* **37** 705–713

[14] Fukumoto, M., Shevrin, D. H. & Roninson, I. B. (1988) Analysis of gene amplification in human tumor cell lines. *Proc. Natl. Acad. Sci. USA* **85** 6846–6850

[15] Roninson, I. B. (1983) Detection and mapping of homologous, repeated and amplified DNA sequences by DNA renaturation in agarose gels. *Nucleic Acids Res.* **11** 5413–5431

[16] Fukumoto, M. & Roninson, I. B. (1986) Detection of amplified sequences in mammalian DNA by in-gel renaturation and SINE hybridization. *Somatic Cell Mol. Genet.* **12** 611–623

[17] Feinberg, A. P. & Vogelstein, B. (1984) A technique for radiolabeling DNA restriction endonuclease fragments to high specific activity. *Anal. Biochem.* **137** 266–267

[18] Wong, A. J., Ruppert, J. M., Eggleston, J., Hamilton, S. R., Baylin, S. B. & Vogelstein, B. (1986) Gene amplification of c-*myc* and N-*myc* in small cell carcinoma of the lung. *Science* **233** 461–464

[19] Brodeur, G. M., Hayes, F. A. Green, A. A., Casper, J. T., Wasson, J., Wallach, S. & Seeger, R. C. (1987) Consistent N-*myc* copy number in simultaneous or consecutive neuroblastoma samples from sixty individual patients. *Cancer Res.* **47** 4248–4253

[20] Yokota, J., Wada, M., Yoshida, T., Noguchi, M., Terasaki, T., Shimosato, Y., Sugimura, T. & Terada, M. (1988) Heterogeneity of lung cancer cells with respect to the amplification and rearrangement of *myc* family oncogenes. *Oncogene* **2** 607–611

[21] Alitalo, K., Schwab, M., Lin, C. C., Varmus, H. E. & Bishop, J. M. (1983) Homogeneously staining chromosomal regions contain amplified copies of an abundantly expressed cellular oncogene (c-*myc*) in malignant neuroendocrine cells from a human colon carcinoma. *Proc. Natl. Acid. Sci. USA* **80** 1707–1711

[22] Maniatis, T., Fritsch, E. F. & Sambrock, J. (1982) In: *Molecular Cloning*, Cold Harbor Laboratories, pp. 109–112

[23] Lawrence, J. B. & Singer, R. H. (1985) Quantitative analysis of *in situ* hybridization methods for the detection of actin gene expression. *Nucleic Acids Res.* **13** 1777–1799

[24] Hosokawa, Y., Ueda, K., Nishi, M., Tsuji, Y., Murata, S., Konishi, E., Kanaizuka, T., Kuzuhara, K., Takeshita, H., Urata, Y., Dobashi, Y., Hirabayashi, K., Yano, J. & Ashihara, T. (1987) Detection of c-*myc* mRNA in HL-

60 cells and human cancer tissues by *in-situ* hybridization. *Oncologia* **20** 101–103 (in Japanese)

[25] Singer, R. H., Lawrence, J. B. & Villnave, C. (1986) Optimization of *in situ* hybridization using isotopic and non-isotopic detection methods. *Biotechniques* **4** 230–250

[26] Schwab, M., Ramsey, G., Alitalo, K., Varmus, H. E., Bishop, J. M., Martinsson, T., Levan, G. & Levan, A. (1985) Amplification and enhanced expression of the c-*myc* oncogene in mouse SEWA tumour cells. *Nature (London)* **315** 345–347

[27] Eick, D., Berger, R., Polack, A. & Bornkamm, G. W. (1987) Transcription of c-*myc* in human mononuclear cells is regulated by an elongation block. *Oncogene* **2** 61–65

[28] Sugimoto, Y., Roninson, I. B. & Tsuruo, T. (1987) Decreased expression of the amplified *mdr*1 gene in revertants of multidrug-resistant human myelogenous leukemia K562 occurs without loss of amplified DNA. *Mol. Cell. Biol.* **7** 4549–4552

[29] Mariani-Costantini, R., Escort, C., Theillet, C., Gentile, A., Merlo, G., Lidereau, R. & Callahan, R. (1988) *In situ* c-*myc* expression and genomic status of the c-*myc* locus in infiltrating ductal carcinoma of the breast. *Cancer Res.* **48** 199–205

[30] Gros, P., Croop, J., Roninson, I., Varshavsky, A. & Housman, D. E. (1986) Isolation and characterization of DNA sequences amplified in multidrug-resistant hamster cells. *Proc. Natl. Acad. Sci. USA* **83** 337–341

[31] Kinzler, K. W., Bigner, S. H., Bigner, D. D., Trent, J. M., Law, M. L., O'Brien, S. J., Wong, A. J. & Vogelstein, B. (1987) Identification of an amplified, highly expressed gene in a human glioma. *Science* **236** 70–73

[32] Roninson, I. B. (1987) Use of in-gel DNA renaturation for detection and cloning of amplitude genes. *Methods Enzymol.* **151** 332–371

11

Oncogenes in human lung cancer

Frederic J. Kaye*[†], **Sarah K. Barksdale***[‡], **J. William Harbour***[‡] and **John D. Minna***[†]

* NCI–Navy Medical Oncology Branch, National Cancer Institute and Naval Hospital
† The Uniformed Services University for the Health Sciences
‡ The Howard Hughes Medical Institute Scholars Program, Bethesda, MD 20814, USA

> *It is definitely no exaggeration to say that the lung cancer of today is the greatest cancer [carcinogenesis] experiment ever carried out in the history of mankind.*
> (K. H. Bauer, 1969)

1. INTRODUCTION

Establishing the detailed pathogenesis of a human neoplastic disease is a formidable task. Such an effort requires the assimilation of clues from diverse fields such as clinical medicine, epidemiology, human genetics, and cellular and molecular carcinogenesis. Further, the long progression time prior to clonal expansion of these tumours and the resulting heterogeneity of the malignant cell phenotype suggest that events underlying the genesis of human cancer must involve the interplay of several distinct mechanisms. In this chapter we review recent work which has focused on the pathogenesis of human lung cancer. The clues we have chosen to pursue come from the identification of tumour-specific genetic alterations which appear either to activate (dominant oncogenes) or to inactivate (recessive or 'tumour-suppressive' oncogenes) a limited number of cellular gene products. Understanding the mechanisms and implications of these genetic changes may in turn provide for a more directed approach to the detection, therapy, and prevention of this cancer.

2. CARCINOGEN (MUTAGEN) EXPOSURE IS A MAJOR RISK FACTOR FOR LUNG CANCER AND SUGGESTS THAT DAMAGE TO CELLULAR DNA WITHIN THE BRONCHIAL EPITHELIUM IS A CRITICAL EVENT

Lung cancer, while once a rare disease, has become the most common malignancy in the industrialized countries, with 140 000 cases projected for 1988 in the US [1].

Numerous environmental exposures have been associated with an increased incidence of this tumour; however, longstanding cigarette smoking remains the major risk factor [2]. Despite extensive research over the past decades and the isolation of several candidate carcinogenic compounds from cigarette smoke [3], no definitive chemical or byproduct has been unequivocally linked to tumorigenesis. However, the identification of covalently bound adducts of human DNA in specimens from 'at risk' subjects has supported the hypothesis of DNA mutagen exposure [4], and a linear correlation between DNA adduct levels in human lung tissue and cumulative cigarette smoking has recently been confirmed [5,6]. In addition, other studies have demonstrated mutagenic substances in tobacco smoke [7,8], and an association between cigarette exposure in young adults and excess non-random chromosomal fragility has been reported [9]. These observations provide a tentative foundation for the hypothesis that somatic damage to cellular DNA is the key event in environmentally induced lung cancer. The possibility of a concomitant, inherited genetic susceptibility to lung cancer (either directly or by altering the effects of carcinogen exposure) has also been proposed [10]. First-degree relatives of patients with lung cancer have a 2.4-fold excess risk for developing lung [11] and other unrelated [12] tumours. Epidemiologic data supporting an inherited predisposition, either in the same patient or within families, however, are difficult to interpret owing to confounding risk factors such as exposure to cigarette smoke, radon gas, or other endemic or occupational hazards. A genetic basis for an increased susceptibility to the damaging effects of carcinogens has long been an attractive hypothesis to explain why some but not the majority of heavy cigarette smokers develop lung cancer. Variations in the level of endogenous aryl hydrocarbon hydroxylase activity was initially speculated to impact on the activation of benzpyrene to its fully carcinogenic form [13]. Recently, genetically determined differences in one of the members of the P450 enzyme family (4-debrisoquine hydroxylase) have also been implicated in excess lung cancer risk [14] and studies are underway to determine whether the inheritance of related alleles for the gene encoding this enzyme are linked with the presence of cigarette-induced lung cancer.

3. GENERATION AND CHARACTERIZATION OF LUNG CANCER CELL LINES

Lung cancer can be segregated into two major clinicopathologic groups: small cell lung cancer (SCLC) and non-SCLC. SCLC is an aggressive tumour which usually presents with widespread metastases and characteristically exhibits features of neuroendocrine differentiation [15,16]. Non-SCLC constitutes 75% of all lung cancer and consists of three major categories: (1) adenocarcinoma (including bronchioloalveolar carcinoma), (2) squamous cell carcinoma, and (3) large cell carcinoma. These lung tumours can be easily distinguished by light microscopy; however, occasionally features of each histologic type can be seen in adjoining regions of the same specimen. SCLC will, initially, respond to systemic chemotherapy, in contrast to non-SCLC tumours which are not generally responsive to systemic chemotherapy or immunotherapy and therefore early detection with surgical excision is the only form of curative therapy for these patients.

Over the past decade there has been a developing interest in the study of the molecular and cellular biology of lung cancer. One major advance that helped these studies was the establishment of a panel of tumour cell lines obtained from biopsy specimens of patients with all histologic types of lung cancer [16]. The cytology and the histology of nude mouse xenografts from these cell lines are representative of the original biopsy material. This panel, which now includes over 100 independent lung cancer lines, has allowed us to search for common cellular and molecular properties of these tumours. In view of the strong evidence for a genetic basis of cancer, our laboratory has taken two approaches to investigating genetic causes of lung cancer. The first approach involves the study of oncogene activation ('dominant oncogenes') in these tumours. A second approach involves the identification of patterns of tumour-specific chromosomal deletions present in the lung cancer cell lines and then using this information as clues to help to identify genes which may function in what has been referred to as 'tumour suppressive' or 'recessive oncogenes'.

4. DOMINANT ONCOGENES

The dominant oncogenes represent cellular genes which have gone awry either by deregulated expression of their protein products (i.e. *myc* gene amplification with overexpression of the mRNA transcript), or by mutational alterations within the transcriptional unit, such as seen with the *ras* or *neu* genes [17,18]. These cellular genes are presumed to subserve critical roles in the growth and differentiation blueprints of normal cells; however, a clear function for any of these gene products has yet to be identified. Clues such as the identification of regions of amino acid sequence similarity with genes already implicated in the process of differentiation (such as *myc* genes) [19], the presence of tyrosine kinase activity (such as *abl* and many others) [20], specific DNA-binding activity (such as *jun* and *fos*) [21], and signal transduction (such as *ras*) hold promise that we may understand in the future how oncogene activation may impact on the initiation or progression of human cancers.

Several dominant oncogenes have been implicated in the pathogenesis of lung cancer. The most extensively studied has been the deregulated expression of the *myc* family genes in SCLC. We use the term 'deregulated expression' to refer to the detection of marked increased levels of steady state mRNA, while there appears to be no structural change in the proteins encoded by these genes. This overexpression can occur with or without gene amplification. *myc* gene amplification in SCLC has been always associated with increased mRNA and protein expression in those specimens tested; however, increased *myc* expression can also be seen in the absence of gene amplification. In contrast to Burkitt's lymphoma and certain plasma cell leukaemias, rearrangement of the *myc* gene by chromosomal translocation has not been reported to be a common mechanism for activation in lung cancer. Preliminary data from our laboratory (Nau, M. N., et al., unpublished data), however, have shown that at least one SCLC line has a rearrangement in the 5' portion of the transcribed L-*myc* gene. Structural analysis of the 5' end of the cDNA shows that a novel, distinct gene has been translocated to this site resulting in a chimaeric mRNA species. It remains to be seen whether this will represent another mechanism for *myc* gene activation in these tumours. In summary, identification of the physiologic signals (*cis* and *trans*) which regulate *myc* gene transcription initiation, elongation of the mRNA transcript, and mRNA and protein stability will be essential to understand the mechanism of *myc* gene activation in human tumours.

The *myc* family of proto-oncogenes consists now of at least four members: c-*myc*, N-*myc*, L-*myc*, and the recently described B-*myc* [22]. Two other *myc*-related genes, P- and R-*myc*, have also been identified in a preliminary report [23]. Of these genes, three of them (c-, N-, and L-*myc*) have been extensively characterized showing that they share striking structural similarities and strongly suggesting that they play a common physiological role. In 1983, Little *et al.*, reported the presence of c-*myc* gene amplification in 7 out of 18 small cell lung cancer cell lines [24] and this finding has now been confirmed from many different laboratories although the incidence in fresh tumours from patients with SCLC appears lower than that observed from derived cell lines. Similarly, both N-*myc* and L-*myc* have also been found to be overexpressed, with or without gene amplification, in distinct subsets of SCLC cell lines [25,26]. In summary, approximately 60% of SCLC cell lines have evidence for deregulation of c-, N-, or L-*myc*, but interestingly we have yet to find a cell line that has co-amplified more than one *myc*-related gene.

Despite these structural similarities, each *myc* family member is distinct, and dramatic differences exist between *myc* genes with respect to tissue and developmental stage specificity [27]. For example, constitutive expression of c-*myc* mRNA with intermittent variations in response to external signals [28] is seen in most cell types, and occasional amplification and overexpression of this gene is observed in many different tissue types. L-*myc* and N-*myc*, in contrast, are expressed in a more restricted set of tissues and overexpression of their gene product appears confined to a more narrow selection of tumour types (Table 1).

Table 1 — Expression of *myc* genes

myc gene	Tissue expression	Deregulated expression in human tumours
c-*myc*	All tissues	SCLC, non-SCLC, lymphoma, leukaemia, colon, breast, prostate, other
N-*myc*	Restricted	SCLC, non-SCLC, neuroblastoma, retinoblastoma, astrocytoma, rhabdomyosarcoma
L-*myc*	Restricted	SCLC, non-SCLC

Our laboratory has been particularly interested in the biologic role of the L-*myc* gene in lung cancer. The L-*myc* gene was first identified in a SCLC cell line on the basis of gene amplification and homology to c-*myc* and N-*myc* [26]. This cellular oncogene is located on human chromosome region 1p32, and while its expression is detected in the newborn murine brain, lung, and kidney, its normal expression appears to be specific for adult pulmonary tissue [27]. For these reasons we have focused much of our attention on this member of the *myc* oncogene family. The structural organization of the human L-*myc* gene shows that this gene consists of three exons and two introns with the major open reading frame (ORF) starting near the 5' end of exon 2, features similar to those observed with c-*myc* and N-*myc*. In contrast to c-myc and N-*myc*, the L-*myc* gene is a complex transcriptional unit, exhibiting high levels of mature mRNA that has undergone alternative splicing of intron 1 and 2 and/or alternative polyadenylation site selection within intron 2 (Fig. 1) [29]. These events predict the possibility of novel *myc* protein products that would

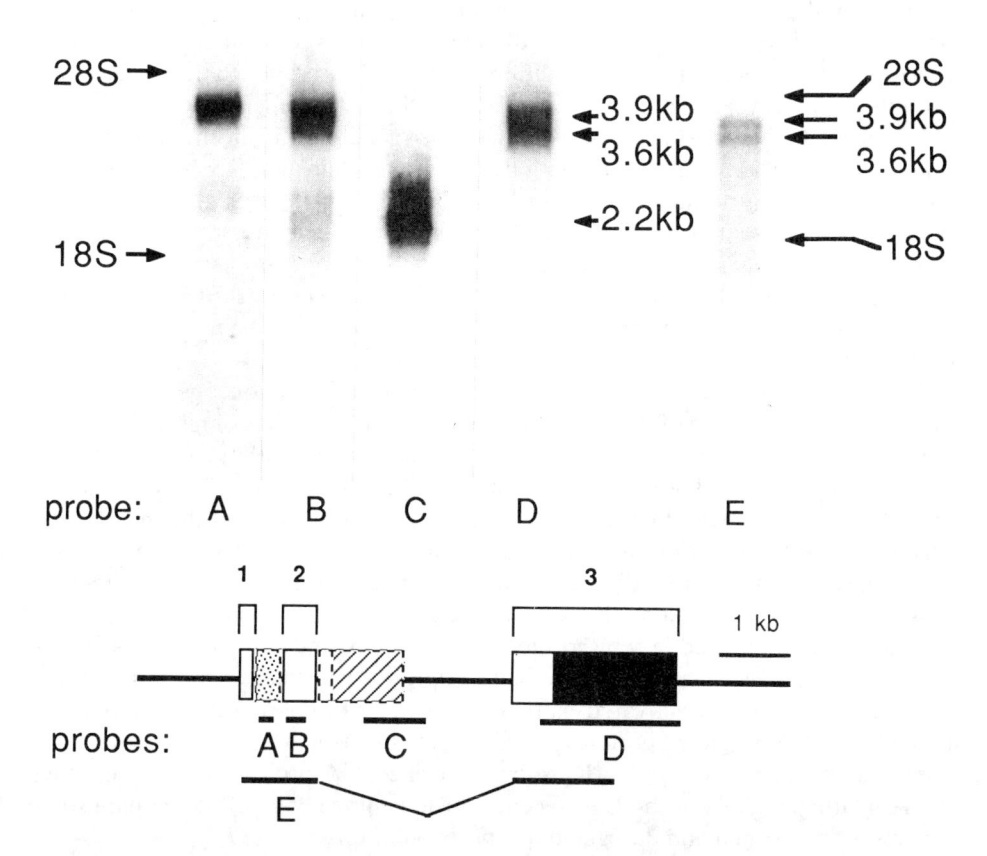

Fig. 1 — Northern analysis using non-overlapping probes along L-*myc* gene reveal distinct mRNA species. Boxes enclosed with solid lines identify exons 1, 2, and 3. The stippled box between exons 1 and 2 represents intron 1 which, if retained, results in a 3.9 kb mRNA transcript or, if spliced, results in a 3.6 kb transcript. The box enclosed with a dashed line adjacent to exon 2 represents the portion of intron 2 retained in 'short-form' L-*myc* mRNA which terminates at one of several alternative polyadenylation signals within intron 2. The area filled in with slanted hatched lines represents the predicted untranslated region of the 'short-form' transcript. The solid filled-in area of exon 3 represents the untranslated region of the full-length L-*myc* transcript. 1 μg poly(A) RNA from an SCLC cell line was size fractionated and transferred to nitrocellulose. Duplicate lanes were then hybridized with probes A (intron 1), B (exon 2), C (intron 2), D (exon 3), and E (a near full-length cDNA insert), demonstrating the different sized mRNAs. Reprinted with permission from American Society for Microbiology [29].

differ in their amino terminal ends or carboxyl terminal ends [30]. Recent data from *in vitro* transcription and translation experiments suggests that a cryptic non-AUG initiation site may be located within intron 1. This start site could result in a protein that contains additional amino terminal residues in frame with the generic exon 2–exon 3 coding domain of *myc*, and might explain the doublet pattern seen on *in vivo* immunoprecipitation experiments [30]. This predicted protein could only be translated from mature mRNAs which have not spliced intron 1 sequences, thus lending significance to the regulation of alternative intron 1 splicing. Alternatively, Saksela *et al.* [31] have reported that the multiple bands seen on *in vivo* immunoprecipitation are only the result of differences in the phosphorylation state of the protein. Further work will be needed to resolve these differences.

Truncated or 'short-form' L-*myc* transcripts, generated from the utilization of one of three polyadenylation sites within intron 2, result in an mRNA species which lacks exon 3 coding domains (Fig. 1). This predicted 'short-form' protein would lack peptide sequences implicated in nuclear targeting, protein dimerization via a leucine 'zipper' domain, and *in vito* transformation of rat embryo fibroblasts [32,33]. Two separate groups have recently found *in vivo* evidence for a 'short-form' cytoplasmic L-*myc* protein [31,34] and characterization of this protein will be an important and exciting study.

Despite the abundance of data on the structure of these oncogenes, until we understand the function of the normal *myc* product and the physiologic mechanisms controlling its expression, the role *myc* genes play in the pathogenesis of human cancer will remain unknown. The activation of the *myc* genes in SCLC and the availability of cell lines which selectively express one or another of the members of this oncogene family offer an ideal model system to study key transcriptional and post-transcriptional factors. To this end we have undertaken a study to identify regulatory *cis* elements along the L-*myc* gene, including the exon 1-intron 1 region that has been implicated in transcriptional pausing [35] and control of translational efficiency [36]. One strategy that we have employed is to harvest protein extracts from cell lines that have activated expression of one of the *myc* family genes and then to search for DNA binding proteins specific for the L-*myc* gene in these extracts. This approach has identified at least a half-dozen specific DNA:protein interactions along the regulatory regions of the L-*myc* gene which we hope will help to explain some aspects of the normal and deregulated expression of this gene [37].

What is the role of *myc* gene activation in lung cancer? There is strong circumstantial evidence implicating the deregulation of the c-*myc* gene in the pathogenesis of Burkitt's lymphoma and some related human plasmacytomas. However, the evidence is less compelling that *myc* genes are necessary for initiation of tumorigenesis in solid tumours such as SCLC. A study designed to address the clinical relevance of *myc* gene activation noted that the presence of c-*myc* activation was associated with a 'large cell' or 'variant' histology of SCLC and was also associated with a slightly shortened survival time in extensive stage patients [38]. No similar correlation in SCLC was observed with N-*myc* or L-*myc* activation and these analyses, in general, suffer from small numbers of cases and selection bias. In conclusion it appears that more information on the role of this gene family in normal cellular physiology will be required before we will be able to understand its role in the tumorigenesis of solid malignancies. Fortunately, research into the *myc* genes is acquiring renewed vigour and we can realistically expect to decipher the mysteries of this oncogene in the near future.

5. OTHER DOMINANT ONCOGENES

In order to induce tumour foci formation within experimental systems (such as transforming rat embryo fibroblast cells), the deregulated expression of a *myc* gene must cooperate with a cytoplasmic oncogene product, such as a *ras* gene which has undergone an activating point mutation [39]. Such activation of *ras* family members by acquired point mutations has been demonstrated in a high frequency of adenocarcinomas of the colon [40] and in approximately 30% of adenocarcinomas of the lung [41]. In addition, studies have also shown that the transfection of a mutated *ras* gene into SCLC cells resulted in a change of phenotype to that resembling non-SCLC [42]. *ras* genes are presumed to facilitate the transfer of information from external signals to cellular machinery; however, as with the *myc* genes, we must wait until more is known of their role in normal physiology to understand their impact on tumorigenesis. Constitutive expression of another group of genes, the *jun* family members and c-*raf*-1, is found in all types of lung cancer. Since it appears that these genes are expressed at high levels in normal pulmonary tissue, it is unclear whether the overexpression of these proto-oncogenes has a role in the initiation or progression of these tumours. We have also been particularly interested in these genes since c-*jun*-A maps to chromosomal band 1p32 which is also the localization of L-*myc* and the src-related oncogene *lck*, and since c-*raf*-1 localizes to chromosomal band 3p25, an area associated with mutational deletions in lung cancer and renal cancers.

Other oncogenes studied in lung cancer include the nuclear oncogenes c-*myb* and p53. *myb* gene amplification and overexpression have been reported in SCLC [43]; however, the significance of this finding is uncertain as c-*myb* expression has been most closely associated with cells of lymphoid lineage. In contrast, evidence for mutational alterations within the p53 gene has been observed in our laboratory in cell lines and primary tissue from lung cancers of different histologic types (Takahashi *et al.*, in press). The pattern of these mutations in several tumours resembles that reported with the retinoblastoma susceptibility gene, suggesting that inactivation of the wild-type p53 gene is occurring in these tumours (see section 6). A 'dominant-negative' mechanism of tumorigenesis has also been suggested by the detection of single point mutations within the open reading frame of the p53 gene. The mutated gene might then induce the neoplastic phenotype by interacting with (and inactivating) the product of the wild-type p53 allele.

6. RECESSIVE ONCOGENES

As mentioned in the beginning of this chapter, there is now considerable evidence that the fundamental defects in malignant cells are genetic in nature. Whether the earliest genetic lesions are dominant or recessive in their effects on cellular growth remains unknown. Evidence for recessively acting cancer genes came initially from cell fusion experiments [44] and from epidemiological studies of familial cancers [45]. Cell fusion experiments were performed in which tumorigenic cells were fused to non-tumorigenic cells (generally rodent cells) to determine the phenotype of the resultant daughter cells. The majority of these studies showed that these somatic cell hybrids had lost the tumorigenic characteristics of the malignant parent cell, and that subsequent reversion to tumorigenicity was associated with the non-random loss of specific chromosomes from the normal cell line [44]. These experiments suggested that some factor that was absent or defective in the tumorigenic parent cell was supplied by the non-tumorigenic parent cell, and thus a recessive defect might be responsible for the malignant phenotype. Further support for a recessive genetic mechanism came from the study of familial cancers such as familial retinoblastoma. Based on statistical data, Knudson hypothesized that a single gene was responsible for initiation of this tumour by mutational inactivation of both alleles of a putative 'retinoblastoma' or 'Rb' gene [45]. Advances in cytogenetics and in the increasing availability of RFLP probes which have been mapped to all the human chromosomes have confirmed this hypothesis and allowed the cloning of the first recessive, 'tumor-suppressor' gene, the *Rb* gene [46–48].

The success of the use of cytogenetics and RFLP analysis to identify the presence of recessive genes in hereditary cancers was not lost on investigators studying adult, non-hereditary cancers. In 1982, Whang-Peng *et al.* described the cytogenetic analysis of 16 SCLC cell lines [49,50]. She observed mild to moderate aneuploidy in all lines; however, the most consistent abnormality seen was a deletional event involving the short arm of chromosome 3 in all the tumour lines studied. This was confirmed by others and extended to non-SCLC as well. The narrowest consensus deletion by cytogenetics was estimated to be between chromosome bands 3p14 and 3p23. With the advent of restriction fragment length polymorphism analysis it became possible to test at the molecular level whether the cytogenetic events represented true DNA loss from tumour cells. Studies by Naylor *et al.* [51], Brauch *et al.* [52], Johnson *et al.* [53], and Kok *et al.* [54] convincingly confirmed the presence of a consistent DNA deletion at 3p present in tumour cells compared with normal DNA in essentially 100% of cases of SCLC and in about 50% of non-SCLC tumours as well (Table 2). These data provided strong circumstantial evidence that

Table 2 — Loss of heterozygosity at chromosome 3p by RFLF analysis in lung cancer

SCLC	58/59 (98%)
Non-SCLC	16/32 (50%)

Data from Naylor *et al.* [51], Brauch *et al.* [52], Kok *et al.* [54], Yokota *et al.* [59], Johnson *et al.* [53].

perhaps a tumour-suppressive or recessive cancer gene, analogous to that seen in paediatric hereditary cancers (such as retinoblastoma), mapped to this chromosome region. This hypothesis would propose that this putative cancer gene would then manifest its phenotype through its absence, presumably by a mutational event such as a deletion. However, the existence of such a gene(s) will have to be proven by isolation of its corresponding cDNA and demonstration of mutations in lung cancer specimens, since the band at 3p14 is reported to be among the most fragile sites in the human genome [55]. The data supporting the existence of such a 3p recessive oncogene includes the following: (1) RFLP evidence for DNA loss in lung cancer, renal cancer, and some, but not all, other human tumours; (2) low or absent level of expression of the aminoacylase-1 gene (assigned to 3p21) in certain small cell lung cancers, a feature resembling the low esterase-D levels observed in some retinoblastoma tumours [56]; (3) the occurrence of familial renal cancer associated with a constitutional chromosome 3:8 translocation. Seizinger et al. [57] and Zbar et al. [58], using RFLP probes, have recently confirmed these original cytogenetic data suggesting the presence of a renal cancer susceptibility locus on chromosome 3p. The availability of multiple kindreds with familial renal cancer syndromes, such as Von Hippel Lindau disease, should facilitate efforts to identify and clone this gene which may also be the locus implicated in lung cancer.

While we were involved in an effort to identify transcribed genes which map to the deleted region on chromosome 3p, we re-reviewed our cytogenetic data and observed a large number of abormalities involving the chromosomal band 13q14 which we recognized as the locus for the retinoblastoma susceptibility gene. These abnormalities included deletions and very often translocations targeted to this chromosomal band. At this time Yokota and his colleagues from the National Cancer Institute in Japan reported a comprehensive RFLP analysis using primary tumour from both SCLC and non-SCLC patients [59]. They confirmed the loss of heterozygosity on the short arm of chromosome 3 in lung cancer, but also noted tumour-specific deletions on the short arm of chromosome 17 and the long arm of chromosome 13. The 13q RFLP probes used in this analysis spanned the RB locus at 13q14 and strongly suggested the possibility that the *RB* gene might be the target of these events. SCLC resembles retinoblastoma in that both malignancies morphologically are small, round cell tumours which in culture grow as non-adherent clusters of cells. Both tumours also exhibit neural features of differentiation and frequently amplify or overexpress *myc*-related genes. For these reasons we undertook a study of the structure and expression of the *Rb* gene in lung cancer [60]. Using the p0.9R and p3.8R complementary DNA (cDNA) probes [46] we found that 1/8 SCLC primary tumours and 4/22 cell lines had evidence for gross structural abormalities within the *Rb* gene (Fig. 2). This frequency of gross deletion events resembles the 20–40% incidence reported for retinoblastoma and the associated mesenchymal tumours. Similarly, we found that in 1/4 pulmonary carcinoid tumours (a tumour which phenotypically resembles SCLC), but in 0/20 non-SCLC cell lines, we could detect structural abnormalities by Southern blot analysis. We then studied RNA expression of this gene (Fig. 3) and found absent *Rb* expression in 60% of SCLC cell lines (including all samples which had detectable DNA abnormalities) and trace or absent expression in 90% of cases. These data are summarized in Table 3.

The obvious question that arises after inspection of these data is, if inactivation of the *Rb* gene is so important in the genesis of SCLC, how do you explain the fact that about 1/3 of all tumours (including retinoblastoma tumours) have detectable *Rb* expression? In the case of retinoblastoma, the answer is that these tumours have subtle mutations that nonetheless result in absent *Rb* protein production [61]. Therefore, in collaboration with Jon Horowitz and his colleagues in Dr Weinberg's laboratory, we were interested in studying expression of the *Rb* protein in SCLC and, in particular, we were interested in studying those SCLC tumours that had detectable *Rb* mRNA. Preliminary results suggest that the majority of these samples

216   Oncogenes in human lung cancer   [Ch. 11

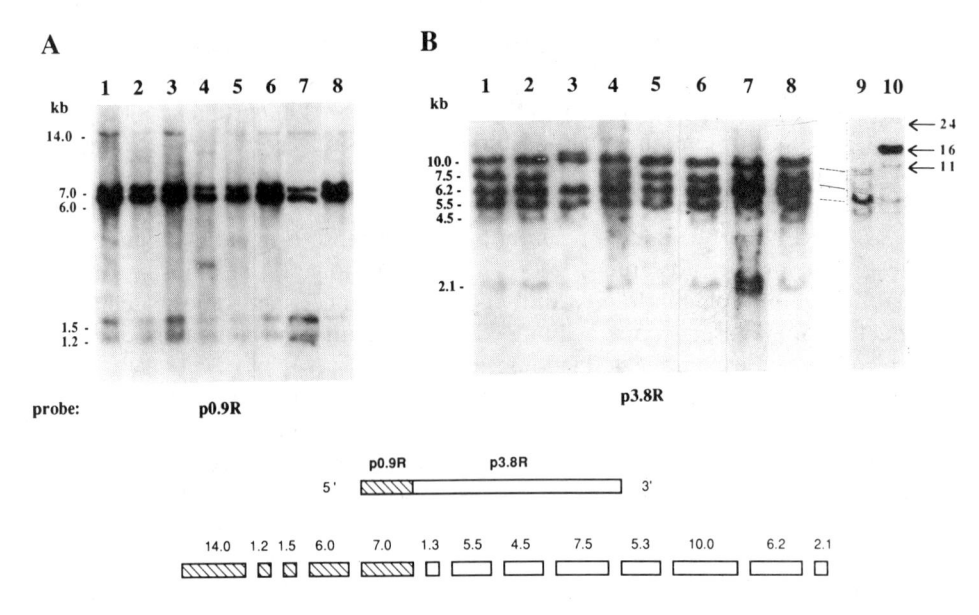

Fig. 2 — Southern analysis of the *Rb* gene in small cell lung cancer. Hind III-digested genomic DNA from 7 SCLC cell lines (lanes 1–7), normal thymus DNA (lane 8), and two primary small cell tumours (lanes 9 and 10) were probed using an *Rb* cDNA fragment. The diagram at the bottom of the figure depicts the previously described *Rb* cDNA clone and the Hind III fragments detected by the p0.9R and the p3.8R probes [46]. Reprinted with permission from *Science* [60].

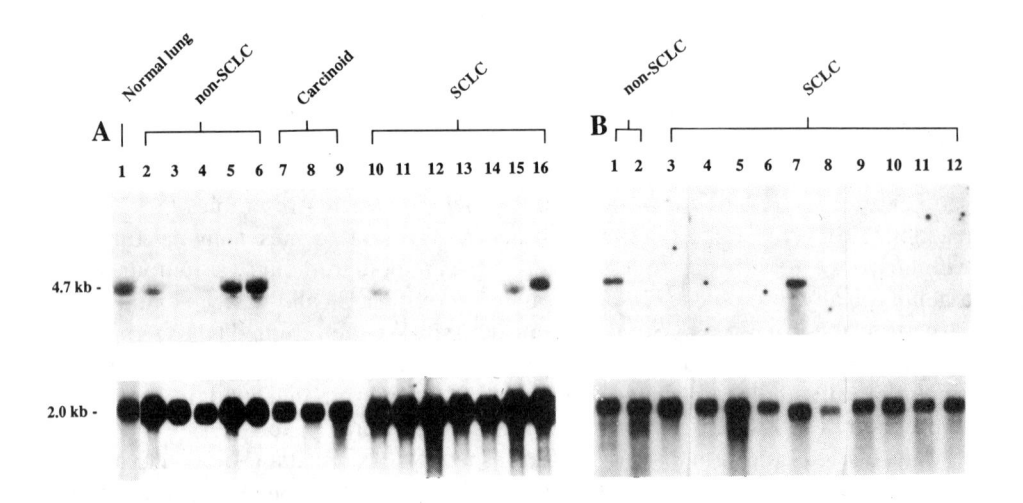

Fig. 3 — Northern analysis of *Rb* gene in lung cancer cell lines. (A) Total RNA (15 μg) was extracted from normal lung tissue (lane 1), 5 non-SCLC lines, 3 carcinoid tumours, and 7 SCLC lines. (B) Poly(A) RNA (2 μg) was extracted from 2 additional non-SCLC lines and 10 additional SCLC lines. RNA samples were then hybridized with the p3.8R probe which detects a 4.7 kb mRNA species. A β-actin probe, which detects a 2.0 kb species, was used as control for quantity and integrity of RNA. Reprinted with permission from *Science* [60].

Table 3.1 — Summary of DNA and RNA status of the *Rb* gene in 50 lung cancer lines

Cell line[a]	DNA changes detected[b]	RNA expression[c]	Cell line (histology)[a]	DNA changes detected[b]	RNA expression[c]
Small cell lung cancer			*Pulmonary carcinoids*		
H187	5' rearrangement	−	H679 (atypical)	3' rearrangement	−
H345	3' rearrangement	−	H720 (atypical)	None	−
H378	3' rearrangement	−	H835 (typical)	None	+
H889	3' arrangemement	−	H727 (typical)	None	+
H82	None	−			
H417	None	−	*Non-small cell lung cancer*		
H510	None	−	H125 (adenosqaumous)	None	+
H524	ND	−	H226 (squamous)	None	+
H526	None	−	H322 (bronchioloalveolar)	None	+
H735	None	−	H358 (bronchioloalveolar)	None	+
H748	ND	−	H460 (large cell)	None	+
H774	None	−	H522 (adenocarcinoma)	None	+
H1284	None	−	H661 (large cell carcinoma)	None	+
H1304	None	−	H810 (large cell)	None	+
H1450	ND	−	H820 (bronchioloalveolar)	None	+
H69	None	tr	H838 (adenocarcinoma)	None	+
N592	None	tr[d]	H1373 (adenocarcinoma)	None	+
H711	ND	tr	H1385 (large cell)	None	+
H847	None	tr	H1404 (bronchioloalveolar)	None	+
H1622	None	tr	H1435 (adenocarcinoma)	None	+
H209	None	+	H1437 (adenocarcinoma)	None	+
H841	None	+	H23 (adenocarcinoma)	None	tr
H1092	None	+[d]	H1355 (adenocarcinoma)	None	tr
H1105	None	+	H596 (adenosquamous)	None	−
H1184	None	+	H1155 (large cell)	None	−
H1436	None	+[d]	H1445 (adenocarcinoma)	None	ND

[a] Full designation of the cell lines includes the prefix 'NCI-'.
[b] See text for discussion of structural abnormalities.
[c] ND = not done; tr = trace.
[d] Truncated mRNA species (4.4 kb).
Reprinted with permission from *Science* [60].

have absent or inactive *Rb* protein (Horowitz, J., *et al.*, in press) and that inactivation of this gene occurs in excess of 90% of all cases studied to date. These data have been extended by Yokota's group where he reported the absence of *Rb* protein in 9 samples of SCLC primary tumour material [62], which reinforces the hypothesis that inactivation of this gene product may be a necessary event in these tumours.

Another oncogene, p53, has recently been proposed to have a similar 'tumour-suppressive' activity. Tumours with homozygous deletions of this gene (suggesting a recessive mechanism) or with 'activating' point mutations on one allele (dominant recessive mechanism) have been reported [63]. Preliminary data demonstrate that 2/3 of lung cancer tumours and cell lines of all histologic types have evidence for detectable mutational events within the p53 locus (Takahashi *et al.*, in press). Further work will be needed to determine whether these alterations occur early or late in the progression of malignancy.

The critical experiment for defining the role of the *Rb* or p53 genes in the pathogenesis of SCLC is to revert the phenotype of tumorigenicity by the reintroduction of a wild-type gene. This experiment has recently been reported in a retinoblastoma and osteosarcoma cell line using a retroviral vector to introduce the *Rb* gene by infection of the recipient cell lines [64]. After treatment, the authors documented that the tumour cells now expressed the exogenous wild-type protein, and that this was associated with a change in their histologic morphology, in their growth characteristics, and loss of the ability to form tumours in nude mice. After several months in culture, however, many of these transformant lines had reverted to the original, fully malignant phenotype and this reversion was associated with acquired loss of the exogenous *Rb* gene product. Efforts to reproduce these findings are ongoing and it will be of great importance to determine whether similar results can be obtained with recipient small cell lung cancer cells. Such functional studies with the *Rb* gene, p53 gene, chromosome 3p loci, or a combination of these will be necessary to prove a definitive role for any of these genes in the pathogenesis of adult lung cancer.

REFERENCES

[1] American Cancer Society (1989) *Cancer Facts and Figures — 1989*, American Cancer Society, Atlanta, GA, pp. 4–6
[2] Fielding, J. E. (1985) Smoking — health effects and control (parts 1 and 2). *New Engl. J. Med.* **313** 491–498, 555–561
[3] Stedman, R. L. (1968) The chemical composition of tobacco and tobacco smoke. *Chem. Rev.* **68** 153–207
[4] Miller, E. C. (1978) Some current perspectives on chemical carcinogenesis in humans and experimental animals. Presidential address. *Cancer Res.* **38** 1479–1496
[5] Gupta, R. C., Sopori, M. L. & Gairola, C. G. (1989) Formation of cigarette smoke-induced adducts in the rat lung and nasal mucosa. *Cancer Res.* **49** 1916–1920
[6] Phillips, D. H., Hewer, A., Martin, C. N., Colin Garner, R. and King, M. M. (1988) Correlation of DNA adducts levels in human lung with cigarette smoking. *Nature (London)*, **336** 790–792
[7] Gairola, C. (1982) Genetic effects of fresh cigarette smoke in *Saccharomyces cerevisiae*. *Mutat. Res.* **102** 123–136
[8] Demarini, D. M. (1983) Genotoxicity of tobacco smoke and tobacco smoke condensate. *Mutant. Res.* **114** 59–89
[9] Kao-Shan, C. S., Fine, R. L., Whang-Peng, J., Lee, E. C. & Chabner, B. A. (1987) Increased fragile sites and sister chromatid exchanges in bone marrow and peripheral blood of young cigarette smokers. *Cancer Res.* **47** 6278–6281
[10] Harris, C. C., Vahakangas, K. & Autrup, H. (1985) Biochemical and molecular epidemiology of cancer risk. In: D. Scartelli, J. Craighead & N. Kaufman (eds), *The Pathologist and the Environment,* International Academy of Pathology, Monograph No. 26, Williams and Wilkins, Baltimore, MD, pp. 140–167
[11] Ooi, W. L., Elston, R. C., Chen, V. W., Bailey-Wilson, J. & Rothschild, H. (1986) Increased familial risk for lung cancer. *J. Natl. Cancer Inst.* **76** 217–222
[12] Lynch, H. T., Kinberling, W. J., Markvicka, S. E., Biscone, K., Lynch, J.,

Whorton, E. & Maillard, J. (1986) Genetics and smoking-associated cancers. A study of 485 families. *Cancer* **57** 1640–1646
[13] Kellerman, G., Shaw, C. R. & Luyten-Kellerman, M. (1973) Aryl hydrocarbon hydroxylase inducibility and bronchogenic carcinoma. *New Engl. J. Med.* **289** 934–937
[14] Ayesh, R., Idle, J. R., Ritchie, J. C., Crothers, M. J. & Hetzel, M. R. (1984) Metabolic oxidation phenotypes as markers for susceptibility to lung cancer. *Nature (London)* **312** 169–170
[15] Minna, J. D., Pass, H., Glaststein, E., & Ihde, D. C. (1989) Cancer of the lung. In: V. T. deVita, S. Hellman & S. Rosenberg (eds) *Principles and Practice of Oncology*, Lippincott, Philadelphia, PA, pp. 591–687
[16] Carney, D. N., Gazdar, A. F., Bepler, G., Guccion, G. C., Marangos, P. J., Moody, T. W., Zweig, M. H. & Minna, J. D. (1985) Establishment and identification of small cell lung cancer lines having classic and variant features. *Cancer Res.* **45** 2913–1923
[17] Tabin, C. J., Bradley, S. M., Bargmann, C. I., Weinberg, R. A., Papageorge, A. G., Scolnick, E. M., Dhar, R., Lowy, D. R. & Chang, E. H. (1982) Mechanism of activation of a human oncogene. *Nature (London)* **300** 143–146
[18] Bargmann, C. I. & Weinberg, R. A. (1988) Oncogene activation of the *neu*-encoded receptor protein by point mutation and deletion. *EMBO J.* **7** 2043–2052
[19] Tapscott, S. J., Davis, R. L., Thayer, M. J., Cheng, P. F., Weintraub, H. & Lasser, A. B. (1988) MyoD1: a nuclear phosphoprotein requiring a *myc* homology region to convert fibroblasts to myoblasts, *Science* **242** 405–409
[20] Reddy, E. P., Smith, M. J. & Srinivasan, A. (1983) Abelson murine leukemia virus genome: structural similarity of its transforming gene product to other *onc* gene products with tyrosine-specific kinase activity. *Proc. Natl. Acad. Sci. USA* **80** 3623–3627
[21] Angel, P., Allegretto, E., Okino, S., Hattori, K., Boyle, W., Hunter, T. & Karin, M. (1988) Oncogene *jun* encodes a sequence-specific trans-activator similar to AP-1. *Nature (London)* **332** 166–171
[22] Ingvarsson, S., Asker, C., Axelson, H., Klein, G. & Sumegi, J. (1988) Structure and expression of B-*myc*, a new member of the *myc* gene family. *Mol. Cell. Biol.* **8** 3168–3174
[23] Alt, F. W., DePinho, R., Zimmerman, K., Legouy, E., Hatton, K., Ferrier, P., Tesfaye, A., Yancopoulos, G. & Nisen, P. (1986) The human *myc* gene family. *Cold Spring Harbor Symp. Quant. Biol* **59** 931–941
[24] Little, C. D., Nau, M. N., Carney, D. N., Gazdar, A. F. & Minna, J. (1983) Amplification and expression of the c-*myc* oncogene in human lung cancer cell lines. *Nature (London)* **306** 194–196
[25] Nau, M. N., Brooks, B. J., Carney, D., Gazdar, A., Battey, J., Sausville, E. & Minna, J. (1986) Human small cell lung cancers show amplification and expression of the N-*myc* gene. *Proc. Natl. Acad. Sci. USA* **83** 1092–1096
[26] Nau, M. N., Brooks, B. J., Battey, J., Sausville, E., Gazdar, A. F., Kirsch, I., McBride, O. W., Bertness, V., Hollis, G. & Minna, J. D. (1985) L-myc: a new *myc*-related gene amplified and expressed in human small cell lung cancer. *Nature (London)* **318** 69–73

[27] Zimmerman, K., Yancopoulos, G., Collum, R. G., Smith, R. K., Kohl, N. E., Denis, K. A., Nau, M. M., Witte, O., Toran-Allerand, D., Gee, C. E., Minna, J. & Alt, F. W. (1986) Differential expression of *myc* family genes during murine development. *Nature (London)* **319** 780–783
[28] Kathy, K., Cochran, B. H., Stiles, C. D. & Leder, P. (1984) Cell-specific regulator of the c-*myc* gene by lymphocyte mitogens and platelet-derived growth factor. *Cell* **35** 603–611
[29] Kaye, F. J., Battey, J., Nau, M., Brooks, B., Seifter, E., DeGreve, J., Birrer, M., Sausville, E. & Minna, J. (1988) Structure and expression of the human L-*myc* gene reveal a complex pattern of alternative mRNA processing. *Mol. Cell. Biol.* **8** 186–195
[30] DeGreve, J., Battey, J., Fedorko, J., Birrer, M., Evan, G., Kaye, F., Sausville, E. & Minna, J. (1988) The human L-*myc* gene encodes multiple phosphoproteins from alternatively processed mRNAs. *Mol. Cell. Biol.* **8** 4381–4388
[31] Saksela, K., Makela, T. P., Evan, G. & Alitalo, K. (1989) Rapid phosphorylation of the L-myc protein induced by phorbal ester tumor promoters and serum. *EMBO J.* **8** 149–157
[32] Stone, J., deLange, T., Ramsey, G., Jakobovits, E., Bishop, J. M., Varmus, H. & Lee, W. (1987) Definition of regions in human c-*myc* that are involved in transformation and nuclear localization. *Mol. Cell. Biol.* **7** 1697–1709
[33] Landshultz, W. H., Johnson, P. F. & McKnight, S. L. (1988) The leucine zipper: a hypothetical structure common to a new class of DNA binding proteins. *Science* **240** 1759–1762
[34] Ikegaki, H., Minna, J. & Kennett, R. (1989) The human L-*myc* gene is expressed as two forms of protein in small cell lung carcinoma cell lines: detection by monoclonal antibodies specific to two *myc* homology box sequences. *Embo J.* **8** 1793–1799
[35] Bentley, D. L. & Groudine, M. (1986) A block to elongation is largely responsible for decreased transcription of c-*myc* in differentiated HL-60 cells. *Nature (London)* **321** 702–706.
[36] Parkin, N., Darveau, A., Nicholson, R. & Sonenberg, N. (1988) *Mol. Cell Biol.* **8** 2875–2881
[37] Barksdale, S., Buchmann, R. & Kaye, F. J. (1989) Characterization of promoter activity and identification of multiple DNA: protein binding sites within the L-*myc* gene. In: P. K. Vogt & R. L. Eriks (eds), *Proc. Fifth Annual Oncogene Meeting, Frederick, MD*, p. 444
[38] Johnson, B. E., Ihde, D. C., Makuch, R. W., Gazdar, A. F., Carney, D. N., Oie, H., Russel, E., Nau, M. & Minna, J. (1987) *Myc* family oncogene amplification in tumor cell lines established from small cell lung cancer patients and its relationship to clinical status and course. *J. Clin. Invest.* **79** 1629–1634
[39] Land, H., Parada, L. F. & Weinberg, R. A. (1983) Cellular oncogenes and multistep carcinogenesis. *Science* **222** 771–773
[40] Forrester, K., Almoguera, C., Han, K., Grizzle, W. E. & Perucho, M. (1987) Detection of high incidence of K-*ras* oncogene during human colon tumorigenesis. *Nature (London)* **327** 398–303
[41] Rodenhuis, S., van de Wetering, M. L., Mooi, W. J., Evers, S., van Zandwijk, N. & Bos, J. (1987) Mutational activation of the K-*ras* oncogene: a possible

pathogenic factor in adenocarcinoma of the lung. *New Engl. J. Med.* **317** 929–935

[42] Mabry, M., Nakagawa, T., Gesell, M., *et al.* (1987) Introduction of Harvey murine sarcoma-virus (Ha-MSV) into human small cell lung cancer (SCLC) is associated with phenotypic changes. *Proc. Am. Assoc. Cancer Res.* **28** 39

[43] Griffin, C. A. & Baylin, S. B. (1985) Expression of the *c-myb* oncogene in human small cell lung carcinoma. *Cancer Res.* **45** 272–275

[44] Klein, G. (1987) The approaching era of the tumor suppressor genes. *Science* **238** 1539–1545

[45] Knudson, A. G. Jr. (1986) Genetics of human cancer. *Ann. Rev. Genet.* **20** 231–251

[46] Friend, S., Bernards, R., Rogelj, S., Weinberg, R., Rapapport, J. H., Albert, D. & Dryja, T. R. (1986) A human DNA segment with properties of the gene that predisposes to retinoblastoma and osteosarcoma. *Science* **323** 643–645

[47] Lee, W.-H., Bookstein, R., Hong, F., Young, L., Shew, J. & Lee, E. (1987) Human retinoblastoma susceptibility gene: cloning, identification, and sequence. *Science* **235** 1394–1396

[48] Fung, Y.-T. K., Murphree, A. L., T'Ang, A., Quian, J., Hinrichs, S. H. & Benedict, W. F. (1987) Structural evidence for the authenticity of the human retinoblastoma gene. *Science* **236** 1657–1659

[49] Whang-Peng, J., Bunn, P., Kao-Shan, C., Lee, E., Carney, D., Gazdar, A. & Minna, J. (1982) A non-random chromosomal abnormality, del 3p(14–23) in human small cell lung cancer (SCLC). *Cancer Genet. Cytogenet.* **6** 119–124

[50] Whang-Peng, J., Kao-Shan, C., Lee, E. C., Bunn, P. A., Carney, D., Gazdar, A. F. & Minna, J. D. (1982) Specific chromosome defect associated with human small cell lung cancer-deletion 3p(14–23). *Science* **215** 181–183

[51] Naylor, S., Johnson, B., Minna, J., & Sakaguchi, A. (1987) Loss of heterozygosity of chromosome 3p markers in small cell lung cancer. *Nature (London)* **329** 451–453

[52] Brauch, H., Johnson, B., Hovis, J., Yano, T., Gazdar, A., Pettingill, O., Graziano, S., Sorenson, G., Poiesz, B., Minna, J., Linehan, M. & Zbar, B. (1987) Molecular analysis of the short arm of chromosome 3 in small cell and non-small cell carcinoma of the lung. *New Engl. J. Med.* **317** 1109–1111

[53] Johnson, B., Sakaguchi, A., Gazdar, A., Minna, J., Burch, D., Marshall, A. & Naylor, S. (1988) Restriction fragment length polymorphism studies show consistent loss of chromosome 3p alleles in small cell lung cancer patients' tumours. *J. Clin. Invest.* **82** 502–509

[54] Kok, K., Osinga, J., Carritt, B., Davis, M., van der Hout, A., van der Veen, A., Landsvater, R., de Leij, L., Berendsen, H., Postmus, P., Poppema, S., & Buys, C. (1987) Deletion of a DNA segment at the chromosomal region 3p21 in all major types of lung cancer. *Nature (London)* **330** 578–580

[55] Smeets, D. F., Scheres, J. M. & Hustinx, T. W. (1986) The most common fragile site in man in 3p14. *Hum. Genet.* **72** 215–220

[56] Sparkes, R. S., Murphree, A. L., Lingua, R. W., Sparkes, M. C. & Field, L. L. (1983) Gene for hereditary retinoblastoma assigned to human chromosome 13 by linkage to esterase D. *Science* **219** 971–973

[57] Seizinger, B., Rouleau, G., Ozelius, L., Lane, A., Farmer, G., Lamiell, J.,

Haines, J., Yuen, J., Collins, D., Majoor-Krakauer, D., Bonner, T., Mathew, C., Rubenstein, A., Halperin, J., McConkie-Rosell, A., Green, J., Trofatter, J., Ponder, B., Eierman, L., Bowmer, M., Schimke, R., Oostra, B., Aronin, N., Smith, D., Drabkin, H., Waziri, M., Hobbs, W., Martuza, R., Connelly, P., Hsia, Y. & Gusella, J. (1988) Von Hippel-Lindau disease maps to the region of chromosome 3 associated with renal cell carcinoma. *Nature (London)* **332** 268–270

[58] Zbar, B., Brauch, H., Talmadge, C. & Linehan, M. (1987) Loss of alleles of loci on the short arm of chromosome 3 in renal cell carcinoma. *Nature (London)* **327** 721–724

[59] Yokota, J., Wada, M., Shimosato, Y., Terada, M. & Sugimura, T. (1987) Loss of heterozygosity on chromosomes 3, 13, 17 in small cell carcinoma and on chromosome 3 in adenocarcinoma of the lung. *Proc. Natl. Acad. Sci. USA* **84** 9252–9254

[60] Harbour, J. W., Lai, S., Whang-Peng, J., Gazdar, A., Minna, J. & Kaye, F. (1988) Abnormalities in structure and expression of the human retinoblastoma gene in SCLC. *Science* **241** 353–357

[61] Dunn, J., Phillips, R. A., Becker, A. J. & Gallei, B. L. (1988) Identification of germline and somatic mutations affecting the retinoblastoma gene. *Science* **241** 1797–1800

[62] Yokota, J., Akiyama, T., Fung, Y., Benedict, W., Namba, Y., Hanaoka, M., Wada, M., Terasaki, T., Shimosato, Y., Sugimura, T. & Terada, M. (1988) Altered expression of the retinoblastoma (*Rb*) gene in small cell carcinoma of the lung. *Oncogene* **3** 471–475

[63] Green, M. R. (1989) When the products of oncogenes anti-oncogenes meet. *Cell* **56** 1–3

[64] Huang, H., Yee, J.-K., Shew, J.-Y., Chen, P.-L., Bookstein, R., Friedman, T., Lee, E.-Y. & Lee, W.-H. (1988) Suppression of the neoplastic phenotype by replacement of the *Rb* gene in human cancer cells. *Science* **242** 1563–1565

12

Neural cell adhesion molecule (NCAM) and human small cell lung cancer

Rob J. A. M. Michalides*, Kitty C. E. C. Moolenaar*, Robby E. Kibbelaar†
and Wolter J. Mooi†
*Division of Tumour Biology and †Department of Pathology, The Netherlands
Cancer Institute, Plesmanlaan 121, 1066 CX Amsterdam, The Netherlands

1. INTRODUCTION

Human lung carcinomas constitute a spectrum of epithelial tumours in which squamous, glandular and neuroendocrine differentiation can be present, alone or in various combinations, and to varying degrees [1,2]. Mainly on the basis of these differentiation patterns as assessed light microscopically, pathological classification systems of lung tumours have been devised, the one most widely used now being the WHO classification of 1981 [3].

Such morphological tumour classifications are of clinical interest, since they allow some prediction of tumour behaviour. The distinction between small cell lung carcinoma (SCLC) and non-SCLC is most important, since SCLC are usually rapidly growing, early metastasizing tumours, which are highly chemo- and radiosensitive, at least initially [4]. For these reasons, SCLC are only rarely surgically resected, and require a different therapeutic approach.

SCLC can usually be distinguished satisfactorily from non-SCLC on the basis of light microscopic morphology [5,6]. In diagnostically difficult cases, additional electron microscopy has been found useful [7,8]. Some cases, however, remain difficult to classify, mainly because lung carcinomas do not constitute a collection of different tumour types; instead, the defined tumour types are part of a morphologically continuous spectrum so that no strict margins can be drawn between subgroups [2].

However, even more importantly, such morphological investigations leave unanswered which factors are responsible for the differences in biological behaviour of different lung tumours. Should these factors be elucidated, a classification of lung tumours based on the assessment of these factors would obviously have a better chance of predicting more precisely the clinical behaviour. To this purpose, we, and others, have raised monoclonal antibodies directed against SCLC membrane preparations. With one of these, designated MoAb 123C3, we performed tissue distribution and clinicopathological correlation studies in a series of surgically resected lung carcinomas. We found that these antibodies actually recognize epitopes present on the neural cell adhesion molecule (NCAM) and established thereby for the first time the presence of NCAM on human SCLC. We further investigated the immunoreactivity of various lung tumour types with monoclonal antibody 735, which cross-reacts with long polysialic acid side-chains present on some, but absent on other, NCAM subtypes. Negativity of some carcinoids with this antibody, but positivity of carcinoids with 123C3, indicates that differential expression of NCAM subtypes may occur in lung carcinomas and may cause differences in lung tumour behaviour, not only between SCLC and non-SCLC, but also between SCLC and carcinoids.

2. HISTOPATHOLOGICAL INVESTIGATIONS WITH 123C3

Monoclonal antibody 123C3 was raised against a membrane preparation of an SCLC specimen and its reactivity on normal tissues was tested [9]. For the endocrine system, positive tissues included pituitary and adrenal glands, thyrocytes and C cells of the thyroid, the parathyroids, Leydig cells of the testis and pancreatic islet cells. In bronchioles and intertestinal epithelium, occasional cells were positive. Wholly or largely non-neuroendocrine epithelial tissues such as the rete testis, mammary epithelium and gastric mucosa were positive in all or a significant proportion of cells, the numbers of positive cells being so large that it could be excluded that all positive cells represented neuroendocrine cells. Neurones were negative or weakly positive; their supportive cells, such as glial cells, Schwann cells and satellite cells, were, however, positive. Smooth muscle cells, cardiac muscle cells, ovarian stromal cells were weakly or strongly positive.

NCAM therefore appears to be present not only on neuroendocrine cells, as to be expected, but also on other cell types.

3. EXAMINATION OF 123C3 REACTIVITY ON LUNG TUMOURS

MoAb 123C3 was tested on a series of 358 resected primary epithelial tumours of the lung [10]; see Table 1. It recognized all 30 small cell carcinomas and carcinoids tested, but also stained 20% of various types of non-small cell lung carcinomas, and almost all bronchial gland tumours investigated (Fig. 1). Staining for 123C3 in squamous carcinomas or adenocarcinomas, when present, was usually focal and weak or moderately strong, whereas staining of SCLC and carcinoids was generally strong and present throughout the tumour. The positivity found in some non-small cell lung carcinomas was considered yet another illustration of the fact that the major types of lung tumours cannot be strictly separated, and that overlaps in differentation states and antigenic make-up exist between different histological types of lung tumours.

The ultrastructure of 32 resected 123C3-positive lung carcinomas was investigated: in only 9 were neuroendocrine granules found; in the others, no morphological sign of neuroendocrine differentiation could be detected.

Table 1 — 123C3 positivity in 358 resected lung tumours

	123C3 positive	123C3 negative
Squamous carcinoma	38	185
Adenocarcinoma	4	67
Adenosquamous carcinoma	4	18
Carcinoid	15	0
Small cell carcinoma	15	0
Small cell–squamous carcinoma	5	1
Pleiomorphic adenoma	1	0
Adenoid cystic carcinoma	3	1
Mucoepidermoid carcinoma	1	0

Fig. 1 — Histological section of small cell lung carcinoma, immunostained for MoAb 123C3. Large, compact fields of tumour cells are strongly positive, while the stroma is negative.

The prognostic value of 123C3 reactivity was evaluated in a series of 278 radically resected non-SCLC, where 123C3 reactivity was compared with postoperative survival. Disease-free survival time was significantly shorter for 123C3-positive tumours: 62% of 123C3-positive non-SCLC had a disease-free survival time of one year, 52% a disease free survival time of two years. 82% of the 123C3-negative non-

SCLC had a one year disease-free survival time, and 70% a two year disease-free survival time [11]. This renders 123C3 a prognostic marker for lung cancer.

4. CHARACTERIZATION OF THE ANTIGEN RECOGNIZED BY MoAb 123C3

We investigated the antigen recognized by 123C3 by immunocompetition, immunoprecipitation and *Staphylococcus aureus* V8 digestion experiments [12] and found, by comparison with a polyvalent rabbit anti-NCAM-serum obtained from Dr E. Bock [13], that 123C3 recognizes the human neural cell adhesion molecule NCAM. Fig. 2 shows a representative example of an experiment demonstrating the similarity in the antigens detected by a polyclonal anti NCAM serum and MoAb 123C3.

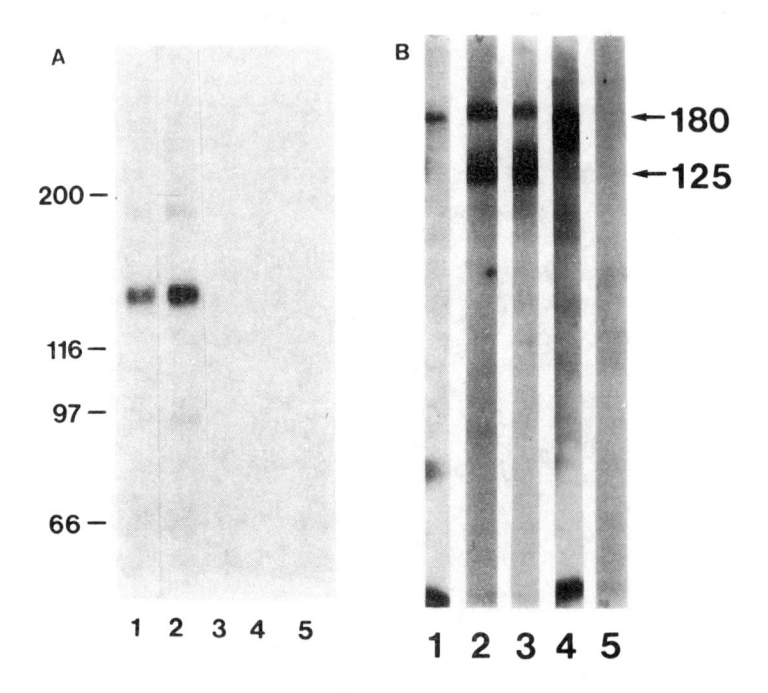

Fig. 2 — Identification of the antigen recognized by 123C3 antiserum. (A) In neuroblastoma CHP212 cells: immunoprecipitation of 123C3 (lane 1), polyclonal rabbit antiserum directed against NCAM (lane 2), normal rabbit serum (lane 3), control mouse IgG (lanes 4 and 5). (B) In SCLC H69 cells: Western blots of H69 (lanes 1, 2 and 3) and T47D, a control human breast tumour cell line (lanes 4 and 5), were immunostained with MoAb 123C3 (lanes 2 and 5), another anti NCAM serum, NKI-nbl-2 (lane 3) and 66IG 10 (lanes 1 and 4), directed against the human transferrin receptor.

5. NEURAL CELL ADHESION MOLECULE

NCAM belongs to a family of cell surface sialoglycoproteins involved in binding of the same types of cells or different cell types to one another [14–17]. NCAM is involved in a number of developmental processes including segregation of cells into discrete regions during development [15, 18], axon guidance, and cellular adhesion [14, 19]. NCAMs are expressed transiently in various embryonic structures. In adult tissues NCAMs are present on neurones, glia and skeletal muscle cells. NCAMs are present on a few tumours such as neuroblastoma, Wilms' tumour, and Ewing sarcoma and, as we have found, on small cell lung cancer.

The NCAM gene is expressed in various isoforms in a developmentally regulated and tissue-specific manner [15, 17]. These different isoforms may have different binding functions. It is therefore quite possible that the NCAM isoforms that we and others have found in particular tumours, and especially in human small cell lung cancer, are in part responsible for the differences in invasive and metastatic potential of these tumours.

Much is known about the structure of the NCAM gene and the different RNA splicing variants that are being generated. Less is known of the biological function of the individual NCAM isoforms, and how these may affect the malignant behaviour of tumour cells. We will discuss these items separately below.

5.1 Structure of the NCAM gene and its products

The major isoforms of NCAM are generated from a single gene by alternative RNA splicing and polyadenylation [16, 20–24]. The NCAM genes in chicken, rodents and man are highly homologous and contain at least 19 exons [22]. Three NCAM mRNAs of 7.4, 6.7 and 4.2 kb are found in chicken [25]; in mouse and man five NCAM mRNAs of 7.4 or 7.2, 6.7, 5.2, 4.3 and 2.9 kb [24, 26] are present. The tissue-specific pattern (brain versus muscle) of these mRNAs and the different RNAs and corresponding NCAM isoforms are represented in Table 2 and Fig. 3, which are a

Table 2 — Expression of NCAM mRNAs and proteins

Sizes of NCAM mRNAs (kb)				Mouse NCAM proteins (kDa); desialylated forms under reducing conditions	
Human		Mouse			
Embryonic brain	Skeletal muscle	Adult brain	Skeletal muscle	Adult brain	Skeletal muscle
7.2	7.2	7.4		180[a]	155[a]
6.7	6.7	6.7	6.7	140[a]	145[a]
5.2[c]	5.2[c,d]	5.2[c]	5.2[c,d]	120[b]	125[b]
	4.3[d]				
	2.9	2.9	2.9[d]		

[a] Transmembrane NCAM.
[b] Non-transmembrane domain NCAM.
[c] Contains SEC domain.
[d] Contains MSD-1 domain.
After Nybroe et al. [17], Gower et al. [24], Barbas et al. [23], and Lipinsky et al. [57].

228   Neural cell adhesion molecule (NCAM)   [Ch. 12

alignment of N-CAM isoforms - boxed areas indicate ORFs

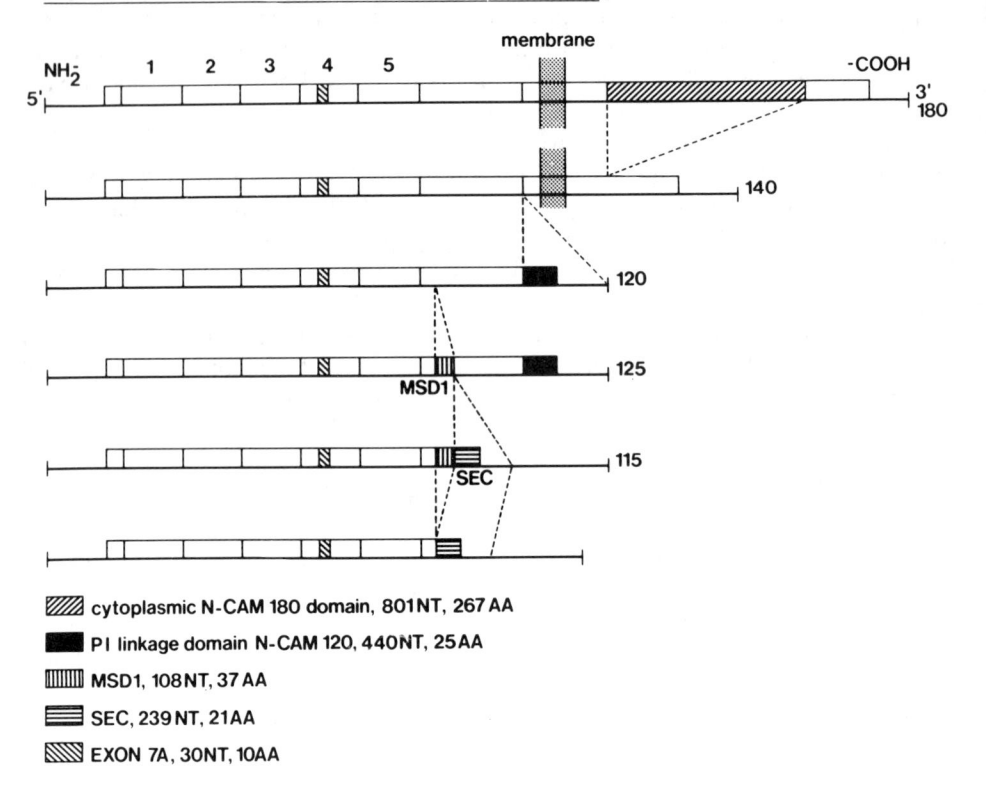

▨ cytoplasmic N-CAM 180 domain, 801NT, 267AA
■ PI linkage domain N-CAM 120, 440NT, 25AA
▦ MSD1, 108NT, 37AA
☰ SEC, 239NT, 21AA
▩ EXON 7A, 30NT, 10AA

Fig. 3 — Alignment of NCAM isoforms. The lines indicate the different mRNAs and the boxed areas represent the open reading frames. Numbers 1–5 refer to the immunoglobulin-like domains in NCAM. Numbers at the 3' end of the RNA refer to the molecular weight in kDa of the corresponding NCAM proteins. Shaded boxes refer to RNA splice variants.

compilation of various studies in different species. All NCAM mRNAs share a common 5' end coding for five immunoglobulin-like domains [16]. The largest NCAM 180 kDa isoform is encoded by the 7.4 or 7.2 kb mRNA [20, 27].

Differential splicing of exon 18 shortens the mRNA from 7.2 to 6.7 kb, which yields an NCAM 140 kDa isoform (NCAM-140) [25, 28, 29]. Splicing of exon 14 to exon 15 generates a 5.2 kb mRNA. This exon 15 codes for 25 amino acids to which a phosphatidyl anchor is attached [22, 25, 30], yielding the non-transmembrane isoform of NCAM-120.

The 5.2 and 4.3 kb muscle mRNAs code for the same isoform; the difference therefore probably arises in the 5' or 3' untranslated regions. The 2.9 kb mRNA codes for NCAM-120 [17] and differs from the 5.2 kb RNA in the 3' untranslated region. A first polyadenylation signal is used to generate the 2.9 kb mRNA; the 5.2 kb message uses a second poly(A) addition signal 2.3 kb further downstream. Both messengers code for NCAM-120 [31].

As a result of further alternative splicing, the 5.2, 4.3 and 2.9 kb mRNAs in muscle cells contain a muscle-specific domain of 108 nucleotides, which is not found in brain cells [21,32]. This MSD-1 domain is inserted at a splice junction between exons 12 and 13 [22] and appears to be encoded by three novel exons. This MSD-1 region contains 36 amino acids. This splice junction site between exons 12 and 13 appears highly unusual; it harbours multiple exon combinations and provides therefore multiple NCAM isoforms [33,34].

Gower et al. [24] described a novel, secreted isoform of NCAM with a specific exon, SEC. The inserted sequence is present as a discrete exon and is specifically associated with a 5.2 mRNA species from brain and skeletal muscle. It incorporates a novel sequence of 21 amino acids and a premature termination signal, leading to a secreted isoform of NCAM-115. This may well correspond to NCAM isoforms identified in conditioned media from neural cells and spinal fluid [13,35,36]. A 5.2 kb mRNA is found in brain cells that does hybridize with this SEC fragment, but not with MSD-1, and may therefore yield the sixth isoform of NCAM as depicted in Fig. 3.

An alternative splicing of exon 7A confers a new level of polypeptide diversity on the NCAM glycoproteins. It results in the insertion of 10 amino acids into the fourth of five extracellular immunoglobulin-like domains [33,37].

These findings indicate that many different NCAM isoforms exist based on alternative RNA splicings. These splicing events include tiny exons of as few as three nucleotides, which are overlooked in regular Northern blot analyses, unless specific probes are being used. They also suggest that many more RNA splice variants of NCAM exist, since the analysis of these hypervariable NCAM gene products has just been started.

5.2 NCAM proteins

The NCAM isoforms are, moreover, subject to several types of post-translational modifications, including glycosylation, sulphatation, phosphorylation and sialylation, adding to the diversity of the NCAM gene product [17,27]. The embryonic form of NCAM contains a high amount of an α-(2,8)-polysialic acid, characteristic of the embryonic form of NCAM [38]. The different forms of NCAM are involved in adhesive binding of cells, and are expressed in a developmental stage and a cell type specific manner. It is therefore obvious that this scheduled appearance of different NCAM isoforms plays a critical role in tissue development. Aberrant expression of NCAM proteins in tumours may therefore be of significance for the tumorigenic behaviour of these cells.

6. BIOLOGY OF NCAM ISOFORMS

Studies of the expression of NCAM during normal embryogenesis, in brain and muscle development, and following disruption of muscle and neurone interactions, indicate a precise regulatory mechanism controlling quantity, isoform ratios and post-translational processes [14,15,39]. They all indicate that NCAM normally has a function in neurone–neurone, neurone–astrocyte and astrocyte–astrocyte adhesion, and is involved in nerve–muscle cell interaction [18,40].

The binding of cells via NCAM can be influenced by the extent of sialylation (embryonic versus adult form of NCAM), by the different isoforms of NCAM expressed on these cells, or by a combination of both.

They will be dealt with separately.

6.1 Embryonic versus adult forms of NCAM

NCAM-mediated cell–cell adhesion occurs by a homophilic binding mechanism in a Ca^{2+}-independent manner [41]. The first three IgG-like domains appear to be involved in NCAM binding. The NCAM binding is heavily influenced by the extent of sialylation of the NCAM protein. The heavily sialylated embryonic form of NCAM contains approximately 3 times more sialic acid than the adult NCAM. This causes a four-fold decrease in the rate of binding [42,43]. Also, different amounts of NCAM at the cell surface affect binding of cells: a two-fold increase in the amount of NCAM associated with reconstituted vesicles results in a 30-fold increase in binding rates [43].

The conversion of E (embryonal) to A (adult) NCAM is expected to change the binding efficiencies of various cells during histogenesis, as observed during development in brain and kidney. A failure of the normal E–A conversion has been observed in genetic mutants of so-called staggerer mice [38], and results in animals with a small and disordered cerebellum which die within one month after birth.

The E form of NCAM is also found, albeit in lower amounts, in adult brain and muscle [44,45], and in the olefactory system of adult mice. The latter may be associated with the continuous generation and growth of olfactory neurones in this system [46].

It appears, in general, that conversion of E to A NCAM isoforms strongly affects the binding between cells. This suggests that cells with E-NCAMs more 'free' to move, which is a prerequisite for invasiveness.

NCAM also influences via its binding a number of diverse intracellular events, such as junctional communication and the function of other cell surface ligands [47]. This is caused by

(1) changes in the number of homophilic bonds between NCAMs on opposing membranes, and
(2) modulation of adhesion as a result of the content of polysialic acid in NCAM.

E-NCAM with high contents of α-(2,8)-polysialic acid impedes cell–cell contacts and could thereby interfere with ligand–receptor interactions. This is relevant for tumour cells carrying E-NCAM: the tumour cells could thereby be 'loosely' attached to one another, allowing invasive growth, whereas the presence of E-NCAM would make them less sticky by reducing the binding affinity of these tumour cells to other substrates, and would render them less responsive to, for example, differentiation-inducing ligands. This, of course, is still hypothetical.

6.2 Different NCAM domains affect binding capacity

The NCAM proteins contain different domains; see Fig. 3. Recently some insight has been gained as to how these domains might affect the binding properties of NCAM: transfection of cells with the chicken NCAM-180 and 140 isoforms resulted in phenotypic alterations. The cells aggregated specifically with each other or with membrane vesicles from chicken brains, exhibited a rounded morphology and spread inefficiently in culture, whereas cells transfected with the NCAM-120 isoform did not show any such alterations [48].

Studies by Pollerberg *et al.* [49], moreover, showed that NCAM-180 has a reduced surface mobility compared with NCAM-140, which was attributed to the binding of NCAM-180 to brain spectrin, a membrane–cytoskeleton linker. NCAM-180 is concentrated at contact areas between adjacent cells, whereas NCAM-140 is dispersed over the cell membrane. These findings suggest that the binding of the different NCAM isoforms, although they all have the same N-terminal binding site, is affected by the cytoplasmic domain of NCAM. This would imply that cells with an NCAM-180 isoform are more tightly bound than cells with NCAM-140, indicating that the ratios of NCAM-180 and 140 would affect the binding capacity. The other NCAM isoforms have not yet been introduced into other cells; but their structures would have the following biological implications.

The secreted form of NCAM, NCAM-115 (see Fig. 3), could interfere with cellular adhesion and perhaps cell–substrate interactions. The latter is possible because NCAM contains a heparin-binding domain involved in neural cell–substrate adhesion [50,51].

The specific exons MSD-1, SEC and 7A (see Fig. 3) may have profound effects on the biological function of the corresponding isoforms: MSD-1 provides an O-linked glycosylation attachment site, which might alter the biological properties, whereas the SEC exon results into a secreted form of NCAM.

The 7A exon appears as a result of postnatal induction of alternative splicing, which closely parallels the conversion of the embryonic to the adult form of NCAM [37]. It is tempting to speculate that this splicing event may also contribute to the change in binding affinity between embryonic and adult NCAM.

The studies mentioned above suggest the assignment of biological functions to the different NCAM isoforms, but most of these assignments remain hypothetical. This can be studied in more detail, since it has now become feasible to generate full-length cDNAs of the different NCAM mRNAs and to study the biological functions of the individual NCAM isoforms.

7. NCAM AND MALIGNANT BEHAVIOUR OF TUMOUR CELLS

Adhesive properties of tumour cells are considered important for metastasis formation. Reduced adhesion between cancer cells is probably one of the factors responsible for the onset of the metastatic process, whereas adhesion between tumour cells and components of the target tissue is essential for invasion into that tissue [52–54]. NCAM is involved in cellular adhesion, and, when present on tumour cells, affects the adhesive properties and thereby invasive properties of tumour cells.

The presence of NCAM on human tumours has been described in Wilms' tumours [55] and in Ewing sarcomas [56,57]. The investigation of 25 Wilms' tumours comprising all different histological types demonstrated homogeneous staining with NCAM antibodies, and with an antibody that recognizes the α-(2,8)-linked polysialic acid on NCAM [58,59]. This embryonic form of NCAM, E-NCAM, is also found in developing kidney, but it is absent in adult kidney. Lipinsky *et al.* [56] described the presence of NCAM in eight out of ten Ewing sarcoma cell lines. The NCAM isoforms found in the Ewing sarcomas were of similar sizes to those in human brain [57] and did not carry polysialic acid chains.

We found that all but one ($N=38$) small cell lung cancers contain NCAM proteins which carry long α-(2,8)-polysialic acid side-chains, recognized by MoAb 735 [59]. This embryonic form of NCAM is present on all but one of the SCLCs examined, whereas only 9 out of 18 typical carcinoids were positive [60] (Table 3).

Table 3 — NCAM expression in neuroendocrine tumours (positive/negative)

	123C3	735
Small cell carcinoma	19/19	37/38
Atypical carcinoid	5/5	4/6
Typical carcinoid	16/16	9/18

Until now, all typical carcinoids tested by us ($N=18$) have been positive for MoAb 123C3. Since SCLC do metastasize much more frequently than carcinoids, and since the higher content of polysialic acid in E-NCAM reduces the binding capacity of NCAM, it is suggestive that differences in metastatic potential of pulmonary neuroendocrine neoplasms coincide with the observed differences in sialylation of NCAM.

NCAM either appears to be (re-)expressed in certain tumours, which are derived from organs that carry NCAM in an earlier developmental stage, or appears as a novel antigen on that cell lineage. In three out of the four tumour systems studied (SCLC, neuroblastoma and Wilms' tumour) in which NCAM was found, NCAM is present in an embryonic, polysialic-acid-rich form. (Re-)expression of NCAM in these tumour cells apparently coincides with the activity of a particular sialyltransferase, specific for α-(2,8)-polysialylation.

This also suggests that polysialylation of NCAM reduced intercellular adhesion of the tumour cells which is the first step in the metastatic process. The neural cell adhesion molecule may therefore provide a direct tool for studying cell adhesion and malignant behaviour of tumour cells.

ACKNOWLEDGEMENTS

Our studies were partly supported by a grant from Centocor Europe B.V., Leiden, The Netherlands.

We thank Yvonne Nijdam for the assistance in the preparation of the manuscript.

REFERENCES

[1] McDowell, E. M., Mclaughlin, J. S., Merenyl, D. K., Kieffer, R. F., Harris, C. C. & Trump, B. F. (1978) The respiratory epithelium. V. Histogenesis of lung carcinomas in the human. *J. Natl. Cancer Inst.* **61** 587–606

[2] Yesner, R. (1981) The dynamic histopathologic spectrum of lung cancer. *Yale J. Biol. Med.* **54** 447–456
[3] World Health Organization (1981) *The World Health Organization Histological Typing of Lung Tumours*, World Health Organization, Geneva
[4] Ihde, D. C. (1984) Current status of therapy for small cell carcinoma of the lung. *Cancer* **54** 2722–2728
[5] Carter, D. (1983) Small-cell carcinoma of the lung. *Am. J. Surg. Pathol.* **7** 787–795
[6] Yesner, R. (1983) Small-cell tumours of the lung. *Am. J. Surg. Pathol.* **7** 775–786
[7] Elema, J. D. & Keuning, M. T. (1985) The ultrastructure of small cell lung carcinoma in bronchial biopsy specimen. *Hum. Pathol.* **16** 1133–1140
[8] Mooi, W. J., Van Zandwijk, N., Dingemans, K. P., Koolen, M. G. J. & Wagenvoort, C. A. (1986) The "grey area" betweeen small cell and non-small cell lung carcinomas: light and electron microscopy versus clinical data in 14 cases. *J. Pathol.* **149** 49–54
[9] Schol, D. J., Mooi, W. J., Van der Gugten, A. A., Wagenaar, S. J. S. C. & Hilgers, J. (1988) Monoclonal antibody 123C3, identifying small cell carcinoma phenotype in lung tumours, recognizes mainly, but not exclusively, endocrine and neuron-supporting normal tissues. *Int. J. Cancer* **2** (Suppl.) 34–40
[10] Mooi, W. J., Wagenaar, S. J. S. C., Schol, D. & Hilgers, J. (1988) Monoclonal antibody 123C3 in lung tumour classification: immunohistology of 358 resected lung tumours. *Mol. Cell Probes* **2** 31–37
[11] Mooi, W. J., Dingemans, K. P., Van Bodegom, P. C., Schol, D. J., Van Zandwijk, N. & Wagenaar, S. S. C. (1989) N-CAM associated with poor prognosis in lung cancer. *Cancer Res.* (submitted)
[12] Moolenaar, C. E. C., Muller, E. J., Schol, D. J., Figdor, C. G., Bock, E., Bitter Suermann, D. & Michalides, R. J. A. M. (1990) Expression of an NCAM related sialoglycoprotein on small cell lung cancer and neuroblastoma cell lines H6g and CHP212. *Cancer Res.* (accepted for publication)
[13] Ibsen, S., Berezin, V., Norgard-Pedersen, B. & Bock, E. (1983) Quantification of the D2 glycoprotein in amniotic fluid and serum from pregnancies with foetal neural tube defects. *J. Neurochem.* **41** 363–366
[14] Edelman, G. M. (1985) Cell adhesion and the molecular process of morphogenesis. *Annu. Rev. Biochem.* **54** 135–136.
[15] Edelman, G. M. (1986) Cell adhesion molecules in the regulation of animal form and pattern formation. *Annu. Rev. Cell Biol.* **2** 81–116
[16] Cunningham, B. A., Hemperley, J. J., Murray, B. A., Prediger, E. A., Brackenbury, R. & Edelman, G. M. (1987) Neural cell adhesion molecule; structure, immunoglobulin-like domains, cell surface modulation and alternative RNA splicing. *Science* **236** 799–806
[17] Nybroe, O., Linnemann, D. & Bock, E. (1988) N-CAM biosynthesis in brain. *Neurochem. Int.* **12** 251–262
[18] Rutishauser, U., Gall, W. E. & Edelman, G. M. (1978) Adhesions among neural cells of the chicken embryo. IV. Role of the cell surface molecule N-CAM in the formation of neurite bundles in cultures of spinal ganglia. *J. Cell Biol.* **79** 382–393

[19] Keane, R. W., Mekta, P. P., Rose, B., Honig, H. S., Loewenstein, W. R. & Rutishauser, U. (1988) Neural differentiation, N-CAM mediated adhesion, and gap functional communication in neuroectoderm. A study *in vitro*. *J. Cell Biol*. **106** 1307–1319

[20] Barthels, D., Santoni, M. J., Wille, W., Ruppert, C., Chaix, J. C., Hirsch, M. R., Fontecilla-Camps, J. C. & Goridis, C. (1987) Isolation and nucleotide sequence of mouse N-CAM cDNA that codes for a Mw 79,000 polypeptide without a membrane-spanning region. *EMBO J*. **6** 907–914.

[21] Dickson, G., Gower, H. J., Barton, H., Prentice, H. M., Moore, S. E., Cox, R. D., Quinn, C. A., Putt, W. & Walsh, F. S. (1987) Human muscle neural cell adhesion molecule (N-CAM); identification of a muscle specific sequence in the extracellular domain. *Cell* **50** 1119–1130

[22] Owens, G. C., Edelman, G. M. & Cunningham, B. A. (1987) Organization of the neural cell adhesion molecule (N-CAM) gene: alternative exon usage as the basis for different membrane associated domains. *Proc. Natl. Acad. Sci. USA* **84** 294–298

[23] Barbas, J. A., Chaix, J. C., Steinmetz, M. & Goridis, C. (1988) Differential splicing and alternative polyadenylation generates distinct N-CAM transcripts and proteins in the mouse. *EMBO J*. **7** 625–632

[24] Gower, H., Barton, C. H., Elsom, V. L., Thompson, J., Moore, S. E., Dickson, G. & Walsh, F. S. (1988) Alternative splicing generates a secreted form of N-CAM in muscle and brain. *Cell* **55** 955–968

[25] Hemperley, J. J., Murray, B. A., Edelman, G. M. & Cunningham, B. A. (1986) Sequence of a cDNA clone encoding the polysialic acid-rich and cytoplasmic domains of the neural cell adhesion molecule N-CAM. *Proc. Natl. Acad. Sci. USA* **83** 3037–3041

[26] Goridis, C., Hirn, M., Santoni, M. J., Gennarini, G., Deagostini-Bazin, H., Jordan, B. R., Kiefer, M. & Steinmetz, M. (1985) Isolation of mouse N-CAM related cDNA. Detection and cloning using monoclonal antibodies. *EMBO J*. **4** 631–635

[27] Hemperley, J. J., Edelman, G. M. & Cunningham, B. A. (1986) cDNA clones of the neural cell adhesion molecule (N-CAM) lacking a membrane-spanning region consistent with evidence for membrane attachment via a phosphatidylinositol intermediate. *Proc. Natl. Acad. Sci. USA* **83** 9822–9826

[28] Murray, B. A., Owens, G. C., Prediger, E. A., Crossin, K. L., Cunningham, B. A. & Edelman, G. M. (1986) Cell surface modulation of the neural cell adhesion molecule resulting from alternative mRNA splicing in a tissue-specific developmental sequence. *J. Cell Biol*. **103** 1431–1439

[29] Santoni, M. J., Barthels, D., Barbas, J. A., Hirsch, M. R., Steinmetz, M., Goridis, C. & Wille, W. (1987) Analysis of cDNA clones that code for the transmembrane forms of the mouse neural cell adhesion molecule (NCAM) and are generated by alternative RNA splicing. *Nucleic Acids Res*. **15** 8621–8641.

[30] Barton, C. H., Dickson, G., Gower, H. J., Rowett, L. H., Putt, W., Elsom, V., Moore, S. E., Goridis, C. & Walsh, F. S. (1988) Complete sequence and in vitro expression of a tissue-specific phosphatidylinositol-linked N-CAM isoform from skeletal muscle. *Development* **104** 165–173

[31] Goridis, C. & Wille, W. (1988) The three size classes of mouse N-CAM proteins

arise from a single gene by a combination of alternative splicing and use of different polyadenylation sites. *Neurochem. Int.* **12**(3) 269–272
[32] Walsh, F. S. (1988) The N-CAM gene is a complex transcriptional unit. *Neurochem. Int.* **12** 263–267
[33] Santoni, M. J., Barthels, D., Vopper, G., Boned, A., Goridis, C. & Wille, W. (1989) Differential exon usage involving an unusual splicing mechanism generates at least eight types of NCAM cDNA in mouse brain. *EMBO J.* **8** 385–392
[34] Prediger, E. A., Hoffman, S., Edelman, G. M. & Cunningham, B., (1988) Four exons encode a 93 base pair insert in three neural cell adhesion molecule mRNAs specific for chicken heart and skeletal muscle. *Proc. Natl. Acad. Sci. USA* **85** 9616–9620
[35] Bock, E. (1987) Determination of the brain proteins N-CAM and GFAP in cerebrospinal fluid. In: E. J. Thompson (ed.) *Advances in CSF Protein Research and Diagnosis*, MTP Press, Lancaster, pp. 89–98
[36] Bock, E., Edvardsen, J. C., Gibson, A., Linneman, D., Lyles, J. M. & Nybroe, O. (1987) Characterisation of soluble forms of N-CAM. *FEBS Lett.* **225** 33–36
[37] Small, S. J., Haines, S. L. & Akeson, R. A. (1988) Polypeptide variation in an N-CAM extracellular immunoglobulin-like fold is developmentally regulated through alternative splicing. *Neuron* **1** 1007–1019
[38] Edelman, G. M. & Chuong, C. M. (1982) Embryonic to adult conversion of neural cell adhesion molecules in normal and staggerer mice. *Proc. Natl. Acad. Sci. USA* **79** 7036–7040
[39] Moore, S. E., Thompson, J., Kirkeness, V., Dickson, J. G. & Walsh, F. S. (1987) Skeletal muscle neural cell adhesion molecule (N-CAM): changes in protein and mRNA species during myogenesis of muscle cell lines. *J. Cell Biol.* **105** 1377–1386
[40] Matsunaga, M., Hatta, K., Nagafuchi, A. & Takeichi, M. (1988) Guidance of optic nerve fibers by N-cadherin adhesion molecules. *Nature (London)* **334** 62–64
[41] Cunningham, B. A., Hoffman, S., Rutishauser, U., Hemperley, J. J. & Edelman, G. M. (1983) Molecular topography of the neural cell adhesion molecule N-CAM; surface orientation and location of sialic acid-rich binding regions. *Proc. Natl. Acad. Sci. USA* **80** 3116–3120
[42] Sadoul, R., Hirn, M., Deagostini-Bazin, H., Rougon, G. & Goridis, C. (1983) Adult and embryonic mouse neural cell adhesion molecules have different binding properties. *Nature (London)* **304** 347–349
[43] Hoffman, S. & Edelman, G. M. (1983) Kinetics of homophilic binding of embryonic and adult forms of the neural cell adhesion molecule. *Proc. Natl. Acad. Sci. USA* **80** 5762–5766
[44] Finne, J., Finne, U., Deagostini-Bazin, H. & Goridis, C. (1983) Occurrence of alpha 2–8 linked poly sialosyl units in a neural cell adhesion molecule. *Biochem. Biophys. Res. Commun.* **112** 482–487
[45] Rothbard, J. B., Brackenbury, R., Cunningham, B. A. & Edelman, G. M. (1982) Differences in the carbohydrate structures of neural cell-adhesion molecules from adult and embryonic chicken brains. *J. Biol. Chem.* **257** 11064–11069
[46] Miragall, F., Kadmon, G., Hussmann, M. & Schachner, M. (1988) Expression

of cell adhesion molecules in the olfactory system of the adult mouse: presence of the embryonic form of N-CAM. *Dev. Biol.* **129** 516–531

[47] Ruthishauser, U., Acheson, A., Hall, A. K., Mann, D. M. & Sunshine, J. (1988) The neural cell adhesion molecule (NCAM) as a regulator of cell–cell interactions. *Science* **240** 53–57

[48] Edelman, G. M., Murray, B. A., Mege, R. M., Cunningham, B. A. & Gallin, W. J. (1987) Cellular expression of liver and neural cell adhesion molecules after transfection with their cDNAs results in specific cell–cell binding. *Proc. Natl. Acad. Sci. USA* **84** 8502–8506.

[49] Pollerberg, E. G., Schachner, M. & Davoust, J. (1986) Differentiation state-dependent surface mobilities of two forms of the neural adhesion molecule. *Nature (London)* **324** 462–465

[50] Cole, G. J. & Glaser, L. (1986) A heparin-binding domain from N-CAM is involved in neural cell–substratum adhesion. *J. Cell Biol.* **102** 403–412

[51] Werz, W. & Schachner, M. (1988) Adhesion of neural cells to extracellular matrix constituents. Involvement of glycosaminoglycans and cell adhesion molecules. *Dev. Brain Res.* **43** 225–234

[52] Roos, E. (1984) Cellular adhesion, invasion and metastasis. *Biochem. Biophys. Acta* **738** 263–284

[53] Liotta, L. A., Rao, C. N. & Barsky, S. H. (1983) Tumour invasion and the extracellular matrix. *Lab. Invest.* **49**(6) 636–649

[54] Nicholson, G. (1987) Tumour cell instability, diversification and progression to the metastatic phenotype: from oncogene to oncofetal expression. *Cancer Res.* **47** 1473–1487

[55] Roth, J., Zuber, Chr., Wagner, Ph., Taatjes, D. J., Weisgerber, C., Heitz, P. U., Goridis, C. & Bitter-Suermann, D. (1988) Reexpression of polysialic acid units of the neural cell adhesion molecule in Wilm's tumours. *Proc. Natl. Acad. Sci. USA* **85** 2999–3003

[56] Lipinsky, M., Braham, K., Philip, J., Wills, J., Philip, Th., Goridis, C., Lenoir, G. & Tursz, T. (1987) Neuroectoderm-associated antigens on Ewing sarcoma cell lines. *Cancer Res.* **47** 183–187

[57] Lipinsky, M., Hirsch, M. R., Deagostini-Bazin, H., Yamada, O., Tursz, Th. & Goridis, C. (1987) Characterization of neural cell adhesion molecules (NCAM) expressed by Ewing and neuroblastoma cell lines. *Int. J. Cancer* **40** 81–86

[58] Roth, J., Taatjes, D. J., Bitter-Suermann, D. & Finne, J. (1987) Polysialic acid units are spatially and temporally expressed in developing postnatal rat kidney. *Proc. Natl. Acad. Sci. USA* **84** 1969–1973

[59] Bitter-Suermann, D. & Roth, J. (1987) Monoclonal antibodies to polysialic acid reveal epitope sharing between invasive pathogenic bacteria, differentiating cells and tumour cells. *Immunol. Res.* **6** 225–237

[60] Kibbelaar, R. E., Moolenaar, C. E. C., Michalides, R. J. A. M., Bitter–Suermann, D., Addis, B. & Mooi, W. J. (1989) Expression of the embryonal neural cell adhesion molecule NCAM in lung carcinoma. Diagnostic usefulness of monoclonal antibody 735 for the distinction between small cell lung cancer and non-small cell lung cancer. *J. Pathol.* **159** 23–28

13

Regulation of cytokine gene expression

Martin Turner and **Marc Feldmann**
Charing Cross Sunley Research Centre, Hammersmith, London W6 8LW, UK

LIST OF ABBREVIATIONS

AIDS, acquired immunodeficiency syndrome; ATL, adult T cell leukaemia; CHX, cycloheximide; CSF, colony-stimulating factor; HIV, human immunodeficiency virus; HTLV, human T lymphotrophic virus; IFN, interferon; IL, interleukin; IRF, interferon response factor; LPS, lipopolysaccharide; LT, lymphotoxin; LTR, long terminal repeat; MHC, major histocompatibility complex; OA, osteoarthritis; PDGF, platelet-derived growth factor; PMA, phorbol-12 myristate-13 acetate; RA, rheumatoid arthritis; TGF, transforming growth factor; TNF, tumour necrosis factor.

1. OVERVIEW

The initiation and amplification of an immune response is regulated in part through the release of soluble proteins (generically referred to as cytokines). These are produced by lymphocytes (lymphokines), or monocytes (monokines) and other non-bone marrow derived cells. During the past seven years the genes for many cytokines have been cloned, often by biotechnology companies hoping to exploit the pharmacological potential of this group of molecules. The availability of large quantities of pure cytokines has led to the realization that these molecules play critical roles in many, often unexpected, biological events. Thus cytokines have been shown to play important roles in development, differentiation and proliferation, not only of immune cells, but also of many (if not all) other cell types.

Cytokines interact in complex ways to bring about their biological effects. This complexity is necessary to impart flexibility on regulation; thus the immune response can be tailored to suit a particular antigenic challenge. Pleiotropism appears to be an important aspect of cytokine function; thus a cytokine identified in one biological system often turns out to be intimately involved in another (apparently) unrelated system. A consequence of pleiotropism appears to be a considerable functional redundancy, but at present the reasons for this are not fully appreciated. Examples of synergism as well as negative cooperativity between cytokines are often encountered, and defining these relationships remains a major goal for workers in this field. A summary of the interrelationships and some of the principal biological effects of the cytokines discussed is provided in Table 1. However, it is not the purpose of this chapter to describe the biological properties of individual cytokines or to catalogue the cells capable of their production; instead, several recent reviews are recommended [1–4].

The cloning of cytokine genes has initiated studies into the molecular mechanisms that regulate their production. Progress in this area has been relatively rapid, partly as a result of the recognition that cytokines are controlled in a manner analogous to other well-studied genes, such as oncogenes. Accordingly, this review contains many references to the control of cellular oncogenes and other cell type specific genes.

2. INTRODUCTION

An intriguing aspect of cytokine gene regulation is tissue specificity; it was originally believed that the production of a particular factor was a cell lineage specific phenomenon. For instance, TNF was believed to be a product of macrophages, but is now known to be produced by a large variety of cell types, including non-haemopoietic cells, although there are quantitative differences. LT, on the other hand, whose coding sequence is only 1.2 kb from TNF in the human genome, appears to be produced exclusively by T and B lymphocytes. The problem of how to impart tissue specificity is highlighted by the case of T cells, since although all T cells are bone marrow derived and develop in the thymus there are many distinct lineages and there are many reports of heterogeneity of cytokine production by T cells, particularly in the mouse [5]. Thus much recent work has demonstrated that cytokines can be expressed by a broad range of tissues. There is increasing evidence that certain cells such as T cells and macrophages can selectively produce cytokines in response to given stimuli and that cytokine production and release are highly regulated processes. However, it is still largely unknown how tissue specificity is conferred. Does each cell type produce cytokines by a different mechanism or are there a few common mechanisms enabling a coordinated response to the initial stimulation?

A further area of importance, of which, unfortunately, very little is known, concerns the mechanism(s) of how cytokine genes are switched off. An understanding of this could have important consequences for chronic inflammatory diseases which are characterized by persistent cytokine production. Disregulation of cytokine gene expression also appears to be the hallmark of some viral diseases and tumours, suggesting that ways of blocking production of or modifying responsiveness to cytokines may be of widespread benefit in treating these diseases.

Cytokine gene expression may in principle be controlled at a number of critical points (Fig. 1). Transcription of the genes appears to be tightly regulated and controlled by a plethora of DNA-binding proteins whose function is either to enhance or to inhibit RNA polymerase II activity. Splicing of immature RNA and poly(A)tailing may also (in principle) be rate limiting for cytokine production;

Table 1 — Principle biological functions mediated by cytokines

Cytokine and abbreviation	Functions
Interleukin 1 (α and β)	Proinflammatory, activates endothelial cells and leukocytes, particularly T cells, induces PGE2 collagenase and synthesis of other cytokines, in particular IL-6 and CSFs, induces fever sleep and stimulates bone resorption, release of corticosteroids, may enhance or inhibit proliferation depending on cell type.
Interleukin 2	Primary growth factor for T lymphocytes, also stimulates macrophages. Appears to be more restricted in target cell type than other cytokines. Effects on oligodendrocytes controversial.
Interleukin 3, interleukin 5, granulocyte–macrophage (GM), macrophage (M), granulocyte (G), colony-stimulating factors (CSFs)	Mediate formation of haemopoietic colonies *in vitro*; IL-3 is multi CSF and stimulates macrophage, neutrophil, megakaryocyte and erythroid colonies, IL-5 stimulates eosinophil colonies: others are more specific as their names imply. Show synergy with other mediators such as IL-1 and IL-6. Stimulate mature cell effector function. Role in steady state haemopoiesis not established.
Interleukin 4	Same as IgG1 induction factor, also important for some T cells. Stimulates macrophages and mast cells and fibroblasts.
Interferons (α,β,γ)	Antiviral, induce class I and class II (γ only) MHC antigens synergistic with TNFs; inhibit collagen synthesis, antiproliferative.
Interleukin 6	Promotes B cell differentiation into plasma cells, stimulates hepatocytes and T cells, but growth inhibitory for some cells.
Tumour necrosis factor TNFα and TNFβ (lymphotoxin)	Similar to IL-1 except some differences noted on immature haemopoietic cells, cytotoxic and growth inhibitory for some cells but stimulates growth of others such as fibroblasts.
Transforming growth factor β	Bifunctional regulator of cell growth; appears to be immunosuppressive; defines a family of at least 10 structurally related cytokines involved in cell growth and differentiation. Includes activins, inhibins bone modifying proteins, decapentaplegic.
Interleukin 8	Formerly monocyte-derived chemotactic factor, represents the prototype of a large number of low molecular weight cytokines. Biological effects are largely unknown but likely to be diverse. Includes macrophage inflammatory proteins, and competence genes KC and JE.

splicing to yield mRNAs coding for different proteins has been documented but little is known concerning its control. Furthermore, as the nucleus of eukaryotic cells represents a discrete compartment separated from the cytoplasm by a membrane,

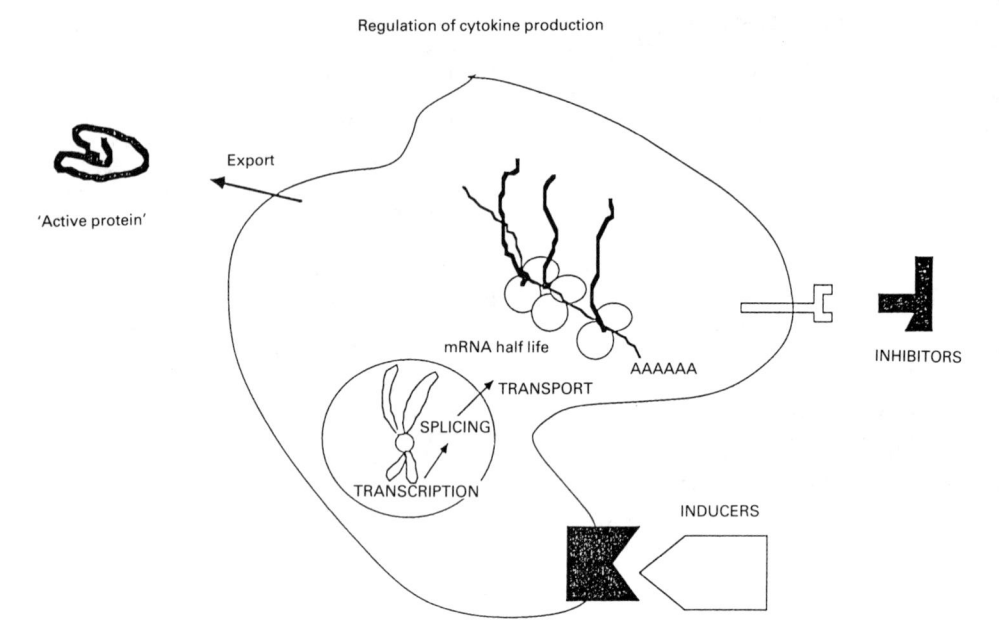

Fig. 1 — Schematic diagram of potential control points for a model cytokine gene.

the rate of transport of mature mRNA into the cytoplasm could represent a potential control point. Control of mRNA stability and the translation of mRNA into protein are among the best-characterized mechanisms controlling cytokine gene expression and there are numerous examples of these phenomena. Cytokine expression may also be controlled at the level of the protein itself, through glycosylation or other covalent modification, which may alter protein stability. The protein may be accumulated within the cell and released as a bolus following stimulation, or may be stored in a latent form requiring activation. Finally, the responsiveness of cells to a first cytokine may be modulated by the presence of numerous other cytokines, or other mediators which act to enhance or inhibit its function.

This chapter is primarily concerned with the regulation of expression of the mRNA for cytokines, and we will start by considering transcription, the process by which mRNA is formed.

3. TRANSCRIPTIONAL CONTROL

Cytokine genes are excellent model systems for the study of transcription, as peripheral blood mononuclear cells represent a good source of non-transformed cells and progression through the cell cycle following stimulation is well documented. Usually, in resting cells cytokine mRNA is undetectable using techniques such as Northern blotting. Following activation of these cells an ordered and transient expression of a number of genes is observed. This may be divided into at least three temporal stages by analogy to the pattern of adenovirus replication (Table 2). It is

Table 2 — Temporal arrangement of genes expressed by activated blood mononuclear cells

Immediate early, peak 1–2 h	Early response, peak 1–6 h	Late response, peak>6 h
c-*fos** (15 min)	IL-1α* (30 min)	c-*myb** (30 min–4 h)
c-*myc** (15–30 min)	IL-β* (3–5 h)	Transferrin receptor*
jun/AP-1*	IL-2* (15 min)	n-*ras*
IRF-1*	IL-5* (5 h)	RANTES*
	TNF* (30 min)	CD2
	LT* (30 min)	Histone H3
	IL-6* (30 min)	PDGF*
	IL-8* (4–5 h)	P53
	TGFβ (>5 h)	

Three categories are envisaged by analogy with patterns of viral gene transcription. The earliest group (immediate early genes) consists of transcriptional activator proteins, while the cytokines fall primarily into the early category. It should be noted that this is somewhat schematic and there is overlap between different groups; for example, the IL-2 receptor appears to bridge the gap between early and late genes. *indicates the presence of multiple repeats of the AUUUA sequence in the 3′ untranslated region of the mature mRNA. Note that genes bearing this sequence can be found in all three temporal response groups. The half-life of each mRNA is indicated in parentheses (note that IL-1β, IL-5, IL-8 and PDGF-A which have the 3′ AU-rich sequence all encode relatively stable mRNAs, $t^{1/2}$>3 h). Half-lives were determined by blocking ongoing transcription with actinomycin D, and assaying for remaining mRNA at different times thereafter by filter hybridization.

assumed that the expression of the 'immediate early' genes is required for the expression of the 'early' genes and so on. Inhibitors of protein synthesis (such as cycloheximide, CHX) have been used as tools to dissect whether protein synthesis is required for the induction of transcription of cytokine genes. It should be noted that interpretation of these studies is complicated by the fact that cytokine mRNA is stabilized following inhibition of protein synthesis (see section 6). Thus large increases in mRNA levels can take place in the absence of transcription. Studies of the cytokine gene expression in macrophage models have yielded paradoxical results, perhaps for precisely the above reason. The use of thioglycollate-elicited murine peritoneal macrophages [6] suggests that the IL-1α and TNF genes are under the control of labile transcriptional repressor proteins, as CHX by itself activated transcription. In contrast, human THP-1 monocytic leukaemia cells treated with CHX alone fail to accumulate significant quantities of IL-1α or IL-1β mRNA [7] and in human U937 cells induction of IL-1β by phorbol ester (PMA) is blocked when CHX is added before or concomitantly with the PMA [8]. Studies in cells isolated from peripheral blood indicate that CHX itself cannot induce transcription of the IL-1 genes and blocks TNF-mediated increases in IL-1 gene transcription, although steady state IL-1α and IL-1β mRNA levels were still increased owing to the ability of CHX to stabilize IL-1 mRNA [9]. Similarly, in non-transformed human monocytes M-CSF transcription is not enhanced by CHX, but steady state mRNA levels are increased 20-fold [10].

The different results using mouse macrophages may be due to their being part of an inflammatory infiltrate and consequently at a different stage of differentiation from human blood monocytes or the cell lines used in other studies. Evidence suggests these 'resting' cells may be partially activated as they express high basal levels of TNF mRNA [11]. Similarly conflicting results have been found in T cells: transcription of IFNγ and IL-2 genes was unaffected by CHX in peripheral T cells purified by rosetting with sheep erythrocytes [12] but blocked in T cell tumour lines [13] or in T cells purified by negative selection using monoclonal antibodies against CD10 [14]. Likewise, viral induction of IFNβ and TNF gene expression was unaffected by CHX in U937 cells but blocked in two transformed B cell lines [15]. Thus some cells must already contain the transcription factors required for cytokine gene expression while in others these need to be translated.

CHX does not seem to increase the rates of ongoing transcription when added to preactivated cells, even during the period when some cytokine mRNAs are being 'superinduced' (a phenomenon where large increases in steady state levels of mRNA are induced by protein synthesis inhibitors) [9,12]. There is some evidence to suggest that CHX may block transcriptional shut-off of some immediate early genes [16,17] and also the TNF gene. Sariban et al. [18] demonstrated that PMA-induced transcription of the TNF gene was easily detectable at 20 min but not at 12 h unless CHX was also present. Although Sariban et al. concluded that CHX increased TNF transcription, their data are also consistent with the notion that CHX delays or prevents transcriptional shut-off. This is further analogous to adenovirus gene expression where the immediate response group contains transcriptional repressors which form a negative feedback loop. These results also suggest that all the necessary components of the transcriptional machinery are rapidly induced or activated.

It should not be assumed that protein synthesis is always required for activation of cytokine gene transcription. Cloning and analysis of the promoter–enhancer regions of cytokines reveals a remarkable plethora of potential regulatory elements. It is known that some of the DNA-binding proteins concerned, e.g. NF-κB, are present in resting cells and can be activated without requirement for protein synthesis, by stimuli such as phorbol esters [19]. It is not clear whether there is a requirement for the presence of all the factors which could potentially bind the promoter–enhancer for the gene to be transcribed or whether only a subset is required for cytokine transcription to a given signal. A recent set of experiments by Sehgal and colleagues highlight the problem [20]. Activation of transcription of the IL-6 gene in fibroblasts is mediated by numerous external stimuli, including IL-1, TNF, PDGF, viral products, LPS and PMA. The effect of CHX on the induction of transcription revealed that with some stimuli such as TNF the level of transcription was unaltered by CHX, while transcription mediated by PDGF was marginally increased. However, IL-1, which was the most potent inducer of IL-6 transcription, was unable to induce IL-6 when protein synthesis was blocked. This may indicate a general property of cytokine genes: their flexible response, being mediated by distinct pathways and perhaps different combinations of transcriptional regulators determined by the nature of the inducing stimulus. Recently DNA sequences of the human and mouse IL-6 genes have been determined [21,22]. Based on conserved elements between the two species potential binding sequences have been identified for SP-1, activator protein 1 (AP-1), NF-kB, serum response factor (SRF), cAMP response element binding protein (CREB), glucocorticoid response element binding protein and interferon regulatory factor 1 (IRF-1), and the more frequently encountered TATA and CCAAT boxes. Whether all of these sites are functional and under what circumstances remains to be determined.

4. TRANS-ACTING FACTORS

Analysis of trans-acting factors in the regulation of cytokine gene expression has been greatly facilitated by their cloning. Recently, a DNA binding protein (IRF-1) recognizing a regulatory element present in the promoter–enhancer of the IFNβ gene has been cloned [23] by screening a cDNA library expressed in λgt11 with double-stranded oligonucleotides corresponding to the regulatory element. Putative binding sites for this factor are also present in genes such as IFNα, TNF, IL-6 and MHC class I, all of which are induced by virus infection. Interestingly, the IRF-1 gene is itself induced by viral stimulation and the kinetics of its expression preceded that of IFNβ, perhaps suggesting a requirement for translation of IRF-1 mRNA. Thus when IRF-1 mRNA was transiently expressed in COS cells by means of an episomal expression vector [24], expression of IFN genes could be detected in the absence of viral stimulation. However, it was noted that constitutive expression of IRF-1 alone was not sufficient to induce as much IFNβ as detected following viral stimulation. This may be explained by the existence of at least one more factor which binds to the IFNβ promoter–enhancer and stimulates transcription. Two sequences termed PRD (positive regulatory domain) I and PRD II have been identified from studies of deletions of the IFNβ promoter [25,26] and the results from DNA footprinting experiments [27,28]. In addition there is evidence for the binding of a protein to the DNA when the gene is in the inactive state. On activation the pattern of digested DNA is altered, consistent with the removal of a (repressor) protein and its replacement with two activator proteins, one of which is probably IRF-1, which appears to bind to PRD I. The other protein has been shown to be NF-κB and this binds to PRD II.

There are some important functional distinctions between the two PRDs within the IFNβ gene promoter [29]. PRD I functions as an inducible element in mouse L cells but as a constitutive element in C127 cells as does PRD II. However, PRD II is a more potent element and more highly virus inducible. In addition, multiple copies of PRD I have no activity when placed downstream of the transcriptional start site, while PRD II functions independently of its position. The viral response element (PRD I) to which IRF-1 may bind could, however, play a dual role in transcriptional regulation as it was found to suppress specifically the function of the SV40 enhancer when sandwiched between the β-globin promoter and the SV40 enhancer [30]. Subsequent studies using structural variations of this viral response element show that this phenomenon correlates with ability to bind IRF-1 itself [31]. Thus IRF-1 may function as a transcriptional activator or silencer depending on the presence of other transcriptional factors. Taken together with the differences between L929 and C127 cells these results lend support to the idea that there may be cell-specific differences in the content or function of transcriptional activators and that transcriptional activators may have the ability to function as inhibitors under certain circumstances. The DNA binding protein recognizing PRD II has recently been shown to be indistinguishable from NF-κB [32,33]. NF-κB binding was shown to be induced by dsRNA and viral stimulation; conversely, treatment of cytoplasmic extracts with deoxycholate (a treatment known to release NF-κB from an inactive complex [34]) led to the appearance of a PRD II binding activity. Furthermore, oligonuclotides corresponding to NF-κB binding sites from different promoters were effectively competed for in gel retardation assays by PRD II sequences. In addition, single base mutations in PRD II known to block IFNβ inducibility [26] bind NF-κB less effectively. Conceptually then, at least, the situation in the nucleus can be likened to the properties of cytokines themselves, with trans-acting factors displaying bifunctional properties depending on the presence and interaction with other agents. The ultimate response is thus determined by the sum of these influences, which may be antagonistic or synergistic. A model of cytokine gene regulation, showing multiple cis-acting sequences, negative regulatory factors and multiple activating proteins, is shown in Fig. 2.

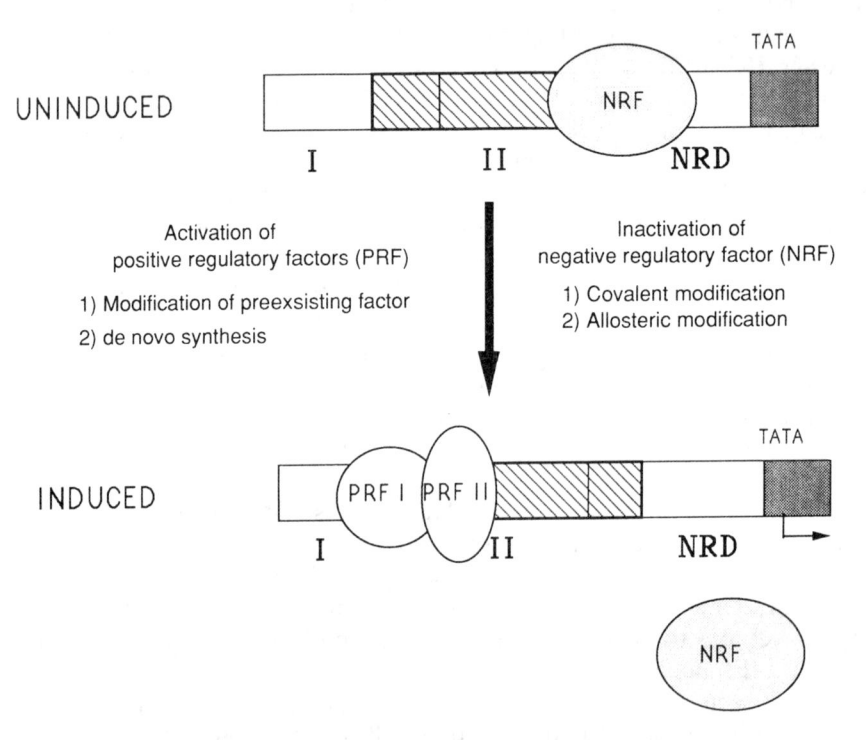

Fig. 2 — Model of cytokine gene demonstrating multiple, sometimes overlapping, cis-acting DNA sequences denoted I, II and NRD (negative regulatory domain). In the inactive or uninduced state a negative regulatory factor (NRF) is bound to the NRD and may prevent the recognition of the promoter by other proteins. On activation the NRF is displaced and multiple positive regulatory factors (PRF I and PRF II) interact with the promoter.

5. INTERACTIONS BETWEEN VIRUSES AND CYTOKINES

The induction of interferons by viruses has been known for many years. However, it has now been shown that other cytokines are also directly virus inducible. These include TNF which has been shown to be induced by Sendai virus, encephalomyocaditis virus, vesicular stomatitis virus, adenovirus, herpes simplex virus II and human immunodeficiency virus [15,35,36]. Interestingly, LT, which is closely related to TNF in both structure and function and located only 1.2 kb upstream from TNF on chromosome 6 in man [37], was also shown to be virus inducible. The LT gene does not contain a viral response element analogous to the IFNβ gene in its 5' region, while the downstream TNF gene does, raising the possibility that these two closely linked genes may share promoter–enhancer elements.

The human T lymphotrophic viruses type I and type II (HTLV I and II) are asociated with various T cell lymphomas and leukaemias, such as adult T cell leukaemia (ATL). These viruses are of particular interest since, following lymphocyte infection, viral transcription responds to the same activation signals as do cytokine genes. Of the two viruses, the trans activator protein of HTLV-1 has been best characterized. This protein (variously referred to as *tat*-1, p40^x, orf IV, tax or X-lor) encodes a 40 kDa nuclear factor which functions to transactivate enhancer sequences within the viral long terminal repeat [38]. It has been proposed that the interaction of *tat*-1 with host cellular genes may be of critical importance for the development of ATL. Transient transfection of T cell tumour lines with *tat*-1 containing expression plasmids resulted in the activation of the endogenous IL-2 receptor (tac protein) and IL-2 genes [39]. Other studies have demonstrated that this is a direct effect on the IL-2 and IL-2R promoters [40,41]. Simultaneous, inappropriate, activation of IL-2 and IL-2R genes may have consequences for autocrine growth of transformed T cells. Infection of CTLL cells (an IL-2 dependent murine T cell line) with retroviruses expressing IL-2 led to a loss of requirement for exogeneous IL-2; in addition, these cells could form tumours in nude mice [42]. The *tat*-2 trans activator from HTLV II appears to have similar properties to *tat*-1 [43]. It has been noted that, although *tat*-1 is a good inducer of the IL-2R α chain promoter, its effects on IL-2 mRNA or IL-2 promoter–CAT constructs were unimpressive [39–41]. It should be noted that IL-2R expression can be induced by a single stimulus in T cells, either PHA or phorbol ester, while IL-2 itself requires a dual stimulus [44, 45]. Thus when Jurkat cells, stably transfected with *tat*-1, were treated with a single stimulus, such as PHA or PMA alone, there was a synergistic effect with *tat*-1 on the IL-2 promoter, which appears to be at least partially resistant to the inhibitory effects of cyclosporin A [46]. Further experiments [47] where Jurkat cells were stably transfected with constructs expressing either sense or antisense *tat*-1 mRNA revealed that GM-CSF but not IL-1α, IL-1β, IL-3, IL-4 or G-CSF mRNA was expressed specifically in the cells containing *tat*-1 sense constructs. Similar to the results with IL-2, a synergistic response between *tat*-1 and mitogenic stimuli was observed for GM-CSF mRNA induction. The lack of IL-1 induction is perhaps surprising as there are several reports of primary as well as established ATL lines' producing IL-1 mRNA [48, 49], which is normally not considered a T cell product; this may suggest that other factors as well as *tat*-1 are involved in the disregulation of cytokine genes. The IL-3 and IL-4 promoters have also been shown to be activated by *tat*-1 in a transient co-transfection assay [50–52], in contrast to the results using the stably transfected Jurkat cells (discrepancies of this type may be due to the assays or differences in Jurkat populations in different laboratories).

That *tat*-1 can also transactivate the c-*fos* promoter linked to the CAT gene as well as the endogenous c-*fos* gene [46] is particularly exciting. This appears to be a specific phenomenon as the c-*myc* promoter was not activated. As c-*fos* is associated with the AP-1/*jun* family of transcription factors [53] for which there appear to be recognition sequences in a number of cytokine genes this observation may be highly relevant to the overproduction of cytokines in ATL.

The HIV retrovirus is associated with acquired immune deficiency syndrome (AIDS) and is cytopathic to CD4+ T cells *in vitro*. HIV-infected T cells produce a number of cytokines including IL-2, TNF and IFNγ, and TGFβ as well as a growth factor for Kaposi's sarcoma cells which appears to be a novel cytokine [55]. In addition, peripheral blood monocytes from AIDS patients have been shown to produce high levels of TNF and to be spontaneously cytotoxic to tumour cell lines [56]. The HIV long terminal repeat (LTR) contains regions homologous to some cytokine promotor–enhancers, as pointed out by Fujita *et al.* [57]. Also, the HIV LTR is capable of responding to the same T cell activation signals as T cell derived cytokines and the IL-2R [58,59]. Common DNA motifs and DNA binding factors may mediate these processes, as appears to be the case for HTLV I and II. A sequence present in the LTR denoted as TAR is responsible for binding of *tat*-3, a virus-encoded nuclear protein which acts as an enhancer of virus gene expression. Sequences present in the HIV LTR responsible for virus induction have been mapped by similar methods as used for HTLV 1. One sequence identified by this technique has homology to the Ig κ light chain gene enhancer and it has been shown that an NF-κB like protein present in both Jurkat cells and macrophages can specifically bind to this region of DNA [59,60]. In addition mutational analysis demonstrated that this sequence was vital for the function of the inducible element in the HIV LTR. Other DNA binding proteins present in H9 cells (a clonal line selected for permissiveness for HIV replication) have been identified by utilizing biotinylated oligonucleotides corresponding to the HIV enhancer region [61]. These were mixed with nuclear extracts and purified by binding to streptavidin, and individual proteins were resolved by two-dimensional gel electrophoresis. This study demonstrated multiple proteins in the molecular weight range 70–86 kDa bound to the HIV enhancer. At least one of these proteins, termed HIVEN 86–A, has been shown to bind to a region in the IL-2R promoter near or at the κB-like element [62,63], and is also active in HTLV-1 infected T cells [63]. It remains to be shown whether sequences present in IL-2 and IFNγ upstream regions are capable of interacting with *tat*-3 or perhaps more likely with *tat*-3 and various cellular factors, as it has recently been shown that the IL-2 gene also contains a functional NF-κB binding site and appears to interact with similar factors to the IL-2R α chain gene [64].

The overproduction of cytokines by HIV-infected cells can account for some of the symptoms observed in AIDS patients. A T cell derived factor has been described that can act as a growth factor for Kaposi's sarcoma cells [55]. Chronic B cell activation and hypergammaglobulinaemia, a common feature of AIDS, has been suggested to be due to overproduction of B cell stimulating factors, in particular IL-6 [65]. Other evidence suggests that cytokines can facilitate viral replication. Factors derived from monocyte-conditioned medium enhance virus replication [66], and it has been shown that both TNF and IL-1 can stimulate the HIV LTR and activate NF-κB [67]. It has also been shown that neutralizing antibodies to TNF block syncytium formation by HIV-infected cells (Anna Vyakarnam, personal communication). It is ironic that some of the reported effects of TNF include antiviral activity [68], since HIV infection induces the production of TNF which in turn stimulates virus production.

6. POST-TRANSCRIPTIONAL CONTROL

Although transcriptional control is clearly important for the rapid induction and repression of cytokine genes, equally complex mechanisms may exist to regulate mRNA stability. A common motif (3' AU sequence, consisting of repeats of AUUUA) in the 3' untranslated domain of some cytokine as well as oncogene and other tissue-specific mRNAs [69,70] has been identified which may play a critical role in determining mRNA stability. This sequence is not found in all cytokine mRNAs (e.g. TGFβ and the acidic and basic fibroblast growth factors). The first evidence that this 3' AU sequence plays a role in destabilizing cytokine mRNA was suggested by Shaw and Kamen [70] who demonstrated that GM-CSF mRNA had a short half-life, of around 30 min. They constructed a chimaeric gene, consisting of the 3' untranslated region of GM-CSF, fused to the normally stable ($\approx$15 h) β-globin coding sequence. When transfected into NIH 3T3 cells the chimaeric β-globin mRNA was rendered unstable while control plasmids gave rise to stable mRNAs. Furthermore, these workers demonstrated that the chimaeric β-globin–GM-CSF mRNA could be stabilized by CHX. Similarly, 3' sequences of the c-*fos* mRNA have been shown to be necessary for its normal metabolism [71,72], and their deletion may contribute to overexpression of c-*fos* [73]. Subsequent measurements of the stabilities of mRNAs bearing the 3' AU-rich sequence have revealed that they do not necessarily coincide with unstable mRNAs. Thus while c-*fos* [72], c-*myc* [74], TNF, IL-6, LT [75], IL-1α [76], M-CSF [10], PDGF-B chain [77] and IL-2 [78] mRNAs (see Table 1) have half-lives estimated at around 1 h or less, IL-1β [76,79], IL-8 (monocyte-derived neutrophil chemotactic factor, see Fig. 3) and PDGF-A [77] and IL-5 [78] are considerably more stable with half-lives greater than 3 h. An example of an experiment determining RNA stability is shown in Fig. 3. It can be seen that TGFβ mRNA decays during the 3 h period but is not stabilized by CHX (TGFβ does not contain the 3' AU-rich sequence), and IL-6 and IL-8 both contain the AU sequence but IL-6 is unstable while IL-8 is not. Furthermore, IL-6 is stabilized by CHX while IL-8 is not. 7B6 mRNA does not contain the 3' AU-rich sequence (S. Ferrari, personal communication) and does not encode a particularly unstable mRNA. The most common interpretation of the ability of CHX to stabilize cytokine mRNA is by the inhibition of the synthesis of a labile protein that degrades cytokine mRNA. However, it should be borne in mind that CHX causes translating ribosomes to stall and accumulate on the mRNA template and thus possibly sterically hinders RNA-degrading enzymes or stabilizes mRNA directly. We have performed some experiments using puro*my*cin, an inhibitor of protein synthesis which leads to the disassembly of the polysome, and have found similar results to those obtained using CHX (unpublished data).

Are steady state levels of cytokine mRNAs containing the 3' AU-rich sequence influenced through changes in mRNA stability? There is a discrepancy between the period of active transcription of some cytokine genes and steady state levels of the respective mRNA. Thus Fig. 4 shows that IL-2 and IFNγ genes continue to be transcribed even after steady state levels of mRNA have decayed to near zero. It is of

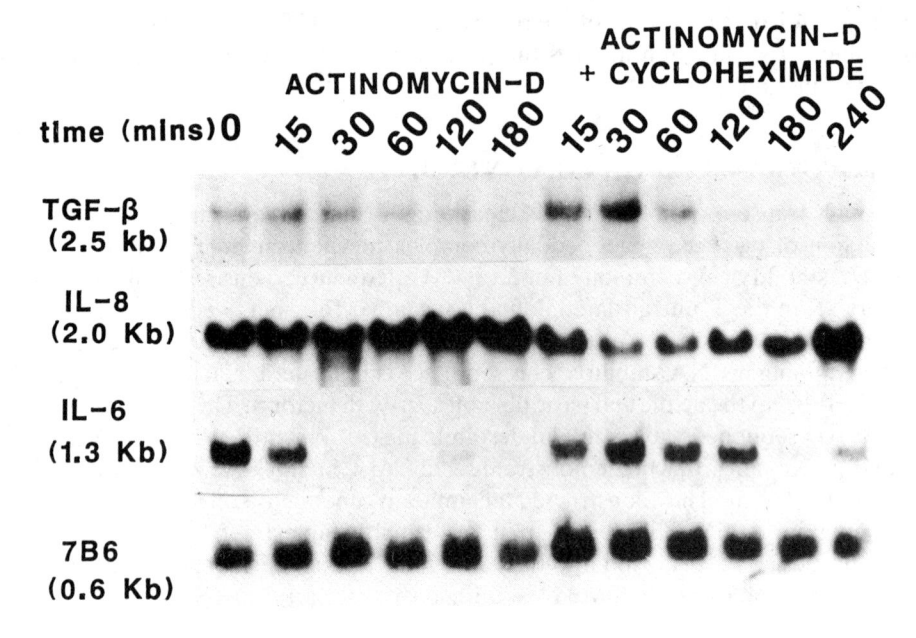

Fig. 3 — Half-lives of different cytokine mRNAs analysed by Northern blotting after actinomycin D chase in the absence or presence of protein synthesis inhibitor cycloheximide. The sizes of the mRNAs are indicated in kb on the left.

interest that transcription of the IL-2 receptor α chain (tac antigen) is more closely correlated with its mRNA concentration in the same experiments. This suggests that post-transcriptional control is important during the later stages of IL-2 and IFNγ gene expression. Several studies of oncogenes bearing the 3' AU-rich sequence suggest that changes in mRNA stability can be mediated by external stimuli. Expression of c-*myc* in the Daudi lymphoblastoid cell line is rapidly down regulated when the cells are exposed to type 1 interferons [80, 81], but the transcription of c-*myc* is unaffected [80,81]; instead, the stability of c-*myc* mRNA is decreased [81]. c-*myc* stability has also been reported to alter in response to stimulation through cell surface immunoglobulin in murine WEHI 231 cells [82]. In another example Thompson *et al.* [83] have provided evidence that c-*myb* mRNA may have different stabilities during its induction and declining phases. Thus the c-*myb* half-life measured at 2 h after serum stimulation of chick embryo fibroblasts was 246 min but at 9 h after stimulation (when c-*myb* levels were declining) it was only 28 min.

Further evidence for selective stabilization comes from studies of IL-2 and TNF gene expression in T cells [14,84]. Stimulation through the CD28 T cell surface antigen results in cyclosporin A resistant T cell proliferation and cytokine production [85]. Nuclear transcription assays reveal that CD28 cross-linking does not affect IL-2 or TNF transcription, but rather increases the stabilities of these mRNAs. Unlike what is believed to occur following CHX treatment, not all mRNAs with the 3' AU sequence are stabilized after CD28 stimulation, raising the possibility of specific

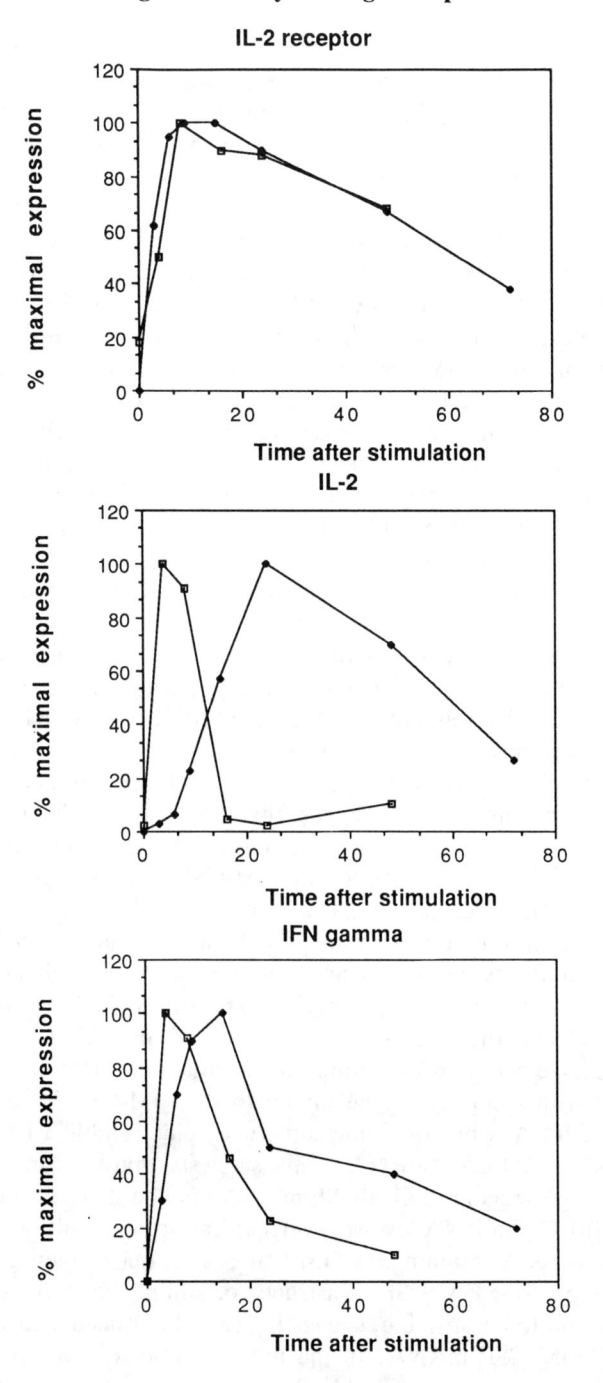

Ch. 13] **Regulation of cytokine gene expression** 249

Fig. 4 — Lack of correlation between steady state RNA levels and nuclear transcription rates for IL-2 and IFNγ genes. The results from the nuclear transcription assay are taken from ref. [12] and recalculated from parts per million into per cent maximal expression; steady state mRNA levels were calculated as per cent maximal expression from densitometer scans of Northern blots (see ref. [116] for original blots).

recognition of different classes of these mRNAs. There are some examples of genes being actively transcribed in the absence of detectable mRNA, such as c-*fos* [86] in human monocytes and GM-CSF in murine monocytes [87]. Induction of GM-CSF mRNA appears to take place in the absence of a change in the rate of transcription, again suggesting a post-transcriptional effect.

The mechanism of control of RNA stability is poorly understood. As already mentioned, many cytokines and oncogenes contain a highly conserved sequence in the 3' untranslated region. What role does this sequence play? It has been suggested that a nuclease may recognize this site, and degrade the mRNA in the cytoplasm. However, as the sequence also occurs in the chromosomal DNA it may be involved in nuclear processing events [88]. As the 3' AU sequence is not always a predictor of a short half-life (Table 1 and Fig. 3) it is likely that its role is not straightforward. There are few studies of the pathway of degradation of 3' AU mRNAs. However, studies of c-*fos* [89] and c-*myc* [90,91] suggest an early shortening and removal of the poly(A) tail to yield a pool of deadenylated mRNA which is rapidly degraded. By analogy with other systems of mRNA degradation such as tubulin [92] and histone [93] and the essential role of functioning ribosomes, the degradation process may be coupled to the translational process. Beutler *et al.* [94] have performed experiments to investigate the nature of the 3' AU-rich sequence required for instability. Instability increases as a function of the number of AU-rich repeats (up to five repeats based on the sequence of TNF 3' AU-rich region were tested). Interestingly, randon AU sequences also rendered the β-globin mRNA unstable but less so than the TNF-derived sequence. They have used this assay to purify a protein by more than 10 000-fold which selectively recognizes the 3' AU-rich sequence.

Two recent experiments [76, 95] suggest that selective stabilization of cytokine mRNAs may contribute to their steady state levels, and that this may be a tissue-specific phenomenon. In the first model the GM-CSF gene is constitutively transcribed in a number of different murine macrophage cell lines, but steady state mRNA is detectable in only one tumour [95]. Measurement of GM-CSF mRNA half-life in this tumour, after blocking transcription with actinomycin D, showed it to be about 2.5 h (5-fold more stable than previously reported). This enhanced stability was not due to a defect in the 3' AU-rich region of GM-CSF mRNA. This was shown when the macrophage cell lines which did not express detectable GM-CSF mRNA were transfected with a reporter gene linked to the GM-CSF 3' sequence. This yielded unstable mRNA, while the same construct yielded stable mRNA in the cell line overexpressing GM-CSF mRNA. This suggests that a trans-acting factor recognizing the 3' AU region of GM-CSF mRNA may be deficient in the cell line which has high GM-CSF mRNA levels. A further important result of this study was that, while the construct containing the GM-CSF 3' AU-rich sequence was stable in the tumour overexpressing GM-CSF, constructs containing the c-*myc* or c-*fos* 3' AU sequence were found to be unstable, suggesting that the trans deficit was a specific phenomenon for GM-CSF mRNA. In the monocytic leukaemia line THP-1 the steady state levels of expression of the IL-1α and IL-1β genes differ by at least an order of magnitude; however, measurements of nuclear transcription by run-on assay suggest that both genes are transcribed at similar rates. This discrepancy is apparently due to the extremely unstable nature of IL-1α mRNA as compared with IL-1β [76]. We have also noted that in other cell lines, in contrast to THP-1, IL-1α is the predominant form of IL-1 mRNA and that this may also be due to post-transcriptional effects (M.T. and M. F., in preparation). Both of these observations are certainly consistent with others showing that 3' AU containing mRNAs are not all unstable, and that mRNAs containing the 3' AU sequence can be found in the immediate early and early as well as the late response groups (see Table 2). Furthermore, the findings [96] that in some instances the stabilizing effect of CHX is tissue specific allows us to speculate that the regulation of turnover of 3' AU containing RNAs is regulated in a tissue-specific fashion, with perhaps multiple trans-acting factors interacting with the 3' AU sequences of different mRNAs. This would confer additional fine tuning to the temporal order of gene expression and allow a more flexible response to external stimuli. The level of complexity could parallel that found at the level of transcription where it is becoming increasingly apparent that a large number of promoter and enhancer binding factors may interact with a single gene. Progress in methods for characterizing cytoplasmic ribonucleoproteins [97] should aid in further definition of the RNA binding proteins involved in this system.

As well as control of mRNA stability, cytokine genes appear to be regulated post-transcriptionally at several other points. Alternative splicing has been described for TGFβ-2 [98], PDGF-A [99], G-CSF [100], M-CSF [101] and IL-7 [102] but the consequences of this differential splicing are largely unknown. Transfection studies untilizing either the 4 kb M-CSF cDNA or the 1.6 kb M-CSF cDNA for example reveal that both give rise to biologically active M-CSF; however, the product of the 1.6 kb cDNA clone (which lacks 298 amino acids) is stably expressed as a transmembrane glycoprotein [103]. Splicing may also be a tissue-specific phenomenon; for example, PDGF-A mRNA generated by endothelial cells lacks 69 bp, encoded on a small exon, specifying the C-terminal region of PDGF and appears to be involved in the determining the secretability of the protein [104].

Untranslated mRNA pools have been described for a number of cytokines including TNF, IL-1 and TGFβ. Beutler et al. [105] showed that thioglycollate-elicited murine peritoneal macrophages contain untranslated TNF mRNA and that macrophages from the C3H/Hej endotoxin resistant strain fail to release TNF despite the presence of relatively high mRNA levels [105,106]. This post-transcriptional block could be overcome by treatment with the combination of LPS and IFNγ [105]. Prostaglandin E_2 or similar agents that act through cyclic AMP also appear to inhibit translation of TNF mRNA partially [107], although their major effect appears to be at the level of transcription. Prostaglandins and cAMP have also been shown to inhibit secretion of bioactive IL-1 without changes in mRNA levels [108]. The same group has reported similar findings for the inhibitory effects of glucorticoids [109]. Most of these studies detailed so far do not distinguish between true translational inhibition or blockage of some downstream event such as protein processing or secretion. Treatment of peripheral blood mononuclear cells with TGFβ leads to the transcription of the genes of IL-1α and IL-1β, TNF and IL-6 and accumulation of mRNA for these cytokines [110]. However, neither IL-1 nor TNF can be detected in conditioned medium by bioassay or ELISA, or by Western blot or ELISA of cell lysates [110]; on the other hand, high levels of IL-6 activity can be detected in all supernatants, suggesting that there is a specific block to translation for IL-1 and TNF and/or that the mRNA is sequestered away from the ribosomes. Interestingly, TGFβ itself is subject to control at the level of transcription in macrophages [111] and T cells [112].

The importance of untranslated mRNA pools is that they can provide the potential for a very rapid response to external stimuli particularly when the translation product is not usually stored in an intracellular vesicle. It is interesting to note that in xenopus oocytes an untranslated, maternally derived mRNA called Vg1, which encodes a cytokine related to TGFβ [113], is localized to the vegetal pole of the cell as demonstrated by *in situ* hybridization [114], presumably by a specific mechanism. This may have relevance to higher eukaryotes as lymphokine release has been shown to be polar, emanating from the region stimulated rather than the entire surface area of the activated cell [115].

7. CYTOKINE EXPRESSION IN AUTOIMMUNITY

Autoimmune sites of inflammation are characterized by the infiltration of immune cells of lymphoid and myeloid origin, some of which can be shown by cell surface marker analysis to be in an activated state. It is therefore reasonable to assume that these cells are actively releasing cytokines which may then mediate some of the pathology of the disease. Work in our laboratory has focused on cytokine production in diseases such as rheumatoid arthritis, Hashimoto's and Graves thyroiditis and non-toxic goitre (which is a non-autoimmune disease and as such represents a control for the other diseases). Using a rapid and sensitive blotting method, which allows simultaneous analysis of a large number of samples with probes for many cytokines, we demonstrated the presence of mRNA for T cell products such as IL-2, IFNγ[116], and LT [117] in biopsy samples from patients with these diseases. However, the demonstration of the active protein in the large quantities that we would have expected given the mRNA data has not been possible using a variety of sensitive techniques [118]. This suggests that a form of post-transcriptional control is in operation, perhaps similar to that mediated by TGFβ on IL-1 and TNF production or that these mediators are turning over very rapidly, while their steady state levels remain very low. High levels of soluble IL-2 receptor have been detected in synovial fluid and may prevent detection of free IL-2 [119]. While the exclusive products of T cells have been difficult to demonstrate, products common to T cells and other cells such as macrophages have been shown to be present at both the mRNA and the protein level. By *in vitro* culture of cells obtained from synovial effusions, IL-1, IL-6 and TNF are present at biologically relevant levels in rheumatoid arthritis patients [117, 120]. On the other hand, high levels of TNF only can be found in culture supernatants of OA patients' cells. This is in agreement with data obtained from immunostaining (C. Q. Chu and M. Field, personal communication) as TNF can be demonstrated at the cartilage–pannus *jun*ction, where it may be participating in tissue remodelling–destruction. The presence of TNF in both RA and OA may be relevant to bone resorption which is mediated by TNF and can be observed in both diseases. Of particular interest concerning the pathogenesis of RA is that cytokine mRNA levels and protein production remain at high levels when joint cells are cultured *in vitro* in the absence of extrinsic stimulation [116,117]. Chronic cytokine production would be expected to be a necessary feature if it is to be considered important in a chronic disease process, such as RA. The establishment of an *in vitro* model for cell interactions and cytokine production occurring during active RA permits new approaches to experimental analysis, and may lead to the development of new therapies. TNF appears to be critical for the prolonged production of IL-1α and IL-1β and IL-6 in cultured RA cells as it has been shown that TNF is a potent inducer of both IL-1α and IL-1β [121] and that neutralizing antibodies to TNF abrogate the chronic IL-1 production [127]. TNF may thus be at the beginning of the cytokine cascade occurring in RA.

The expression of the IL-1 genes can be shopwn to be different in cells from RA patients and peripheral blood cells activated to express IL-1 by a variety of stimuli. Thus, IL-1β is the predominant IL-1 mRNA species in blood and is estimated to be about 20-fold more abundant than IL-1α mRNA [121]. In RA cells roughly equivalent levels of IL-1α and IL-1β mRNA were present [117]. Preliminary experiments suggest that mRNA stabilities are similar to those found in blood and that relative transcription rates can account for these differences. The RA cells often consist primarily of macrophages (which are typically high IL-1β producers) (F. Brennan, personal communication); however, the IL-1α producing cell population has not yet been defined. Further studies are in progress to dissect the cytokine gene cascade occurring in RA.

8. CONCLUSIONS AND PERSPECTIVES

As we have described, cytokine gene expression is controlled at multiple levels, both transcriptionally and post transcriptionally. Cytokines appear to be controlled in a manner closely analogous to other cell cycle dependent genes such as oncogenes. Although the definition of promoters–enhancers for the majority of cytokine genes is in its infancy, it can be assumed from sequence homologies that there exist multiple cis-acting elements responsible for the coordinated regulation of cytokine transcription. In the case of the IFNβ and IL-2 promoters these are becoming well defined and trans-acting factors are becoming identified and in some cases cloned. It is likely that, given the complexity of inducibility and tissue specificity of expression of cytokine genes, the promoter–enhancers of these genes may prove to be considerably more complex than the current paradigms of gene regulation predict (a case in point being the IL-6 gene [21,22]). Further mechanisms of transcriptional control may apply to the cytokines such as attenuation, a phenomenon where RNA polymerase II pauses, usually within an intron, and appears to require interactions with other proteins in order to continue elongation. Such mechanisms have been described as operative for the c-*myc* [122] and c-*myb* [123] genes in lymphocytes and may potentially also operate for cytokines. Antisense transcription also provides a potential mechanism to regulate the effective concentrations of mRNA within a cell and it is noted that a preliminary report of this phenomenon has been described for the IL-2 gene.

Interactions between viruses and cytokines should also yield further insights into how cytokine genes are regulated and could have important consequences for human diseases. The trans-activating proteins of HTLV-1 and HIV appear to be able to activate a number of cytokine promoters and may well represent examples of promiscuous transcription factors. Following on from the observations that an NF-κB like protein interacted with the HIV LTR, it has now become apparent NF-κB interacts with the promoters for IL-2Rα chain, IL-2 and IFNβ, and NF-κB like recognition sites can be found in the upstream sequences of IL-6, TNF and IL-1, although their ability to bind NF-κB has not been formally demonstrated. Other proteins which interact with κB-like elements, such as HIVEN 86A [61,62], H2TF1 [124] and AP-3 [125], may also be able to regulate multiple genes. Other viruses have been shown to transactivate cytokine promoters; these include the adenovirus E1A protein and the bovine papilloma virus E2 protein [50]. In addition to this the EBV trans activator BZLF-1 has been shown to recognize the consensus AP-1 binding site [126]; as previously noted, this sequence is present in a number of cytokine genes, and thus BZLF-1 may have the capacity to transactivate cellular cytokine genes. The induction of particular cytokines by viruses may represent a mechanism for bypassing normal immune responses, and could contribute to viral transformation and/or replication or to the chronic cytokine production in autoimmune diseases.

ACKNOWLEDGMENTS

We would like to thank all of our colleagues for sharing information and helpful and stimulating discussions during the preparation of this manuscript.

REFERENCES

[1] Sporn, M. B. & Roberts, A. B. (1988) Peptide growth factors are multifunctional. *Nature (London)* **332** 217
[2] Oppenheim, J. J., Kovacs, E. J., Matsushima, K. & Durum, S. K. (1986) There is more than one interleukin 1. *Immunol. Today* **7** 45–56
[3] Beutler, B. & Cerami, A. (1988) The history, properties and biological effects of cachectin. *Biochemistry* **27** 7575–7582
[4] Haworth, C. & Feldmann, M. (1989) Lymphokines and disease. In: M. Feldmann, J. Lamb & M. J. Owen (eds) *Human T Cells and Disease,* Wiley, New York, in press
[5] Cherwinski, H. M., Schumacher, J. H., Brown, K. D. & Mossman, T. R. (1987) Two types of mouse helper T cell clone III. Further differences in lymphokine synthesis between Th1 and Th2 clones revealed by RNA hybridisation, functionally monospecific bioassays and monoclonal antibodies. *J. Exp. Med.* **166** 1229
[6] Collart, M. A., Belin, D., Vassalli, J.-D., de Kossodo, S. & Vassalli, P. (1986) γ-interferon enhances macrophage transcription of the tumour necrosis factor/ cachectin, interleukin-1, and urokinase genes, which are controlled by short-lived repressors. *J. Exp. Med.* **164** 2113–2118
[7] Fenton, M. J., Clark, B. D., Collins, K. L., Webb, A. C., Rich, A. & Auron, P. E. (1987) Transcriptional regulation of the human prointerleukin 1β gene. *J. Immunol.* **138** 3972–3979
[8] Nishida, T., Takano, M., Kawakami, T., Nishino, N., Nakai, S. & Hirai, Y. (1988) The transcription of the interleukin 1β gene in induced with PMA and inhibited with dexamethasone in U937 cells. *Biochem. Biophys. Res. Commun.* **156** 269–274

[9] Turner, M., Chantry, D. H., Buchan, G., Barrett, K. & Feldmann, M. (1989) Transcriptional and post-transcriptional regulation of human IL-1α and β genes. *J. Immunol.* (in press)

[10] Horiguchi, J., Sariban, E. & Kufe, D. (1988) Transcriptional and posttranscriptional regulation of CSF-1 gene expression in human monocytes. *Mol. Cell. Biol.* **8** 3951-3954

[11] Collart, M. A., Belin, D., Vassalli, J.-D. & Vassalli, P. (1987) Modulations of functional activity in differentiated macrophages are accompanied by early and transient increase or decrease in c-*fos* gene transcription. *J. Immunol.* **139** 949-955

[12] Kronke, M., Leonard, W. J., Depper, J. M. & Greene, W. C. (1985) Sequential expression of genes involved in human T lymphocyte growth and differentiation. *J. Exp. Med.* **161** 1593-1598

[13] Shaw, J., Meerovitch, K., Elliot, J. F., Bleackley, R. C. & Paetkau, V. (1987) Induction, suppression and superinduction of lymphokine mRNA in T lymphocytes. *Mol. Immunol.* **24** 409-419

[14] Lindsten, T., June, C. H., Ledbetter, J. A., Stella, G. & Thompson, C. B. (1989) Regulation of lymphokine mRNA stability by a cell surface mediated T cell activation pathway. *Science* **244** 339-343

[15] Goldfeld, A. E. & Maniatis, T. (1989) Coordinate viral induction of tumour necrosis factor α and interferon B in human B cells and monocytes. *Proc. Natl. Acad. Sci. USA* **86** 1490-1494

[16] Lau, L. F. & Nathans, D. (1987) Expression of a set of growth related immediate early genes in BALB/c 3T3 cells: coordinated regulation with c-*fos* or c-*myc*. *Proc. Natl. Acad. Sci. USA* **84** 1182-1186

[17] Almendral, J. M., Sommer, D., MacDonald-Bravo, H., Burckhardt, J., Perera, J. & Bravo, R. (1988) Complexity of the early genetic response to growth factors in mouse fibroblasts. *Mol. Cell. Biol.* **8** 2140-2148

[18] Sariban, E., Imamra, K., Luebbers, R. & Kufe, D. (1988) Transcriptional and posttranscriptional regulation of tumor necrosis factor gene expression in human monocytes. *J. Clin. Invest.* **81** 1506-1510

[19] Sen, R. & Baltimore, D. (1986) Inducibility of κ immunoglobin enhancer protein, NF-κB by a posttranslational mechanism. *Cell* **47** 921-928

[20] Walther, Z., May, L. T. & Sehgal, P. B. (1988) Transcriptional regulation of the interferon-$β_2$/B cell differentiation factor BSF-2/hepatocyte-stimulating factor gene in human fibroblasts by other cytokines. *J. Immunol.* **140** 974-977

[21] Yasukawa, K., Hirano, T., Watanabe, Y., Muratani, K., Matsuda, T., Nakai, S. & Kishimoto, T. (1987) Structure and expression of the human B cell stimulatory factor (IL-6) gene. *EMBO J.* **6** 2939-2945

[22] Tanabe, O., Akira, S., Kamiya, T., Wong, G., Hirano, T. & Kishimoto, T. (1988) Genomic structure of the murine IL-6 gene. *J. Immunol.* **141** 3875-3881

[23] Miyamoto, M., Fujita, T., Kimura, Y., Maruyama, M., Harada, H., Sudo, Y., Miyata, T. & Taniguchi, T. (1988) Regulated expression of a gene encoding a nuclear factor, IRF-1, that specifically binds to IFN-β gene regulatory elements. *Cell* **54** 903-913

[24] Fujita, T., Kimura, Y., Miyamoto, M., Barsoumian, E. L. & Taniguchi, T. (1989) Induction of endogenous IFN-α and IFN-β gene by a regulatory transcription factor, IRF-1. *Nature (London)* **337** 270-272

[25] Goodbourn, S., Burstein, H. & Maniatis, T. (1986) The human β-interferon gene enhancer is under negative control. *Cell* **45** 601–610
[26] Goodbourn, S. & Maniatis, T. (1988) Overlapping positive and negative regulatory domains of the human β-interferon gene. *Proc. Natl. Acad. Sci. USA* **85** 1447–1451
[27] Zinn, K. & Maniatis, T. (1986) Detection of factors that interact with the human β-interferon regulatory region *in vivo* by DNAase I footprinting. *Cell* **45** 611–618
[28] Keller, A. D. & Maniatis, T. (1988) Identification of an inducible factor that binds to a positive regulatory element of the human β-interferon gene. *Proc. Natl. Acad. Sci. USA* **85** 3309–3313
[29] Fan, C.-M. & Maniatis, T. (1989) Two different virus-inducible elements are required for human β-interferon gene regulation. *EMBO J.* **8** 101–110
[30] Kuhl, D., de la Fuente, J., Chatuvedi, M., Parimoo, S., Ryals, J., Meyer, F. & Weissmann, C. (1987) Reversible silencing of enhancers by sequences derived from the human IFNa promoter. *Cell* **50** 1057–1069
[31] Fujita, T., Sakakibara, J., Sudo, Y., Miyamoto, M., Kimura, Y. & Taniguchi, T. (1988) Evidence for a nuclear factor(s), IRF-1, mediating induction and silencing properties to human IFN-β gene regulatory elements. *EMBO J.* **7** 3397–3405
[32] Visvanathan, K. V. & Goodbourn, S. (1989) Double stranded RNA activates binding of NF-κB to an inducible element in the human β-interferon promoter. *EMBO J.* **8** 1129–1138
[33] Lenardo, M. J., Ming-Fan, C., Maniatis, T. & Baltimore, D. (1989) The involvement of NF-κB in β-interferon gene regulation reveals its role as a widely inducible mediator of signal transduction. *Cell* **57** 287–294
[34] Baeuerle, P. A. & Baltimore, D. (1988) IκB: a specific inhibitor of the NF-kB transcription factor. *Science* **242** 540–546
[35] Berent, S. L., Torczynski, R. M. & Bollon, A. P. (1986) Sendai virus induces high levels of tumour necrosis factor mRNA in human peripheral blood leukocytes. *Nucleic Acids Res.* **14** 8997–9015
[36] Wong, G. H. W. & Goeddel, D. V. (1986) Tumour necrosis factor α and β inhibit virus replication and synergise with interferons. *Nature (London)* **323** 819–822
[37] Pennica, D. & Goeddel, D. V. (1987) Cloning and characterisation of the genes for murine and human tumour necrosis factors. In: D. R. Webb & D. V. Goeddel (eds) *Lymphokines*, Vol. 13, Academic Press, New York
[38] Felber, B. K., Paskalis, H., Kleinman-Ewing, C., Wong-Stall, F. & Pavlakis, G. (1985) The pX protein of HTLV-1 is a transcriptional activator of its long terminal repeats. *Science* **229** 675–678
[39] Inoue, J., Seiki, M., Taniguchi, T., Tsuru, S. & Yoshida, M. (1986) Induction of interleukin-2 receptor gene expression by p40^x encoded by human T-cell leukemia virus type 1. *EMBO J.* **5** 2883–2888
[40] Maruyama, M., Shibuya, H., Harada, H., Hatakeyama, M., Seiki, M., Fujita, T., Inoue, J., Yoshida, M. & Taniguchi, T. (1987) Evidence for aberrant activation of the interleukin-2 autocrine loop by HTLV-1-encoded p40^x and T3/Ti complex triggering. *Cell* **48** 343–350

[41] Siekevitz, M., Feinberg, M. B., Holbrook, N., Wong-Staal, F. & Greene, W. C. (1987) Activation of interleukin 2 and interleukin 2 receptor (Tac) promoter expression by the trans-activator (tat) gene product of human T cell leukemia virus, type 1. *Proc. Natl. Acad. Sci. USA* **84** 5389–5393

[42] Yamada, G., Kitamura, Y., Sonoda, H., Harada, H., Taki, S., Mulligan, R. C., Osawa, H., Diamantstein, T., Yokoyama, S. & Taniguchi, T. (1987) Retroviral expression of the human IL-2 gene in murine T cell line results in cell growth autonomy and tumorigenicity. *EMBO J.* **6** 2705–2709

[43] Greene, W. C., Leonard, W. J., Wano, Y., Svetlik, P. B., Peffer, N. J., Sodroski, J. G., Rosen, C. A., Goh, W. C. & Hasletine, W. A. (1986) Trans-activator gene of HTLV-II induces IL-2 receptor and IL-2 cellular gene expression. *Science* **232** 877–880

[44] Wiskocil, R., Weiss, A., Imboden, J., Kamin-Lewis, R. Stobo, J. (1985) Activation of a human T cell line: a two stimulus requirement in the pretranslational events involved in the co-ordinate expression of interleukin 2 and γ-interferon genes. *J. Immunol.* **134** 1599–1603

[45] Kumagai, N., Benedict, S. H., Mille, G. B. & Gelfand, E. W. (1987) Requirements for the simultaneous presence of phorbol esters and calcium ionophores in the expression of human T lymphocyte proliferation-related genes. *J. Immunol.* **139** 1393–1399

[46] Siekevitz, M., Josephs, S. F., Dukovich, M., Peffer, N., Wong-Staal, F. & Greene, W. C. (1987) Activation of the HIV-1 LTR by T cell mitogens and the trans-activator protein of HTLV-1. *Science* **238** 1575–1578

[47] Wano, Y., Feinberg, M., Hosking, J. B., Bogerd, H. & Greene, W. C. (1988) Stable expression of the tax gene of type 1 human T-cell leukemia virus in human T cells activates specific cellular genes involved in growth. *Proc. Natl. Acad. Sci. USA* **85** 9733–9737

[48] Yamashita, U., Shirakawa, F. & Nakamura, H. (1987) Production of interleukin 1 by adult T cell leukemia (ATL) cell lines. *J. Immunol.* **138** 3284–3289

[49] Wano, Y., Hattori, T., Matsuoka, M., Takatsuki, K., Chua, A. O., Gubler, U. & Greene, W. C. (1987) Interleukin 1 gene expression in adult T Cell leukemia. *J. Clin. Invest.* **80** 911–916

[50] Miyatake, S., Seiki, M., DeWaal Malefijt, R., Heike, T., Fujisawa, J., Takebe, Y., Nishida, J., Shlomai, J., Yokota, T., Yoshida, M., Arai, K. & Arai, N. (1988) Activation of T cell-derived lymphokine genes in T cells and fibroblasts: effects of human T cell leukemia virus type 1 $p40^x$ protein and bovine papilloma virus encoded E2 protein. *Nucleic Acids Res.* **16** 6547–6566

[51] Miyatake, S., Seiki, M., Yoshida, M. & Arai, K.-I. (1988) T-cell activation signals and human T-cell leukemia virus type I-encoded p40x protein activate the mouse granulocyte-macrophage colony-stimulation factor gene through a common DNA element. *Mol. Cell. Biol.* **8** 5581–5587

[52] Arai, N., Nomura, D., Villaret, D., DeWaal Malefijt, R., Seiki, M., Yoshida, M., Minoshima, S., Fukuyama, R., Maekawa, M., Kudoh, J., Shimizu, N., Yokota, K., Aba, Yokota, T., Takebe, Y. & Arai, K. (1989) Complete nucleotide sequence of the chromosomal gene for human IL-4 and its expression. *J. Immunol.* **142** 274–282

[53] Fujii, M., Sassone-Corsi, P. & Verma, I. (1988) c-*fos* promoter trans-

activation by the tax protein of human T-cell leukemia virus type 1. *Proc. Natl. Acad. Sci. USA* **85** 8526–8530

[54] Curran, T. & Franza, B. R., Jr. (1989) Fos and Jun: the AP-1 connection. *Cell* **55** 395–397

[55] Nakamura, S., Salahuddin, S. Z., Biberfeld, P., Ensoli, B., Markham, P. D., Wong-Staal, F. & Gallo, R. C. (1988) Kaposi's sarcoma cells: long term culture with growth factor from retrovirus-infected CD4+ T cells. *Science* **242** 426–433

[56] Wright, S., Jewett, A., Mitsuyasu, R. & Bonavida, B. (1988) Spontaneous cytotoxicity and tumour necrosis factor production by peripheral blood monocytes from aids patients. *J. Immunol.* **141** 99–104

[57] Fujita, T., Shibuya, H., Ohashi, T., Yamanishi, K. & Taniguchi, T. (1986) Regulation of human interleukin-2 gene: functional DNA sequences in the 5' flanking region for the gene expression in activated T lymphocytes. *Cell* **46** 401–407

[58] Tong-Starksen, S., Luciw, P. A. & Peterlin, B. M. (1987) Human immunodeficiency virus long terminal repeat responds to T-cell activation signals. *Proc. Natl. Acad. Sci. USA* **84** 6845–6849

[59] Nabel, G. & Baltimore, D. (1987) An inducible transcription factor activates expression of human immunodeficiency virus in T cells. *Nature (London)* **326** 711–713

[60] Griffin, G. E., Leung, K., Folks, T. M., Kunkel, S. & Nabel, G. (1989) Activation of HIV gene expression during monocyte differentiation by induction of NF-κB. *Nature (London)* **339** 70–73

[61] Franza, B. R., Josephs, S. F., Gilman, M. Z., Ryan, W. & Clarkson, B. (1987) Characterization of cellular proteins recognizing the HIV enhancer using a microscale DNA-affinity precipitation assay. *Nature (London)* **330** 391–395

[62] Bohnlein, E., Lowenthal, J. W., Siekevitz, M., Ballard, D. W., Franza, B. R. & Greene, W. C. (1988) The same inducible nuclear proteins regulates mitogen activation of both the interleukin-2 receptor-alpha gene and type 1 HIV. *Cell* **53** 827–836

[63] Ballard, D. W., Bohnlein, E., Lowenthal, J. W., Wano, Y., Franza, B. R. & Greene W. C. (1988) HLTV-1 Tax induces cellular proteins that activate the kB element in the IL-2 receptor α gene. *Science* **241** 1652–1655

[64] Hoyos, B., Ballard, D., Bohnlein, E., Siekevitz, M. & Greene, W. C. (1989) Kappa-B specific binding proteins: role in the regulation of IL-2 gene expression. *Science* **244** 457–460

[65] Nakajima, K., Martinez, O., Hirano, T., Breen, E., Nishanian, P., Gonzalez, J. F., Fahey, J. & Kishimoto, T. (1989) Induction of IL-6 production by HIV. *J. Immunol.* **142** 531–536

[66] Clouse, K. A., Powell, D., Washington, I., Poli, G., Strebel, K., Farrar, W., Barstad, P., Kovacs, J., Fauci, A. & Folks, T. M. (1989) Monokine regulation of human immunodeficiency virus-1 expression in a chronically infected human T cell clone. *J. Immunol.* **142** 431–438

[67] Osborn, L., Kunkel, S. & Nabel, G. (1989) TNF and IL-1 stimulate the immunodeficiency virus enhancer by activation of NF-κB. *Proc. Natl. Acad. Sci. USA* **86** 2336–2340

[68] Wong, G. H. W., Krowka, J. F., Stites, D. P. & Goeddel, D. V. (1988) *In vitro* anti-human immunodeficiency virus activities of tumor necrosis factor-α and interferon-γ. *J. Immunol.* **140** 120–124

[69] Caput, D., Beutler, B., Hartog, K., Thayer, R., Brown-Shimer, S. & Cerami, A. (1986) Identification of a common nucleotide sequence in the 3'-untranslated region of mRNA molecules specifying inflammatory mediators. *Proc. Natl. Acad. Sci. USA* **83** 1670–1674

[70] Shaw, G. & Kamen, R. (1986) A conserved AU sequence from the 3' untranslated region of GM-CSF mRNA mediates selective mRNA degradation. *Cell* **46** 659

[71] Treisman, R. (1985) Transient accumulation of c-*fos* RNA following serum stimulation requires a conserved 5' element and c-*fos* 3' sequences. *Cell* **42** 889–902

[72] Rahmsdorf, H. J., Schonthal, A., Angel, P., Litfin, M., Ruther, U. & Herrlich, P. (1987) Posttranscriptional regulation of c-*fos* mRNA expression. *Nucleic Acids Res.* **15** 1643–1659

[73] Lee, W. M. F., Lin, C. & Curran, T. (1988) Activation of the transforming potential of the human fos proto-oncogene requires message stabilisation and results in increased amounts of partially modified fos protein. *Mol. Cell. Biol.* **8** 5521–5527

[74] Reed, J. C., Alpers, J. D. & Nowell, P. C. (1987) Expression of c-*myc* proto-oncogene in normal human lymphocytes. *J. Clin. Invest.* **80** 101–106

[75] Turner, M. & Feldmann, M. (1988) Comparison of patterns of expression of tumour necrosis factor, lymphotoxin and interleukin 6 mRNA. *Biochem. Biophys. Res. Commun.* **153** 1144–1151

[76] Turner, M., Chantry, D. & Feldmann, M. (1988) Post-transcriptional control of IL-1 gene expression in the acute monocyte leukemia line THP-1. *Biochem. Biophys. Res. Commun.* **156** 830–839

[77] Majesky, M. W., Benditt, E. P. & Schwartz, S. M. (1988) Expression and developmental control of platelet-derived growth factor A-chain and B-chain/Sis genes in rat aortic smooth muscle cells. *Proc. Natl. Acad. Sci. USA* **85** 1524–1528

[78] Shaw, J., Meerovitch, K., Bleakly, R. C. & Paetkau, V. (1988) Mechanisms regulating the level of IL-2 mRNA in T lymphocytes. *J. Immunol.* 2243–2248

[79] Lee, S. W., Tsou, A.-P., Chan, H., Thomas, J., Petrie, K., Eugui, E. M. & Allison, A. C. (1988) Glucocorticoids selectively inhibit the transcription of the interleukin 1β gene and decrease the stability of interleukin 1β mRNA. *Immunology* **85** 1204–1208

[80] Knight, E. Jr., Anton, E. D., Fahey, B. K., Friedland, B. K. & Jonak, G. J. (1985) Interferon regulates c-*myc* gene expression in Daudi cells at the post-transcriptional level. *Proc. Natl. Acad. Sci. USA* **82** 1151–1154

[81] Dani, C., Mechti, N., Piechaczyk, M., Lebleu, B., Jeannteur, P. & Blanchard, J. M. (1985) Increased rate of degradation of c-*myc* mRNA in interferon-treated Daudi cells. *Proc. Natl. Acad. Sci. USA* **82** 4896–4899

[82] Levine, R. A., McCormack, J. E., Buckler, A. & Sonenshein, G. E. (1986) Transcriptional and posttranscriptional control of c-*myc* gene expression in WEHI 231 cells. *Mol. Cell. Biol.* **6** 4112–4116

[83] Thompson, C. B., Challoner, P. B., Neiman, P. E. & Groudine, M. (1986) Expression of the c-*myb* proto-oncogene during cellular proliferation. *Nature (London)* **319** 374–380

[84] June, C. H., Jackson, K. M., Ledbetter, J. A., Leiden, J. M., Lindsten, T. & Thompson, C. B. (1989) Two distinct mechanisms of interleukin 2 gene expression in human T lymphocytes. *J. Autoimmun.* **2** suppl. 55–65

[85] June, C. H., Ledbeter, J. A., Gillespie, M. M., Lindsten, T. & Thompson, C. B. (1987) T-cell proliferation involving the CD28 pathway is associated with cyclosporine-resistant interleukin 2 gene expression. *Mol. Cell. Biol.* **7** 4472–4481

[86] Sariban, E., Luebbers, R. & Kufe, D. (1988) Transcriptional and posttranscriptional control of c-*fos* gene expression in human monocytes. *Mol. Cell. Biol.* **8** 340–346

[87] Thorens, B., Mermod, J. J. & Vassali, P. (1987) Phagocytosis and inflammatory stimuli induce GM-CSF mRNA in macrophages through posttranscriptional regulation. *Cell* **48** 671–679

[88] Reeves, R., Elton, T. S., Nissen, M. S., Lehn, D. & Johnson, K. R. (1987) Posttranscriptional gene regulation and specific binding of the nonhistone protein HMG-1 by the 3' untranslated region of bovine interleukin 2 cDNA. *Proc. Natl. Acad. Sci. USA* **84** 6531–6535

[89] Wilson, T. & Treisman, R. (1988) Removal of poly(A) and consequent degradation of c-*fos* mRNA facilitated by 3' AU rich sequences *Nature (London)* **336** 396–399

[90] Brewer, G. & Ross, J. (1988) Poly(A) shortening and degradation of the 3' A+U-rich sequences of human c-*myc* mRNA in a cell-free system. *Mol. Cell. Biol.* **8** 1697–1708

[91] Swartwout, S. and Kinniburgh, A. J. (1989) c-*myc* degradation in growing and differentiating cells: possible alternate pathways. *Mol. Cell. Biol.* **9** 288–295

[92] Yen, T. J., Machlin, P. S. & Cleveland, D. W. (1988) Autoregulated instability of β-tubulin mRNAs by recognition of the nascent amino terminus of β-tubulin. *Nature (London)* **334** 580–585

[93] Graves, R. A., Pandley, N. B., Chodchoy, N. & Marzluff, W. F. (1987) Translation is required for regulation of histone mRNA degradation. *Cell* **48** 615–626

[94] Beutler, B., Thompson, P., Keyes, J., Hagerty, K. & Crawford, D. (1988) Assay of a ribonuclease that preferentially hydrolyses mRNAs containing cytokine-derived UA-rich instability sequences. *Biochem. Biophys. Res. Commun.* **152** 973–980

[95] Schuler, G. D. & Cole, M. D. (1988) GM-CSF and oncogene mRNA stabilities are independently regulated in *trans* in a mouse monocytic tumor. *Cell* **55** 1115–1122

[96] Dani, C. H., Blanchard, J. M., Piechaczyk, M., El-Sabouty, S., Marty, L. & Jeanteur, P. H. (1984) Extreme instability of *myc* mRNA in normal and transformed human cells. *Proc. Natl. Acad. Sci. USA* **81** 7046–7050

[97] Dreyfuss, G. (1986) Structure and function of nuclear and cytoplasmic ribonucleoprotein particles. *Annu. Rev. Cell. Biol.* **2** 459–498

[98] Webb, N., Madisen, L., Rose, T. M. & Purchio, A. F. (1988) Structural and sequence analysis of TGFβ2 cDNA clones predicts two different precursor proteins produced by alternate mRNA splicing. *DNA* **7** 493–497

[99] Rorsman, F., Bywater, M., Knott, T., Scott, J. & Betsholtz, C. (1988) Structural characterisation of the human platelet-derived growth factor A-chain cDNA and gene: alternative exon usage predicts two different precursor proteins. *Mol. Cell. Biol.* **8** 571–577

[100] Nagata, S., Tsuchiya, M., Asano, S., Yamamoto, O., Hirata, Y., Kubota, N., Oheda, M., Nomura, H. & Yamazaki, T. (1986) The chromosomal gene structure and two mRNAs for human granulocyte colony-stimulating factor. *EMBO J.* **5** 575–581

[101] Ladner, M. B., Martin, G. A., Noble, J. A., Nikoloff, D. M., Tal, R., Kawasaki, E. S. & White, T. J. (1987) Human CSF-1: gene structure and alternative splicing of mRNA precursors. *EMBO J.* **6** 2693–2698

[102] Namen, A. E., Lupton, S., Hjerrild, K., Wignall, J., Mochizuki, D. Y., Schmierer, A., Mosley, B., March, C. J., Urdal, D., Gillis, S., Cosman, D. & Goodwin, R. G. (1988) Stimulation of B-cell progenitors by cloned murine interleukin 7. *Nature (London)* **333** 571–573

[103] Rettenmier, C. W. & Roussel, M. F. (1988) Differential processing of colony-stimulating factor 1 precursors encoded by two human cDNAs. *Mol. Cell. Biol.* **8** 5026–5034

[104] Collins, T., Bonthron, D. T. & Orkin, S. H. (1987) Alternative RNA splicing affects function of encoded platelet-derived growth factor A chain. *Nature (London)* **328** 621–624

[105] Beutler, B., Krochin, N., Milsark, I. W., Leudke, C. & Cerami, A. (1986) Control of cachectin (tumour necrosis factor) synthesis: mechanisms of endotoxin resistance. *Science* **232** 977–980

[106] Beutler, B., Tkacenko, V., Milsark, I., Krochin, N. & Cerami, A. (1986) Effect of γ interferon on cachectin expression by mononuclear phagocytes. *J. Exp. Med.* **164** 1791–1796

[107] Kunkel, S. L., Spengler, M., May, M. A., Spengler, R., Larrick, J. & Remick, D. (1988) Prostaglandin E-2 regulates macrophage-derived tumour necrosis factor gene expression. *J. Biol. Chem.* **263** 5380–5384

[108] Knudsen, P. J., Dinarello, C. A. & Strom, T. B. (1986) Prostaglandins posttranscriptionally inhibit monocyte expression of interleukin 1 activity by increasing intracellular cyclic adenosine monophosphate. *J. Immunol.* **137** 3189–3194

[109] Knudsen, P. J., Dinarello, C. A. & Strom, T. B. (1987) Glucocorticoids inhibit transcriptional and post-transcriptional expression of IL-1 in U937 cells. *J. Immunol.* **139** 4129–4134

[110] Chantry, D., Turner, M., Abney, E. & Feldmann, M. (1989) Modulation of cytokine production by transforming growth factor beta. *J. Immunol.* **142** 4295–4300

[111] Assoian, R. K., Fleurdelys, B. E., Stevenson, H. C., Miller, P. J., Madtes, D. K., Raines, E. W., Ross, R. & Sporn, M. B. (1987) Expression and secretion

of type β transforming growth factor by activated human macrophages. *Proc. Natl. Acad. Sci. USA* **84** 6020–6024

[112] Kehrl, J. H., Wakefield, L. M., Roberts, A. B., Jakowlew, S., Alvarez-Mon, M., Derynck, R., Sporn, M. B. & Fauci, A. S. (1986) Production of transforming growth factor β by human T lymphocytes and its potential role in the regulation of T cell growth. *J. Exp. Med.* **163** 1037–1054

[113] Weeks, D. L. & Melton, D. A. (1987) A maternal mRNA localised to the vegetal hemisphere in xenopus eggs codes for a growth factor related to TGF-β. *Cell* **51** 861–867

[114] Melton, D. A. (1987) Translocation of a localized maternal mRNA to the vegetal pole of xenopus oocytes. *Nature (London)* **238** 80–82

[115] Poo, W.-J., Conrad, L. & Janeway, C. A. Jr. (1988) Receptor-directed focusing of lymphokine release by helper T cells. *Nature (London)* **332** 378–380

[116] Buchan, G., Barrett, K., Fujita, T., Taniguchi, T., Maini, R. N. & Feldmann, M. (1988) Detection of activated T cell products in the rheumatoid joint using cDNA probes to interleukin-2, IL-2 receptor and interferon gamma. *Clin. Exp. Immunol.* **71** 295–301

[117] Buchan, G., Barrett, K., Turner, M., Chantry, D., Maini, R. & Feldmann, M. (1988) Interleukin-1 and tumour necrosis factor mRNA expression in rheumatoid arthritis: prolonged production of IL-1α. *Clin. Exp. Immunol.* **73** 449–455

[118] Brennan, F. M., Chantry, D., Jackson, A. M., Maini, R. N. & Feldmann, M. (1989) Cytokine production in culture by cells isolated from the synovial membrane. *J. Autoimmun.* **2** suppl 177–186

[119] Symons, J. A., Wood, N. C., Di Giovine, F. S., Duff, G. W. (1989) Soluble K-2 receptor in rheumatoid arthritis. Correlation with disease activity, K-1 and K-2 inhibition. *J. Immunol.* **141** 2612–2617

[120] Hirano, T., Matsuda, T., Turner, M., Miyasaka, N., Buchan, G., Tang, B., Sato, K., Shimizu, M., Maini, R. N., Feldmann, M. & Kishimoto, T. (1988) Excessive production of interleukin 6/B cell stimulatory factor-2 in rheumatoid arthritis. *Eur. J. Immunol.* **18** 1797–1801

[121] Turner, M., Buchan, G., Barrett, K. & Feldmann, M. (1986) Tumour necrosis factor induction of IL-1. *Br. J. Rheumatol.* **25** 110

[122] Lindsten, T., June, C. H. & Thompson, C. B. (1988) Multiple mechanisms regulate c-*myc* gene expression during normal T cell activation. *EMBO J.* **7** 2787–2794

[123] Bender, T. P., Thompson, C. B. & Kuehl, W. M. (1987) Differential expression of c-*myb* mRNA in murine B lymphomas by a block to transcription elongation. *Science* **273** 1473

[124] Baldwin, A. S. Jr. & Sharp, P. A. (1987) Binding of a nuclear factor to a regulatory sequence in the promoter of the mouse H-2K^b class I major histocompatability gene. *Mol. Cell. Biol.* **7** 305–313

[125] Chiu, R. M., Imagawa, M., Imbra, R. J., Bockoven, J. R. & Karin, M. (1987) Multiple cis and trans acting elements mediate the transcriptional response to phorbol esters. *Nature (London)* **329** 648–651

[126] Farrell, P. J., Rowe, D. T., Rooney, C. M. & Kouzarides, T. (1989) Epstein–Barr virus BZLF-1 transactivator specifically binds to a consensus AP-1 site and is related to c-*fos EMBO J* **8** 127–132

[127] Brennan, F. M., Chantry, D., Jackson, A. M., Maini, R. N. & Feldmann, M. (1989) Inhibitory effect of TNFα antibodies on synovial cell Interleukin-1 production in rheumatoid arthritis. *Lancet* **ii**, 244–247.

14

The molecular biology of multidrug resistance

Lela Veinot-Drebot and Victor Ling
The Ontario Cancer Institute, Department of Medical Biophysics, University of Toronto, Ontario M4X 1K9, Canada

1. INTRODUCTION

The development of drug resistance is likely to be a major cause of failure in cancer chemotherapy. Some tumours are resistant from the outset to many of the most active antineoplastic agents. Others that are initially responsive to a combination of anticancer drugs frequently recur, and are non-responsive to the same drug combination or to different drug combinations. Drug-resistant cell lines and transplantable tumours have been used as tools to study mechanisms underlying resistance to drugs used in cancer chemotherapy.

Multidrug resistance (MDR) is a selectable phenotype of cultured mammalian cells and transplantable tumours. Multidrug resistance is characterized by the simultaneous acquisition of cross-resistance to structurally and functionally unrelated anticancer drugs. Although MDR is usually associated with the overproduction of P-glycoprotein, alterations in topoisomerases [103], glutathione S-transferases [7, 123] or glutathione content [63] have also been postulated as mechanisms via which a drug resistance phenotype may be mediated.

This chapter focuses on the molecular biology of P-glycoprotein and our current understanding of what role this membrane molecule plays in the multidrug resistance phenotype. Recent reviews on multidrug resistance have included those by Endicott and Ling [36] and Bradley et al. [18a] which discuss P-glycoprotein from biochemical and cellular perspectives (for additional reviews, see [46,51,92,108,156]).

2. THE MULTIDRUG RESISTANCE PHENOTYPE

Animal and human cell lines selected for resistance to a single cytotoxic agent, such as colchicine, adriamyicon, actinomycin D or taxol, often display cross-resistance to unrelated drugs such as vinca alkaloids, anthracyclines, epidophyllotoxins, dactinomycin, melphalan, nitrogen mustard and mitomycin C [8,9,14,16,28,64,74,82,91, 113,118]. This phenotype has commonly been referred to as the multidrug resistance (MDR) phenotype.

One of the most interesting and yet still poorly understood aspects of multidrug resistance is that the pattern of cross-resistance can differ from one MDR cell line to another; this is true even for MDR cell lines selected from the same parent using the same selecting drug (for review, see [108]). The basis of the resistance phenotype in MDR cells [10,15,29,30,40,69,74,81,124,125] appears to result from a net decrease in the intracellular concentration of the drugs involved compared with the drug-sensitive parent. Studies have shown that this reduced accumulation results from the enhanced activity of an energy-dependent, broad specificity, drug efflux system [30, 67,69,125,126].

Overproduction of a 170 kDa protein, P-glycoprotein, in the plasma membrane is the most consistent change detected in independently derived MDR cell lines. The level of P-glycoprotein correlates well with the level of drug resistance and the decrease in intracellular drug accumulation in many MDR cell lines. Monoclonal antibodies prepared against P-glycoprotein from Chinese hamster ovary cells also cross-react with P-glycoprotein from other species and have been used to confirm that many MDR cell lines have increased levels of P-glycoprotein [8]. Analysis of P-glycoprotein DNA sequences shows that P-glycoprotein is a highly conserved protein and it may function as an energy-dependent drug efflux pump.

MDR cells are often hypersensitive (collaterally sensitive) to a broad range of compounds such as local anaesthetics, calcium channel blockers, some non-ionic detergents and steroid hormones (for reviews, see [18a,36]). Although the mechanism underlying this cytotoxic effect is not understood, disruption of the plasma membrane is thought to be involved. In support of this hypothesis, the gross ultrastructure of the cell membranes of highly resistant human and hamster MDR cell lines was shown to be distinctly different from those of the correponding drug-sensitive parental cell lines [4].

Multidrug resistance can be partially or completely reversed by a diverse group of compounds termed 'chemosensitizers' [18a,36]. Compounds involved in collateral sensitivity, such as local anaesthetics and non-ionic detergents, when added at low concentrations (non-cytoxic doses) in combination with cytotoxic drugs (e.g. vinblastine, adriamycin), can enhance their cytotoxicity. Other compounds that are also able to circumvent multidrug resistance can be grouped as (i) non-cytotoxic analogues of drugs involved in MDR (e.g. analogues and anthracyclines and vinca alkaloids) [70,71,127], (ii) calmodulin inhibitors and calcium channel blockers [10,12,19,23,40,68,76,105,109,128,143–145] and (iii) a diverse group of unrelated agents, such as chloroquin, progesterone and cyclosporins [23,48,96,104,121,129, 142,146].

The mechanism(s) underlying reversal of drug resistance are not understood since compounds with apparently different effects on cell physiology are able to interfere with multidrug resistance [75,76]. Many of the compounds which do circumvent multidrug resistance also increase drug accumulation and retention in MDR cells (for review, see [18a]); this effect strengthens the correlation between multidrug resistance and reduced drug accumulation. In the case of verapamil, it has been shown to bind specifically to membrane vesicles prepared from MDR cells and to inhibit vinblastine-analogue photoaffinity labelling of P-glycoprotein [23]. Thus, in some cases drug-reversing agents may act directly on P-glycoprotein interfering with the drug efflux pathway, perhaps as competitive inhibitors. Although many compounds involved in drug reversal are calcium channel blockers or calmodulin inhibitors, studies have shown that calcium-dependent processes are not directly involved in drug efflux [19,23,76,106,143].

In summary, there are several distinguishing features associated with the MDR phenotype: (i) variable patterns of cross-resistance to unrelated cytotoxic compounds, (ii) a net decrease in the intracellular concentration of drug, (iii) collateral sensitivity, (iv) the reversal by 'chemosensitizers' and (v) level of drug resistance correlates with P-glycoprotein content.

3. P-GLYCOPROTEIN AS AN ENERGY-DEPENDENT DRUG EFFLUX PUMP

Genomic and cDNA sequences encoding P-glycoprotein have been isolated and characterized. From the DNA sequence encoding P-glycoprotein from mouse [54,58,66] hamster [36,55,107,111] and human [20,21,149,151] a model of P-glycoprotein as a membrane-associated, energy-dependent, drug efflux pump has been postulated [20,45,54]. The predicted structure of P-glycoprotein (c. 1280 amino acids in length) is a highly conserved tandemly duplicated molecule joined by a 'linker' sequence of approximately 60 amino acids. Each half of the molecule consists of six potential transmembrane regions and two putative nucleotide-binding regions. Fig. 1 depicts the proposed structure of a single P-glycoprotein molecule relative to the plasma membrane. The potential transmembrane domains of P-glycoprotein are proposed to form a channel or pore through the plasma membrane. The orientation of P-glycoprotein relative to the membrane was determined using monoclonal antibodies that localized the C-terminus to the cytoplasm [73]. This model of P-glycoprotein has provided a framework with which it is possible to interpret data on the mechanistic basis of multidrug resistance.

Much of the proposed structure and function of P-glycoprotein is based on the remarkable similarity between the predicted amino acid sequence of P-glycoprotein and a diverse group of bacterial transport proteins [20, 45, 54]. Some of these bacterial proteins include permeases such as the *Salmonella typhimurium* histidine transport subunit, HisP, the *S. typhimurium* oligopeptide permease subunit, OppD, and the *Escherichia coli* maltose transport subunit, MalK. Others are efflux proteins such as those encoded by *Rhizobium meliloti* ndvA, required for the transport of β-1,2 glycan, *Bordetella pertussis* cyaB, involved in the efflux of calmodulin-sensitive adenylate cyclase, and an *E. coli* haemolysin secretion protein, HlyB [45, 47,131] (for review, see [2,79,84]).

The highest sequence similarity was found between the 66 kDa HlyB protein and the C-terminal portion of P-glycoprotein. For example, the predicted amino acid sequence encoded by the hamster *pgp2* gene showed 46.9% identity and a further 27.6% conserved amino acid similarity with the 228 possible amino acid matches of the HlyB protein. Although the highest similarity of the different prokaryotic transport proteins of P-glycoprotein is found in the regions that form a predicted nucleotide-binding domain, the similarity to HlyB and other bacterial transport proteins extends beyond this domain. The extensive conservation of sequences and

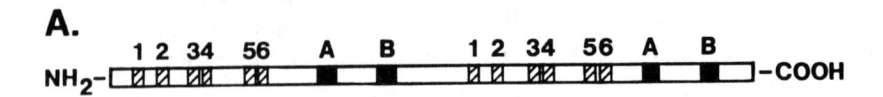

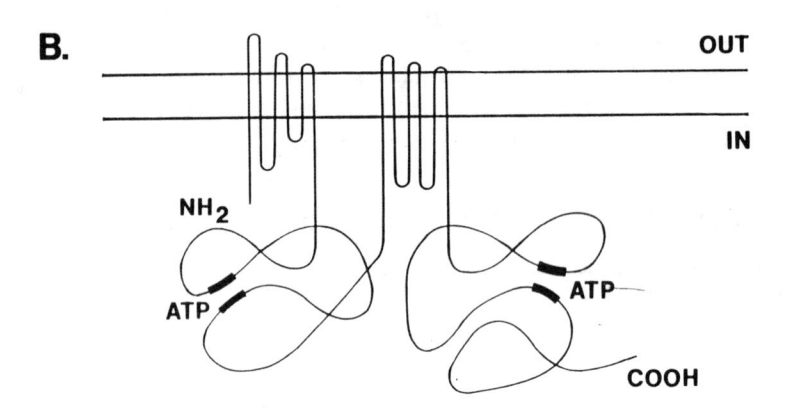

Fig. 1 — (A) A model of P-glycoprotein. The illustration shows that the primary sequence of P-glycoprotein is composed of two tandomly duplicated halves. The proposed transmembrane regions are marked by numbered boxes. The consensus sequences that form the potential nucleotide-binding fold common to P-glycoprotein and the bacterial periplasmic transport systems are marked by solid boxes lettered A and B (taken from ref. [35]). (B) Proposed orientation of P-glycoprotein within the plasma membrane lipid bilayer (redrawn from ref. [54]).

structural similarity between P-glycoprotein and bacterial transport proteins suggests that P-glycoprotein activity involves both nucleotide binding and transport. Indeed, human P-glycoprotein has been shown to bind ATP [24,95] and ATPase activity was recently found in purified preparations of P-glycoprotein [61,62]. Although transport of haemolysin in *E. coli* requires the activity of several proteins including HlyB [44], there is evidence to suggest that P-glycoprotein may interact directly with drugs. *In vitro* studies using photoaffinity-labelled vinblastine and competition with different compounds suggest that drugs bind to P-glycoprotein although it is not yet known whether this binding is necessary for active drug efflux *in vivo* (for review, see [36]).

Sequence similarity amongst functionally distinct eukaryotic and prokaryotic transport proteins would indicate that the P-glycoprotein gene family may belong to a much larger supergene family encoding membrane-associated transport proteins. In support of this hypothesis, two eukaryotic genes, predicted to encode transport proteins, were recently shown to have regions of similarity to the wide range of bacterial transport proteins described above as well as to P-glycoprotein [34,93]. The brown and white genes of *Drosphila melanogaster* are predicted to encode subunits of a hypothesized pteridine precursor permease [34,93,100]. Alignment of either of these genes to bacterial transport proteins (see section 3) showed that the highest amino acid sequence similarity for both of the *Drosophila* sequences was within the consensus ATP binding site [34,93]. In addition, the hydrophobicity and hydrophilicity patterns of both of these *Drosophila* sequences showed that each contains six potential transmembrane regions [34,100]. From molecular and genetic characterization of the transport of pigment precursors across *Drosophila* cell membranes it has been hypothesized that the formation of heteromultimeric transport structures by the putative permeases encoded by the white, brown and scarlet genes could provide the basis of mutiple transport systems [34]. As will be discussed below, P-glycoprotein in mammalian cells is encoded by a small multigene family. It remains to be established whether mammalian P-glycoprotein isoforms interact with each other in a manner analogous to the pteridine permeases in *Drosophila* (see Footnote 1).

P-glycoprotein is post-translationally modified. As much as 40 kDa of its molecular weight is due to *N*-linked glycosylation. Inhibition of P-glycoprotein glycosylation has been found not to affect the MDR phenotype [11,80]. P-glycoprotein is also phosphorylated, but it is less clear what role this modification may play in P-glycoprotein activity although there is evidence that phosphorylation may modulate P-glycoprotein function (for review, see [36]). It was found in an MDR human leukaemia cell line, K562/ADM, that agents which circumvent multidrug resistance, such as verapamil, significantly increased the phosphorylation of P-glycoprotein [60]. It has been suggested by *in vivo* and *in vitro* studies that P-glycoprotein may be a substrate for different protein kinases including protein kinases A and C [60,87,99,113]. Further *in vivo* studies will be required to determine the importance of phosphorylation in modulating P-glycoprotein activity.

The model of P-glycoprotein as an energy-dependent, drug efflux pump correlates well with the phenotype of multidrug resistance and strongly suggests that P-glycoprotein plays a key role in multidrug resistance. In addition, the high sequence similarity between P-glycoprotein molecules of different organisms (see section 4) would indicate that the function of this protein in drug-sensitive cells may be conserved.

4. THE P-GLYCOPROTEIN GENE FAMILY

4.1 Classification and organization of the P-glycoprotein gene family

Evidence showing that P-glycoprotein is encoded by a multigene family has come from several independent investigations. Genomic and cDNA probes homologous to P-glycoprotein sequences were shown by Southern analysis to cross-hybridize to multiple fragments of genomic DNA isolated from human and rodent drug-sensitive cells. In independently derived MDR cell lines, these sequences were found to be either amplified equally or differentially amplified (see section 6). The first cDNA clone, pCHP1, used in such an analysis was isolated from a CHO MDR cell line, CH^{R}B30, by screening a λ gt11 cDNA expression library with P-glycoprotein-specific monoclonal antibodies [73]. pCHP1 consists of 630 bp and spans five exons in the 3' terminal region of the hamster *pgp2* gene (Fig. 2) [98]. This probe also cross-hybridizes to other members in the hamster gene family as well as to those from mouse and human [73]. Evidence of a gene family was also obtained by sequencing independent cDNA clones homologous to P-glycoprotein sequences isolated from the same species. The coding sequences of these different cDNA clones were found

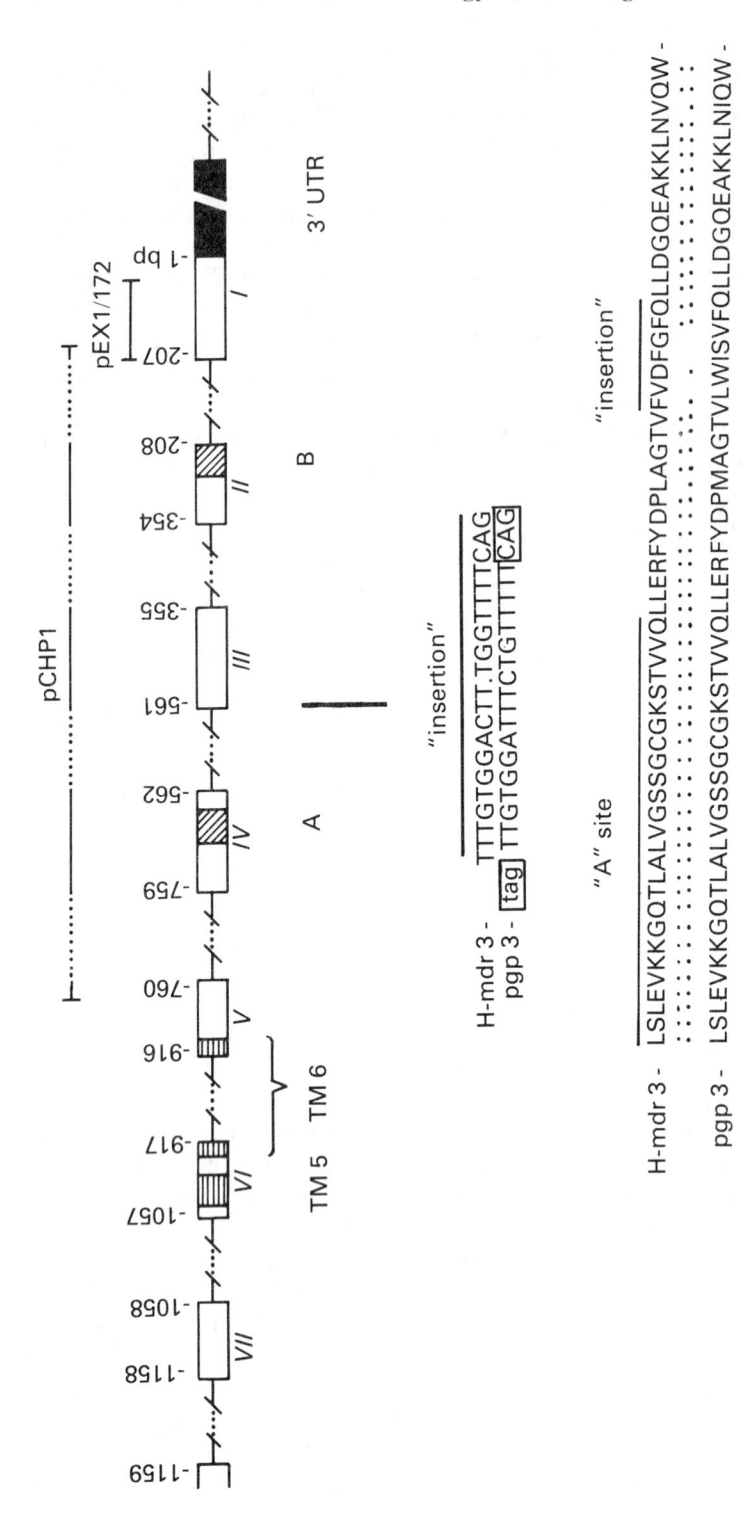

Fig. 2 — Organization of hamster P-glycoprotein genes. The upper illustration shows the 3′ exon–intron organization of hamster P-glycoprotein genes by comparison of the genomic nucleotide sequences with the hamster *pgp1* and *pgp2* cDNA sequences [35] and conformity to splice junction consensus sequences [101]. The locations of pCHP1 cDNA and the pEX1/172 probe are shown. The potential transmembrane regions TM5 and TM6 (▨), the putative nucleotide-binding domains (▥), and the 3′ untranslated region (UTR: ■) are shown. The coding nucleotide positions where the introns separate the exons are indicated above the exons. The nucleotide sequence shows the similarity between the putative 21-nucleotide insertion found in the hamster *pgp3* gene and the 21-nucleotide insertion found in human *mdr3* gene (H-*mdr3*). The lower illustration shows the predicted amino acid similarity between these two regions as well as the adjacent 'A' nucleotide-binding domain. This figure was adapted from refs [98,149].

to be highly similar to each other whereas the 3' non-coding sequences were found to be significantly different from each other [98].

The number of members in the P-glycoprotein gene family of humans and rodents was recently determined using a single-exon DNA probe, pEXI/172 (Fig. 2) [98]. The sequence of this probe, constructed from the hamster *pgp*2 gene, represents a highly conserved sequence (>90% similarity) amongst the P-glycoprotein genes identified thus far from hamster, mouse and human [98]. Sequences in the pEXI/172 probe are only moderately similar to those from the same region within the 5' duplicated half of P-glycoprotein genes from rodent and human cells and consequently do not cross-hybridize to this probe [98]. Thus, the number of restriction fragments detected by this probe was expected to represent the number of members in the P-glycoprotein gene family. By Southern analysis it was shown that the hamster and mouse P-glycoprotein gene families each consist of three members, but only two gene members are found in human. In an independent but similar study the size of the mouse P-glycoprotein gene family was recently confirmed [27]. In addition to the human *mdr*1 gene, two other human DNA sequences have been independently described, pMDR2 and *mdr*3, and were subsequently shown to be encoded by the *mdr*3 gene [112,149].

Members of the P-glycoprotein gene family encode highly similar gene products. For instance, the C-terminal cytoplasmic domains encoded by the hamster *pgp*1 and *pgp*2 genes, except for 13 amino acids, are identical to each other [35]. P-glycoprotein genes from different organisms can be classified into one of three classes based on the similarity of the 3' non-coding sequences with those of the hamster *pgp* genes (Table 1). The 3' non-coding sequences of the hamster *pgp* genes are distinct [35],

Table 1 — Relationships of the three hamster *pgp* genes to the mouse and human *mdr* genes based on direct comparisons of their 3' untranslated regions

Species	Gene associated with the following P-glycoprotein gene class [a]		
	I	II	III
Hamster	*pgp*1 (pL34)[b]	*pgp*2 (pL20)[b]	*pgp*3 (λP13)[c]
Mouse	*mdr*3[d]	*mdr*1 (λDR11)[e]	*mdr*2 (λDR29)[f]
Human	*mdr*1[g]		mdr3[h]

Taken from ref. [98].
[a]The name of the cDNA or genomic clone representing the gene is in parenthesis. Expression of genes designated by asterisks confers the MDR phenotype.
[b]Endicott *et al.* [35].
[c]Ng *et al.* [98].
[d]Gros, personal communication.
[e]Gros *et al.* [54].
[f]Gros *et al.* [58].
[g]Chen *et al.* [20].
[h]van der Bliek *et al.* [150,151].

and represent gene-specific sequences. The fact that P-glycoprotein genes from mouse and human contain 3' non-coding sequences similar to those from hamster suggests that these genes are related to one another. To maintain consistency in this chapter, the gene class is presented wherever specific gene members are discussed.

The geneology of the hamster P-glycoprotein gene family was recently proposed [98] (Fig. 3). Briefly, the high amino acid sequence similarity between P-glycoprotein

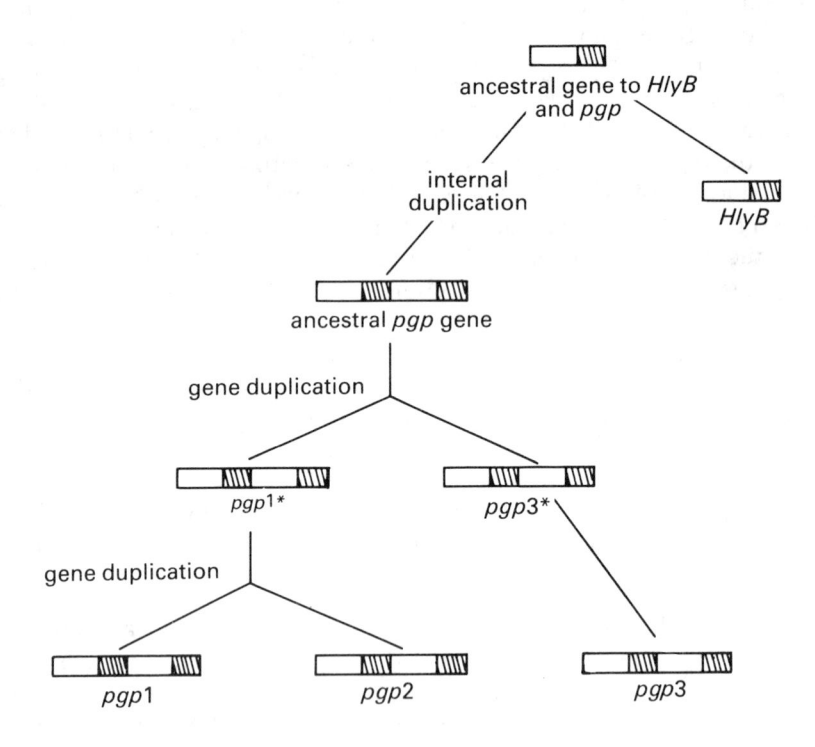

Fig. 3 — Proposed geneology for the hamster P-glycoprotein gene family (taken from ref. [98]).

and the *E. coli* HlyB protein (see section 3) suggests that both genes evolved from a common ancestral gene and at some point the ancestral P-glycoprotein gene became a duplicated sequence. It is proposed that such a duplication event occurred prior to the formation of multiple P-glycoprotein-like genes as in any one species the 5' and 3' halves of P-glycoprotein genes share significantly less similarity with each other than with the equivalent halves of other gene members [58,151]. In addition, a significant similarity is found between the intron sequences of the hamster *pgp*1 and *pgp*2 genes (class I and II respectively), suggesting that these two genes arose from a recent gene duplication (Fig. 3) [35,98]. Sequence analysis of human *mdr*3 cDNA (class III) and hamster *pgp*3 genomic sequences (class III) indicates these genes have diverged from the other members [35,151]; the same may also be true for the mouse *mdr*2 gene (class III) [58].

Additional insight into the evolution of the P-glycoprotein gene family has been obtained by determining the intron–exon arrangement of these genes (Fig. 2) [98, R. Zastawny and V. Ling, unpublished results]. The organization of the complete 3' halves and part of the 5' halves of the three hamster *pgp* genes suggests that all members of this family have a similar intron–exon arrangement (Fig. 2). It is predicted that each hamster P-glycoprotein gene consists of at least 26 exons [98]. This similarity supports the hypothesis that all P-glycoprotein genes arose from a single ancestral P-glycoprotein-like gene which duplicated during the course of speciation (Fig. 3). Determining the intron–exon arrangement of P-glycoprotein genes has also shown that the putative nucleotide-binding domains and many of the potential transmembrane domains are contained on separate exons [98, R. Zastawny and V. Ling, unpublished results] (see Footnote 2).

4.2 P-glycoprotein genes in different species

How widely distributed are P-glycoprotein genes? Hamster P-glycoprotein DNA sequences, pEXI/172 (see above), have been shown to cross-hybridize to genomic DNA isolated from diverse organisms, such as fish, chicken and *Drosophila melanogaster* (L. Veinot-Drebot, P. L. Davies and V. Ling, unpublished results). Thus, it would seem that P-glycoprotein arose early in evolution and has been highly conserved in many organisms. It is likely that P-glycoprotein function is of fundamental importance.

The number of P-glycoprotein gene members differs between human and rodents. Southern analysis of rhesus monkey and orangutan DNA, using the pEXI/172 probe, suggests that all primates may only contain two P-glycoprotein gene members [98]. It is not known how closely related the gene members in either of these primates are to the human *mdr*1 or *mdr*3 genes (classes I and III respectively), but they are probably very similar. Using this same approach, evidence has also been obtained to suggest that both chicken and rabbit may each contain only two P-glycoprotein genes. On the other hand, it was found that bovine species may contain three gene members and porcine species as many as five (L. Veinot-Drebot and V. Ling, unpublished results). If each of the gene members in these organisms encodes functional P-glycoprotein, it is possible that the size of the P-glycoprotein gene family reflects species-specific differences in P-glycoprotein isoforms. Alternatively, some of these gene members may be found to be pseudogenes. However, based on these studies, it would appear that the normal function of P-glycoprotein may require a minimum of two P-glycoprotein genes. Continued efforts to characterize P-glycoprotein genes in different organisms should provide insight into P-glycoprotein biology as well as providing a lineage of the evolution of P-glycoprotein genes.

Is there evidence that lower eukaryotes also acquire multidrug resistance? Pleiotropic drug resistance is a selectable phenotype of *Saccharomyces cerevisiae* mediated by mutations in the *PDR*1 regulatory protein. *Pdr*1 mutant cells are resistant to very different metabolic inhibitors, such as cycloheximide, oligomycin and venturicidin [6]. Although the sequence of the *PDR*1 gene product is not similar to P-glycoprotein, the expression of P-glycoprotein and perhaps other transport protein genes may be regulated by this gene product (see Footnote 1).

Chloroquine-resistant malaria parasite strains (*Plasmodium falciparum*) were recently shown to be cross-resistant to other drugs such as mefloquine and quinine; drug resistance was found to be correlated with an increased rate of drug efflux [78].

In addition, this drug-resistant phenotype was reversed by verapamil and other compounds [85]. These findings suggest that multidrug resistance and P-glycoprotein may be associated with other human diseases in addition to cancer (see Footnote 3).

5. DNA TRANSFECTION STUDIES

Several studies have demonstrated that transfection with high molecular weight genomic DNA from MDR cells results in expression of the MDR phenotype in previously drug-sensitive cells [32,33,56,119,137]. In an initial study, drug-sensitive mouse LTA cells were transfected with DNA from the colchicine-selected Chinese hamster CH^{R}C5 cell line [33]. After a three-step selection with colchicine, independent multidrug resistant clones were shown by Southern analysis to contain multiple copies of the hamster *pgp*1 (class I) gene [33, K. Deuchars and V. Ling, unpublished results]. Amplification of the endogenous mouse P-glycoprotein genes was not observed. In addition, by using monoclonal antibodies that discriminated between P-glycoprotein encoded by mouse and hamster genes, it was demonstrated by Western analysis that the transformed clones overproduced hamster P-glycoprotein. Moreover, genes that normally flank the hamster P-glycoprotein genes and are co-amplified in CH^{R}C5 cells (see section 6) were apparently not co-transfected or expressed in the mouse transfectants. Taken together, transfection studies with genomic DNA support the conclusion that P-glycoprotein is directly causative of multidrug resistance.

The isolation and characterization of cDNAs encoding P-glycoprotein has provided insight into the molecular basis of multidrug resistance. Drug-sensitive mammalian cells tranfected with cDNA encoding P-glycoprotein acquire multidrug resistance. This provides the best evidence thus far that P-glycoprotein plays a causative role in multidrug resistance. For example, drug-sensitive mouse NIH 3T3 cells acquired multidrug resistance after transfection with a full-length, wild-type mouse *mdr*1 cDNA (class II). In this study, the transfectants were selected for resistance to adriamycin and a correlation was found between the level of P-glycoprotein mRNA, the copy number of the transfected cDNA and the level of drug resistance. These transfectants were also cross-resistant to daunorubicin, colchicine, vinblastine and actinomycin D [26].

Table 2 summarizes some of the studies which have shown that expression of cDNA encoding P-glycoprotein is directly associated with multidrug resistance (for others, see [21,26,57,59,102,147]). The main points of Table 2 and of this section are (i) increased expression of a cDNA from a single member of the P-glycoprotein gene family is sufficient for expression of cross-resistance, (ii) different patterns of cross-resistance can be obtained by transfecting cells with a cDNA encoding P-glycoprotein cloned from a drug-sensitive cell line, (iii) infection of a P-glycoprotein cDNA under the control of a retrovirus expression vector is sufficient for cross-resistance without prior selection with drugs, (iv) cDNA encoding P-glycoprotein from one species can confer multidrug resistance onto another species, (v) collateral sensitivity and drug reversal are associated with overexpression of cDNA encoding P-glycoprotein and (vi) only specific P-glycoprotein isoforms may be directly associated with multidrug resistance.

Collateral sensitivity is associated with cells transfected with a cDNA encoding P-

Table 2 — Full-length cDNAs encoding P-glycoprotein confer MDR: summary of cDNA transfection studies.

cDNA	Vector construct	Transfected cell line	Selection scheme	Cross resistance
Mouse *mdr*1[a] (class II)	pDREX4[b]	Hamster LR73	Adriamycin / Colchicine	Adriamycin Colchicine Vincristine
	pBAmdr[c]	Mouse NIH 3T3	G418	Adriamycin Colchicine Vinblastine
	pHmdr[d]	Mouse NIH 3T3	See notes	Adriamycin Colchicine Vinblastine
Human *mdr*1 (class I)	pGMDR[e]	Mouse NIH 3T3 / Human KB	Colchicine	Adriamycin Colchicine Vinblastine
	pHaMDR1/A[f]	Mouse NIH 3T3 / Human KB	See notes	Adriamycin Colchicine Vinblastine
	pHaMDR1/A[f]	Madin-Darby Canine kidney (MDCK)	Colchicine	Not reported
	pUCFVXmdr1(gs)[g]	Human KB3-1	Vinblastine / Colchicine	Adriamycin Colchicine Vinblastine

a All mouse cDNA transfection studies reported thus far have been done using a full-length mouse *mdr*1 cDNA clone (class I), λDR11, obtained from a drug-sensitive mouse pre-B-cell line [57].

b Mouse *mdr*1 cDNA was introduced into the expression vector p91023B. Transfected Chinese hamster ovary cells, LR73, were selected with either adriamycin or colchicine. Transfected cells showed a 10- to 20-fold increase in resistance to colchicine, adriamycin and vincristine when compared with LR73 control cells [57].

c Mouse *mdr*1 cDNA was introduced into the expression vector pBA [26]. Mouse NIH 3T3 cells co-transfected with pBAmdr and an expression vector containing the neomycin resistance gene were selected using G418. These transfected cells were significantly more resistant to adriamycin when compared with control cells, but less resistant when compared with transfected cells selected directly with adriamycin. Resistance was correlated with the steady state level of *mdr* mRNA. Co-transfected cells were resistant to daunorubicin and actinomycin D as well as to those drugs listed in the table [26].

d Mouse *mdr*1 cDNA was introduced into the retrovirus expression vector pH [58]. Infected mouse NIH 3T3 cells were selected for resistance to either colchicine or adriamycin. Infectants were also obtained by growth in the absence of drug selection. Unselected infectants were less resistant to colchicine, vinblastine and adriamycin than were colchicine-selected clones [59].

e A full-length human *mdr*1 cDNA (class I) was constructed using three overlapping cDNA clones isolated from the MDR human KB-C2.5 carcinoma cell line; the cDNA clone was introduced into the expression vector pGV16 [147]. Transfected NIH 3T3 and human KB cells were first enriched by selection in medium containing G418; pooled G418-resistant cells were selected for resistance to colchicine. Variable patterns of cross-resistance were shown for mouse transfected cells; 6- to 15-fold, 3- to 10-fold and 1.5- to 3-fold increases in resistance to colchicine, vinblastine and adriamycin, respectively when compared with control cells [147]. Transfected human KB cells, like mouse transfectants, were more resistant to colchicine than to either adriamycin or vinblastine [147].

f Human *mdr*1 cDNA [147] was introduced into the retrovirus expression vector pCO1 [102]. After exposure to pHaMDR1/A, infected human KB and mouse 3T3 cells were grown in the presence of the cytotoxic drugs listed in the table. Infected KB cells were also shown to be resistant to puromycin, VP16 and actinomycin D [102]. Infected MDCK cells were selected for resistance to colchicine; these clones were subsequently used to show that expression of *mdr*1 cDNA was sufficient for the expression of P-glycoprotein in a polarized manner on the apper surface of cells [102].

g A full-length human *mdr*1 cDNA clone, λHDRV1, obtained from a cDNA library from vinblastine-selected KB-V1 cells was introduced into the retrovirus expression vector pUCFVXMΔH3. This construct encoded the wild-type glycine residue at position 185 and was used to construct additional cDNA clones, one of which encoded the colchicine-specific valine residue at position 185 (see text) [20]. Human KB-3-1 transfectants were selected with either vinblastine or colchicine. Cells expressing the colchicine-specific valine residue at position 185 grew in 3- to 6-fold higher concentrations of colchicine than did other transfectants [20]. Agents used to show cross-resistance, and listed in the table, are not listed by the level of resistance that transfected cells were obtained.

glycoprotein. Mouse NIH 3T3 cells transfected with a wild-type mouse *mdr*1 cDNA (class II) and selected with adriamycin were shown to display a modest increase in sensitivity to verapamil relative to the parent drug-sensitive cell line. In addition, verapamil in combination with adriamycin abolished the MDR phenotype in these transfected cells [26]. However, it should be noted that these studies provide only indirect evidence that collateral sensitivity is directly mediated by P-glycoprotein. These properties were examined in transfected cells that were subsequently stepwise selected for increased resistance to adriamycin [26]. Thus, it is not entirely clear whether overexpression of a P-glycoprotein cDNA *per se* is sufficient to mediate collateral sensitivity or whether additional factors acquired during selection are also involved.

Retrovirus expression vectors that express cDNA encoding P-glycoprotein have been used to show that cross-resistance can occur without first selecting cells resistant to a drug. Mouse NIH 3T3 cells expressing a wild-type mouse *mdr*1 cDNA (class II) were found to display a low level of resistance to colchicine, doxorubicin and vinblastine without first selecting cells resistant to any one of these three drugs [59]. Similar results have also been obtained by transfecting mouse NIH 3T3 cells with a plasmid vector containing a mouse *mdr*1 cDNA (class II) and selecting transfectants using the co-transfected neomycin resistance gene [26]. Thus, these studies would indicate that increased expression of P-glycoprotein genes is sufficient for the MDR phenotype and that, if more than expression of P-glycoprotein genes is required for multidrug resistance, such factors are not limiting under non-selective conditions. It has not been reported whether MDR cells generated in this fashion are also collateral sensitive.

One enigmatic aspect of multidrug resistance is the variable patterns of cross-resistance observed amongst different MDR cell lines. As described in section 2, MDR cells are often more resistant to the agent used in the selection than to other drugs, and the overall pattern of cross-resistance can vary greatly from one cell line to another. Transfection studies using cDNAs have been useful to address the molecular basis of this variability.

One study has shown that a mutation in P-glycoprotein can affect the drug resistance pattern. MDR cells preferentially resistant to colchicine were obtained during the selection of a series of stepwise-selected, colchicine-resistant cells from the drug-sensitive human KB/HeLa cell line [1,112,118,119]. *mdr*1 (class I) from one of these mutant cell lines, KB-C2.5, was shown to confer preferential resistance to colchicine when used to transfect either mouse NIH 3T3 cells or human KB cells [147]. Comparing the seqeunce of this cDNA with a *mdr*1 cDNA (class I) from the vinblastine-selected MDR KB-V1 human cell line revealed a mutation at codon 185 that is predicted to cause a glycine-to-valine substitution. Subsequent studies indicated that this mutation was sufficient to confer preferential resistance to colchicine [21]. Based on the predicted structure of P-glycoprotein and these data (Fig. 1) [20,45,54], the glycine-to-valine substitution is located within the first hydrophobic region, on the cytoplasmic side of the membrane. It has been postulated that this may be a drug-binding region [21].

Although it is still unclear what functional or structural changes occur in P-glycoprotein as a result of such a mutation at codon 185, this study supports the hypothesis that P-glycoprotein is directly involved in multidrug resistance. It is also the first example to show that different patterns of cross-resistance in independently derived MDR cell lines could be caused by an overproduction of variant P-glycoprotein molecules. In addition, these results would indicate that although individual P-glycoprotein isoforms are overall highly conserved subtle differences in the primary sequence could account for different functions of individual wild-type P-glycoprotein isoforms. Identification and characterization of other mutations in P-glycoprotein isoforms may provide further insight into the mechanism underlying the variability of cross-resistance patterns.

In addition to mutations, other factors involving P-glycoprotein gene expression may contribute to the pattern of cross-resistance. It has been hypothesized that unbalanced P-glycoprotein gene expression may contribute to the pattern of cross-resistance. In one study of a CHO MDR cell line, CH^{R}C5, it was shown by Northern analysis that both the *pgp*1 and *pgp*2 genes (classes I and II, respectively) were simultaneously expressed [35]. Cells transfected with a cDNA encoding P-glycoprotein also show variable patterns of cross-resistance. As an example, a variable pattern of cross-resistance to doxorubicin and vinblastine was found amongst mouse NIH 3T3 cells infected with a retroviral expression vector containing a full-length, wild-type mouse *mdr*1 cDNA (class II) and selected with colchicine [59]. The level of cDNA-derived *mdr*1 mRNA from independent colonies varied by 10-fold [59]. Thus, the absolute cellular level of *mdr*1 mRNA (class II) (presumably this represents the level of P-glycoprotein) may, in part, contribute to the overall pattern of cross-resistance [59].

A variable pattern of cross-resistance to vinblastine and doxorubicin was also reported for mouse NIH 3T3 colchicine-selected transfectants expressing an *mdr*1 cDNA (class I) derived from the human KB-C2.5 MDR cell line (shown to express a mutated *mdr*1 gene; see above) [147]. Using P-glycoprotein gene-specific probes for Northern analysis, only the mouse *mdr*1 gene (class II) appears to be expressed in drug-sensitive (untransfected) mouse NIH 3T3 cells [27]. Thus the pattern of cross-resistance may also be affected by the ratio of different P-glycoprotein isoforms. It is not known whether P-glycoprotein isoforms interact with each other. Alternatively, the expression of other genes may contribute to the MDR phenotype [147]. Additional insight into the basis of cross-resistance might be obtained from transfection studies using P-glycoprotein cDNAs encoding different isoforms and by quantitating the individual levels of P-glycoprotein isoforms expressed.

Evidence that different P-glycoprotein isoforms may be overproduced in MDR cells was obtained by the stepwise selection of resistant cells from drug-sensitive mouse J774.2 cells using colchicine, vinblastine or taxol as the selecting agent [52, 53]. In these studies two P-glycoprotein precursors (120 and 125 kDa) were shown tio be overproduced in different sublines. Of three sublines, all equally resistant to vinblastine, the one subline which overproduced the 120 kDa precursor species contained 50% less P-glycoprotein than did the sublines which produced the 125 kDa precursor species. Thus, it was speculated that the P-glycoprotein represented by the 120 kDa precursor may be more efficient for vinblastine efflux compared with the 125 kDa form. An apparent 'switching' from the expression of the 125 kDa precursor to the 120 kDa precursor was observed during the course of selection for increased vinblastine resistance. The mechanism of this switching is as yet unclear. The different molecular weight precursors may represent P-glycoprotein isoforms expressed by different gene members, but this has not yet been established. Such information will be required to investigate the functional significance of these two P-glycoprotein species identified in this system.

Although different P-glycoprotein gene members may be directly involved in multidrug resistance in rodent cells, this may not be the case for humans. Of the two members in the human P-glycoprotein gene family, only *mdr*1 cDNA (class I) has been shown to express the MDR phenotype in transfected cells [21,119,147]. Mouse *mdr*1 cDNA (class II) also confers MDR onto drug-sensitive cells [57,59]. Expression of transfected hamster *pgp*1 genomic DNA (class I) in mouse cells (see section 5) and the simultaneous expression of the hanster *pgp*1 and *pgp*2 genes (classes I and II respectively) in the CHO MDR cell line, CH^{R}C5, are both correlated with multidrug resistance [31, K. Deuchars, P. Juranka and J. Endicott, unpublished results]. However, expression of the hamster *pgp*3 gene (class III) has not been shown to be correlated with drug resistance [31, K. Deuchars and V. Ling, unpublished results]. Expression of the human *mdr*3 gene (class III) is not detectable in the MDR cell lines examined thus far [112,150]. Likewise, melanoma cells transfected with a full-length human *mdr*3 cDNA (class III) are not resistant to vinicristine [151]. However, it could not be determined in this study whether these transfectants expressed a functional P-glycoprotein encoded by the cDNA [151]. Additional transfection studies are needed to define better the role of different P-glycoprotein isoforms in multidrug resistance.

In summary, drug-resistant clones generated by DNA-mediated transfection of particular P-glycoprotein gene family members have many of the characteristics of multidrug resistant cell lines produced by stepwise selection *in vivo*. Thus, P-glycoprotein is capable of conveying the characteristics of multidrug resistant cell lines to previously drug-sensitive cells.

6. AMPLIFICATION OF P-GLYCOPROTEIN GENES

Members of the P-glycoprotein gene family in hamster, mouse and human are physically linked to one another. The P-glycoprotein gene family maps to chromosome region 1q26 in hamster [72,138], to chromosome 5 in mouse [86] and to chromosome region 7q21–31 in human [41,141]. In hamster and human cells, the *pgp*1 (*mdr*1) and *pgp*3 (*mdr*3) genes map within 500 kb of each other [17,149]. The size of the mouse *mdr*1 gene is 68 kb and the human *mdr*1 gene has been shown to encompass approximately 100 kb of genomic DNA [50,154,159].

Karotypic abnormalities normally detected in cells containing amplified DNA are often detected in drug-resistant cells (for review, see [132,133]). In MDR cells, amplified DNA sequences are seen as either (i) one or more homogeneous staining regions (HSR) that map to the same locus as the P-glycoprotein gene family or translocated to other chromosomal sites or (ii) double minute chromosomes. More recently, large circular DNA molecules or episomes have been found to be associated with gene amplification [114]. Episomes containing the human *mdr*1 gene (class I) have been detected in the vinblastine-resistant KB-V1 human carcinoma cell line [114]. The karyotypic properties of MDR cells vary from one cell line to another and may depend, in part, on the level of drug resistance sought in the selection (for review, see [18a]).

Development of multidrug resistance in hamster, mouse and human cell lines is usually associated with P-glycoprotein gene amplification [31, 55, 107, 117, 120]. The level of drug resistance often correlates with the level of P-glycoprotein gene amplification [107, 117]; exceptions to this are discussed in section 7. In addition, P-glycoprotein gene members have been shown to be differentially amplified and expressed in independently derived MDR cell lines. The hamster *pgp*1 (class I) and the human *mdr*1 (class I) genes are amplified and overexpressed in most MDR cell lines of the appropriate species and the degree of overexpression corresponds to relative resistance [31, 86, 120]. Likewise, in the multidrug resistant mouse ECH^R cell line, development of MDR is associated with amplification of only a subset of the DNA fragments recognized by a P-glycoprotein cDNA probe (pCHP1) [107]. These results supports the hyopthesis that the overproduction of specific isoforms may be sufficient for multidrug resistance.

Amplified DNA sequences encoding P-glycoprotein have been cloned from human and rodent MDR cell lines. The technique of DNA renaturation in agarose gels [110] was used to clone amplified DNA sequences from independently derived Chinese hamster MDR cell lines [111]. The sequence isolated, pDR1.1, was used to clone an additional 120 kb of apparently contiguous amplified genomic DNA from one of these hamster cell lines (LZ) [55]. In a subsequent study, amplified sequences from sublines of human KB carcinoma cells, stepwise selected for increasing resistance to colchicine, vinblastine or adriamycin, were cloned and shown to encode P-glycoprotein [112]. Two clones, pMDR1 and pMDR2, were obtained using a DNA probe, pDR4.7, isolated in the study described above [112]. Clone pMDR1 was shown to correspond to the human *mdr*1 gene (class I) and was used in a subsequent study to clone *mdr*1 cDNA from the colchicine-selected human MDR cell line KB-C2.5 [20]. Clone pMDR2 was later shown to correspond to a human *mdr*3 cDNA clone (class III) [149].

By Southern analysis, clones pMDR1 and pMDR2 were found to hybridize to amplified fragments in several independently derived highly drug-resistant human sublines [112]. In addition, by Northern analysis, the steady-state level of the *mdr*1 transcript (class I) (c. 4.5 kb) in these sublines was shown to correlate with the level of gene amplification and drug resistance. On the other hand, the pMDR2 probe (class III) did not detect an RNA transcript in either the MDR sublines or the drug-sensitive parental subline [112]. In another study, the level of *mdr*1 (class I) expression in other independently derived MDR cell lines was shown to be higher than expected from the level of gene amplification in these cells [150]. However, for these cell lines, as for the others described above, expression of the co-amplified *mdr*3 gene (class III) was not detectable. Thus, these studies showed that increased expression of the *mdr*1 gene (class I) is intimately associated with multidrug resistance, but that an equivalent increase in expression of the human *mdr*3 gene (class III) is not required for MDR.

Sequences linked to P-glycoprotein genes are frequently co-amplified in multi-drug resistant cells. Genes overexpressed in the colchicine-selected Chinese hamster CH^{R}C5 MDR cell line were identified by differentially screening cDNA libraries from the drug-resistant cell line and the AuxB1 drug-sensitive parental cell line [31,

153]. This analysis produced six different classes of cDNAs that did not cross-hybridize with each other. *In situ* hybridization and Southern analysis of genomic DNA, resolved by pulse-field electrophoresis, showed these co-amplified genes were physically linked to P-glycoprotein genes. In addition, these genes form part of an amplicon in CH^{R}C5 cells that is at least 1100 kb in size [72,153]. Recently, genes represented by these cDNA clones were shown to be conserved in mouse and human and are co-amplified in MDR cell lines derived from either of these species [17,130, 150].

Only two of the genes represented by the six different cDNAs described above have been extensively characterized. Members from one of these classes of cDNAs were shown to correspond to the hamster P-glycoprotein genes [150]. Members from another class were shown to encode sorcin (V19), a cytosolic calcium-binding protein homologous to calpain [152]. In earlier studies it was shown in independently derived MDR cell lines that both sorcin and P-glycoprotein were overproduced, suggesting that sorcin may also be involved in MDR [88, 89]. However, cloned cDNA transfection studies demonstrated that increased expression of sorcin is not required for MDR.

Genes co-amplified in the CH^{R}C5 hamster MDR cell line have been shown to be differentially amplified and expressed in other MDR cell lines of hamster and human origin [31,150,153]. Indeed, not all of the sequences amplified in CH^{R}C5 cells were co-amplified in other MDR cell lines and some of the amplified coding sequences appeared to be underexpressed. Only DNA encoding P-glycoprotein was consistently amplified and overexpressed. In independently derived human MDR cell lines only the human *mdr*1 gene (class I) was shown to be overexpressed although the *mdr*3 gene (class III) and flanking genes were also co-amplified [150]. Thus, sequences linked to P-glycoprotein may be fortuitously co-amplified in MDR cells. Although the cDNA transfection experiments described above clearly indicate that increased P-glycoprotein gene expression alone is sufficient to elicit MDR, it is possible that overexpression of co-amplified flanking genes may contribute to the complexity and variability of multidrug resistance. Co-transfection and expression of these co-amplified genes with those encoding the P-glycoprotein isoforms would be necessary to explore this hypothesis fully.

7. EXPRESSION OF P-GLYPROTEIN GENES

7.1 Expression in tissues

P-glycoprotein has been detected in numerous tissues and organs using both P-glycoprotein-specific monoclonal antibodies and DNA probes. Distribution of P-glycoprotein in tissues has been studied in human [42,43,90,122,134,135,139], hamster [5, 94, Georges *et al.*, 162], and mouse [27]. Although it has been difficult to detect reproducibly very low levels of P-glycoprotein, expression studies have shown that the level of P-glycoprotein varies over a wide range amongst different tissues.

In humans, P-glycoprotein has been detected consistently in the adult adrenal gland, liver, kidney, and also in the pancreas, jejunum and colon [43,65,134]. Recently, P-glycoprotein was also detected in the capillary endothelium of the brain, testes and skin [22]. The physiological role of P-glycoprotein in these diverse cell types is as yet unknown. In the case of intestinal epithelium cells, however, it has been speculated that P-glycoprotein, as a membrane transport protein, may be involved in a detoxification or protection mechanism to protect against plant alkaloids and other xenobiotics ingested in the diet [77]. The presence of P-glycoprotein in the capillaries of brain and testes is particularly interesting in the context of chemotherapy because these organs are known to be pharmacological sanctuaries for metastatic cancer. Thus, it may be postulated that P-glycoprotein is involved in the transport of molecules including cytotoxic agents out of capillary endothelium cells into the capillary lumen. P-glycoprotein has also been shown by immunohistochemical analysis to be polarized in the kidney, intestine, liver and pancreas. In the liver, P-glycoprotein is found on the biliary surface of hepatocytes, while in the intestinal mucosa and in the proximal renal tubule P-glycoprotein is localized to the apical surface of lining epithelial cells, adjacent to the lumen [139]. The localized distribution of P-glycoprotein in these organs is also suggestive of a membrane transport function for P-glycoprotein.

High levels of P-glycoprotein mRNA are found in carcinogen-induced, preneoplastic, rat liver nodules which are resistant to a large number of xenobiotics [25,37,140]. However, these cells also overproduce the anionic isozyme of glutathione S-transferase (GSTπ), a phase II enzyme (drug conjugating), normally found at low levels in liver and which may be involved in cellular detoxification [25] (for review, see [97]). Thus it remains to be established in these cells how P-glycoprotein and GSTπ may contribute to drug resistance. Only one MDR cell line, the human breast cancer Adr^R MCF7 cell line, has also been shown to overproduce both P-glycoprotein and GSTπ [7,38,39,115,116].

7.2 Regulation of gene expression

At present, few studies have reported DNA sequences involved in the regulation of P-glycoprotein gene expression. The promoter and transcription initiation sites have been reported for the human *mdr1* gene (class I) and the mouse *mdr*1 gene (class II) [159]. A 0.43 kb genomic fragment, 5' to the *mdr*1 gene, was recently shown to contain sequences sufficient for promoter activity *in vivo* [148]. Within this sequence is a consensus CAAT box and two GC box-like sequences; however, it does not contain a TATA-like sequence. Transcription of the human *mdr*1 gene (class I) is initiated predominantly from two sites [148]. Transcription is initiated from an unique site for the mouse *mdr*1 gene (class II) [159]. Transcription in normal human tissues is initiated predominantly from the downstream initiation site. On the other hand, transcription from both sites has been detected in some MDR cell lines [148]. Some human tumours with high levels of *mdr*1 mRNA have been shown to initiate transcription from both of these start sites whereas others have been shown to use only the downstream site [49]. The significance of these multiple transcription sites is not clear, but it suggests that different mechanisms may regulate P-glycoprotein gene expression [49].

P-glycoprotein gene-specific probes were used for Northern analysis to show that P-glycoprotein genes are differentially expressed in normal tissues of mouse [49]. For example, *mdr*1 mRNA (class II) was easily detected in the adrenal gland whereas *mdr*2 mRNA (class III) was found at relatively lower levels and *mdr*3 mRNA (class I) could not be detected. On the other hand, mouse *mdr*3 mRNA (class I) was found to be very abundant in the intestine, but neither *mdr*1 (class II) nor *mdr*2 (class III) genes were found to be expressed (for additional examples, see [27]). The hamster P-glycoprotein gene family has also been shown to be differentially expressed in normal tissues although there may be differences between these two organisms in terms of the relative level of gene expression (K. Deuchars and V. Ling, unpublished results). Expression of the human *mdr*3 gene (class III) has only been detected in liver thus far, indicating that expression of the human P-glycoprotein gene family is also differentially expressed [27] (see Footnote 4).

P-glycoprotein gene expression may be hormonally regulated in some tissues. Using P-glycoprotein gene-specific probes for Northern analysis, low levels of *mdr*1 mRNA (class II) are detected in the non-gravid mouse uterus [27]. On the other hand, high levels of *mdr*1 mRNA (class II) are detected in the gravid mouse uterus [27]. Neither the *mdr*2 gene (class III) nor the *mdr*3 gene (class I) appeared to be expressed in either the gravid or the non-gravid uterus [27]. In another study, *in situ* hybridization of the gravid mouse uterus, using P-glycoprotein-specific DNA probes, demonstrated that increased levels of mRNA encoding P-glycoprotein were restricted to the secretory epithelial cells of the endometrium [27]. Immunohistochemistry using antiserum directed against P-glycoprotein showed that P-glycoprotein is localized to the luminal surface of these secretory epithelial cells; no evidence for P-glycoprotein gene expression was obtained in the non-gravid uterus in this particular study [3]. P-glycoprotein has also been detected in the hamster gravid uterus by immunohistochemistry using P-glycoprotein-specific monoclonal antibodies, but not in the non-gravid uterus (G. Bradley and V. Ling, unpublished results). High levels of P-glycoprotein are also found in the secretory epithelium of human placenta [135].

P-glycoprotein gene expression may also be regulated at the level of RNA processing. Human *mdr*3 mRNA (class III) has been shown to be alternatively spliced in normal cells. In one study, three different cDNA clones derived from human *mdr*3 mRNA (class III) were isolated (and characterized) from two different libraries; one library was constructed from mRNA of normal liver and the other was derived from mRNA of a drug-sensitive human liver cell line, HepG2 [149]. The more abundant *mdr*3 cDNA aligned with a full length human *mdr*1 cDNA sequence (class I) without significant insertions or deletions. However, the second *mdr*3 cDNA clone (class III) contained a large deletion (129 bp in the case of the cDNA from the library constructed from liver and 141 bp for clones isolated in the other library) spanning the fifth putative transmembrane domain encoded by the 3' half of the gene. It is not known whether such aberrant transcripts are translated *in vivo* or whether this gene product is involved in the normal function of P-glycoprotein.

Perhaps the most interesting cDNA variant found in this particular study was a human *mdr*3 cDNA clone (class III) containing a 21 bp insertion between the domains encoding the putative nucleotide binding site in the carboxy terminal half of P-glycoprotein [149]. The seven additional amino acids resulting from this insertion are predicted to lie within the cytoplasmic portion of P-glycoprotein. The hamster *pgp*3 gene (class III) was also recently shown to contain a 21 bp sequence highly similar (similarity score 90%) to the insertion in the human *mdr*3 cDNA clone and located in the same region of the gene, immediately 5' to exon III in the hamster gene (Fig. 2) [98]. Although a hamster *pgp*3 cDNA containing this 21 bp sequence has not been identified thus far, the consensus splice recognition sequence immediately 5' to this sequence suggests that the hamster *pgp*3 mRNA (class III) may also be alternatively spliced (class III) [98]. It should be noted, however, that despite the high sequence similarity at the nucleotide level only 29% perfect identify and 14% conserved similarity is found between the seven putative amino acids encoded by each of these 21 bp regions. As shown in Fig. 2, these seven amino acids are within a highly conserved region of all P-glycoprotein molecules. The significance of this low similarity at the amino acid level is not clear. However, the identification of a hamster *pgp*3 cDNA (class III) containing such an insertion should verify if hamster *pgp*3 mRNAs are alternatively spliced.

Overproduction of P-glycoprotein in MDR cells is usually correlated with a corresponding increase in the steady state level of P-glycoprotein mRNA. However, especially in cultured human MDR cell lines, this correlation is not always found. For example, the human KB-8-5 MDR cell line exhibits a low level of resistance to colchicine without any detectable change in the P-glycoprotein gene copy number or steady state level of mRNA [120]. Alternatively, SKOV3 human ovarian MDR cell lines resistant to low levels of either vinblastine or vincristine have an increased steady state level of P-glycoprotein mRNA without any detectable increase in the gene copy number [18b]. In both of these examples DNA amplification was only observed after these cell lines were used to select mutants expressing higher levels of drug resistance. Although overproduction of P-glycoprotein in cells expressing low levels of drug resistance may occur in the absence of gene amplification, high levels of P-glycoprotein are most commonly achieved in cell lines via gene amplification. In another study, two revertant cell lines of the human K562/ADM MDR cell line, selected for resistance with adriamycin, maintained the same copy number of the amplified *mdr*1 gene (class I), but the steady-state level of *mdr*1 mRNA was greatly decreased [136]. Taken together, these studies would indicate that transcriptional modification of P-glycoprotein genes can occur in MDR cells. Further characterization of P-glycoprotein gene expression in MDR cells may provide additional insight into the regulation of these genes in drug-sensitive cells.

Additional properties of P-glycoprotein gene expression have been described. Mature mRNA transcripts encoded by different P-glycoprotein gene members have different molecular weights in sensitive and MDR cells. Using gene-specific probes, it has been shown by Northern analysis that different lengths of 3' non-coding sequences partially explain the approximately 400 nucleotide difference between the human *mdr*1 (class I) transcript (4.5 kb) and the human *mdr*3 (class III) transcript (4.1 kb) [149]. A similar situation exists between the mRNA transcripts encoded by the hamster *pgp*1 and *pgp*2 genes (classes I and II respectively) [35]; transcripts encoded by mouse P-glycoprotein genes also have different molecular weights. For mouse, it has been shown by Northern analysis, using gene-specific probes, that two mature transcripts (5 and 6 kb) are encoded by the *mdr*3 gene (class I). Sequence analysis of cDNA derived from these transcripts indicates that alternative polyadenylation is responsible for these different sizes [27]; hamster *pgp*1 and *pgp*2 mRNAs (classes I and II) have also been shown to be alternatively polyadenylated [35].

In summary, many aspects of P-glycoprotein gene expression have been described and it is clear from these studies that multiple factors regulate the expression of the P-glycoprotein gene family. In general, drug-sensitive cell lines express low to undetectable levels of P-glycoprotein whereas it is easily detected in many normal tissues. This observation would indicate that the level of P-glycoprotein gene expression and the specific gene members expressed are dependent on tissue-specific fators. Thus, future studies of the regulation of P-glycoprotein gene expression will need to consider how different cell systems effect this regulation.

8. P-GLYCOPROTEIN GENE EXPRESSION IN TUMOURS AND RESPONSE TO CHEMOTHERAPY

It has been well established that cultured cells overproducing P-glycoprotein acquire multidrug resistance. Efforts have been made to determine whether the MDR phenotype is also expressed in human tumour cells and whether this phenotype correlates with chemotherapy failure. Numerous studies show that many human cancers, prior to chemotherapy or following chemotherapy failure, produce high levels of P-glycoprotein. Using immunoblot analysis, the initial study by Bell *et al.* [13] detected increased P-glycoprotein in some advanced non-responsive ovarian cancers. Subsequently, using a sequential series of biopsy specimens from two patients with acute non-lymphoblastic leukaemia, Ma *et al.* [83] detected a correlation between disease progression and the proportion of blast cells staining positive for P-glycoprotein. In another survey of adult patients with sarcoma, Gerlach *et al.* [44] detected P-glycoprotein in approximately 20% of the biopsy specimens; however, positive samples were also found in the group that had not received chemotherapy. More recently, using a human *mdr*1 cDNA probe (class I) for Northern and slot-blot analyses, Goldstein *et al.* [49] surveyed over 400 human cancers of various types. This large survey extended the findings made by Gerlach *et al.* which showed that some tumours have high levels of P-glycoprotein even prior to chemotherapy [44]. Tumours resistant to chemotherapy from the outset may express an 'intrinsically resistant' MDR phenotype. Untreated tumours derived from tissues that normally contain high levels of P-glycoprotein frequently also contain similar high levels of P-glycoprotein. However, for these tumours the level of P-glycoprotein may only be a differentiated phenotype and not necessarily the cause of chemotherapy failure.

It has not been directly demonstrated whether overproduction of P-glycoprotein is causative of chemotherapy failure. Molecular studies using cell lines indicate that only specific P-glycoprotein isoforms are directly involved in MDR. Thus, clinical studies need not only to detect P-glycoprotein but also to identify the P-glycoprotein isoform(s) produced by tumour cells to establish an accurate correlation between P-glycoprotein and chemotherapy failure. Epitopes of P-glycoprotein recognized by three monoclonal antibodies, C219, C494 and C32 [73], were recently mapped; from this analysis it was shown that this particular panel of monoclonal antibodies can be used to distinguish P-glycoprotein isoforms in human and rodent cells (Georges *et al.*, [162]). An immunohistochemical analysis of tumour samples which can identify individual P-glycoprotein isoforms may provide new insight into clinical studies and has great potential for diagnostic screenings. In addition, a similar analysis of normal tissues may provide insight into the normal function of individual P-glycoprotein isoforms.

9. CONCLUDING REMARKS

Multidrug resistance is a phenotype most commonly associated with the overproduction of P-glycoprotein, but expression of other genes may contribute to this phenotype. Although independently derived MDR cell lines often differ in their cross-resistance pattern, DNA-mediated transfection studies have shown that P-glycoprotein gene expression is associated with this variability.

P-glycoprotein is detected in many normal tissues and organs. The level of P-glycoprotein amongst these cells varies greatly and there is evidence that the P-glycoprotein gene family is differentially regulated. It is hypothesized that P-glycoprotein is a protein associated with the transport of drugs and possibly tissue-specific factors. As the normal function of P-glycoprotein becomes better understood, more will also be learned about the molecular basis of multidrug resistance.

ACKNOWLEDGMENTS

The authors would like to thank their colleagues at the Ontario Cancer Institute for helpful discussions and for critical reading of this manuscript. Studies conducted in the authors' laboratory were supported by Public Health Service grant CA37130 from the National Institutes of Health, USA, and a grant from the National Cancer Institute of Canada.

RFERENCES

[1] Akiyama, S.-i., Fojo, A., Hanover, J. A., Pastan, I. & Gottesman, M. M. (1985) *Somatic Cell Mol. Genet.* **11** 117–126
[2] Ames, G. F.-L. (1986) *Annu. Rev. Biochem.* **55** 397–425
[3] Arceci, R. J., Croop, J. M., Horwitz, S. B. & Housman, D. (1988) *Proc. Natl. Acad. Sci. USA* **85** 4350–4354
[4] Arsenault, L., Ling, V. & Kartner, N. (1988) *Biochim. Biophys. Acta* **938** 315–321
[5] Baas, F. & Borst, P. (1988) *FEBS Lett.* **229** 329–332.
[6] Balzi, E., Chen, W., Ulaszewski, S., Capieaux, E. & Goffeau, A. (1987) *J. Biol., Chem.* **262** 16871–16879
[7] Batist, G., Tulpule, A., Sinha, B. K., Katki, A. G., Myers, C. E. & Cowan, K. H. (1986) *J. Biol. Chem.* **261** 15544–15549
[8] Bech-Hansen, N. T., Till, J. E. & Ling, V. (1976) *J. Cell Physiol.* **88** 23–31
[9] Beck, W. T. (1983) *Cancer Treat. Rep.* **67** 875–882
[10] Beck, W. T. (1984) *Adv. Enzyme Regul.* **22** 207–227
[11] Beck, W. T. & Cirtain, M. C. (1982) *Cancer Res.* **42** 184–189
[12] Beck, W. T., Cirtain, M. C., Look, A. T. & Ashmun, R. A. (1986) *Cancer Res.* **46** 778–784
[13] Bell, D. R., Gerlach, J. H., Kartner, N., Buick, R. N. & Ling, V. (1985) *J. Clin. Oncol.* **3** 311–315
[14] Biedler, J. L., Chang, T., Meyers, M. B., Peterson, R. H. F. & Spengler, B. A. (1983) *Cancer Treat. Rep.* **67** 859–867
[15] Biedler, J. L. & Peterson, R. H. F. (1981) In: A. C. Sartorelli, J. S. Lazo & J.

R. Bertino (eds) *Molecular Actions and Targets for Cancer Chemotherapeutic Agents*, Academic Press, New York, pp.453–482
[16] Biedler, J. L. & Riehm, H. (1970) *Cancer Res.* **30** 1174–1184
[17] Borst, P. & van der Bliek, A. M. (1988) In: I. B. Roninson (ed.) *Molecular and Cellular Biology of Multidrug Resistance in Tumor Cells*, CRC, Boca Raton, FL, chapter 12
[18a] Bradley, G., Juranka, P. F. & Ling, V. (1988) *Biochim. Biophys. Acta* **948** 87–128
[18b] Bradley, G., Naik, M. & Ling, V. (1989) *Cancer Res.* **49** 2790–2796
[19] Cano-Gauci, D. F. & Riordan, J. R. (1987) *Biochem. Pharmacol.* **36** 2115–2123
[20] Chen, C.-j., Chin, J. E., Ueda, K., Clark, D. P., Pastan, I., Gottesman, M. M. & Roninson, I. B. (1986) *Cell* **47** 381–389
[21] Choi, K., Chen, C.-j., Kriegler, M. & Roninson, I. B. (1988) *Cell* **53** 519–529
[22] Cordon-Cardo, C., O'Brien, J. P., Casals, D., Rittman-Grauer, L., Biedler, J. L., Melamed, M. R. & Bertino, J. R. (1989) *Proc. Natl. Acad. Sci. USA* **86** 695–698
[23] Cornwell, M. M., Gottesman, M. M. & Pastan I. (1986) *J. Biol. Chem.* **261** 7921–7928
[24] Cornwell, M. M., Tsuruo, T., Gottesman, M. M. & Pastan, I. (1987) *FASEB J.* **1** 51–54
[25] Cowan, K. H., Batist, G., Tulpule, A., Sinha, B. K. & Myers, C. E. (1986) *Proc. Natl. Acad. Sci. USA* **83** 9328–9332
[26] Croop, J. M., Guild, B. C., Gros, P. & Housman, D. E. (1987) *Cancer Res.* **47** 5982–5988
[27] Croop, J. M., Raymond, M., Haber, D., Devault, A., Arceci, R. J., Gros, P. & Housman, D. E. (1989) *Mol. Cell. Biol.* **9** 1346–1350
[28] Dalton, W. S., Durie, B. G. M., Alberts, D. S., Gerlach, J. H. & Cress, A. E. (1986) *Cancer Res.* **46** 5125–5130
[29] Dano, K. (1972) *Cancer Chemother. Rep.* **56** 701–708
[30] Dano, K. (1973) *Biochim. Biophys. Acta* **323** 466–483
[31] de Bruijn, M. H. L., van der Bliek, A. M., Biedler, J. L. & Borst, P. (1986) *Mol. Cell. Biol.* **6** 4717–4722
[32] Debenham, P. G., Kartner, N., Siminovitch, L., Riordan, J. R. & Ling, V. (1982) *Mol. Cell. Biol.* **2** 881–889
[33] Deuchars, K. L., Du, R.-P., Naik, M., Evernden-Porelle, D., Kartner, N., van der Bliek, A. M. & Ling, V. (1987) *Mol. Cell. Biol.* **7** 718–724
[34] Dreesen, T. D., Johnson, D. H. & Henikoff, S. (1988) *Mol. Cell. Biol.* **8** 5206–5215
[35] Endicott, J. A., Juranka, P. F., Sarangi, F., Gerlach, J. H., Deuchars, K. L. & Ling, V. (1987) *Mol. Cell. Biol.* **7** 4075–4081
[36] Endicott, J. A. & Ling, V. (1989) *Annu. Rev. Biochem.* **58** 137–171
[37] Fairchild, C. R., Ivy, S. P., Rushmore, T., Lee, G. & Koo, P. et al. (1987) *Proc. Natl. Acad. Sci. USA* **84** 7701–7705
[38] Farber, E., Parker, S. & Gruenstein, M. (1976) *Cancer Res* **36** 3879–3887
[39] Fine, R. L., Patel, J. & Chabner, B. A. (1988) *Proc. Natl. Acad. Sci. USA* **85** 582–586

[40] Fojo, A., Akiyama, S., Gottesman, M. M. & Pastan, I. (1985) *Cancer Res.* **45** 3002–3007
[41] Fojo, A., Lebo, R., Shimizu, N., Chin, J. E., Roninson, I. B., Merlino, G. T., Gottesman, M. M. & Pastan, I. (1986) *Somat. Cell Mol. Genet.* **12** 415–420
[42] Fojo, A. T., Shen, D.-W., Mickley, L. A., Pastan, I. & Gottesman, M. M. (1987) *J. Clin. Oncol.* **5** 1922–1927
[43] Fojo, A. T., Ueda, K., Slamon, D. J., Poplack, D. G., Gottesman, M. M. & Pastan, I. (1987) *Proc. Natl. Acad. Sci. USA* **84** 265–269
[44] Gerlach, J. H., Bell, D. R., Karakousis, C., Slocum, H. K., Kartner, N., Rustum, Y. M., Ling, V. & Baker, R. M. (1987) *J. Clin. Oncol.* **5** 1452–1460
[45] Gerlach, J. H., Endicott, J. A., Juranka, P. F., Henderson, G., Sarangi, F., Deuchars, K. L. & Ling, V. (1986) *Nature (London)* **324** 485–489
[46] Gerlach, J. H., Kartner, N., Bell, D. R. & Ling, V. (1986) *Cancer Surv.* **5** 25–46
[47] Glaser, P., Sakamoto, H., Bellalou, J., Ullmann, A. & Danchin, A. (1988) *EMBO J.* **7** 3997–4004
[48] Goldberg, H., Ling, V., Wong, P.-Y. & Skorecki, K. (1988) *Biochem. Biophys. Res. Commun.* **152** 552–558
[49] Goldstein, L. J., Galski, H., Fojo, A. et al. (1989) *J. Natl. Cancer Inst.* **81** 116–124
[50] Gottesman, M. M. & Pastan, I. (1988) *J. Biol. Chem.* **263** 12163–12166
[51] Gottesman, M. M. & Pastan, I. (1988) *Trends Pharmacol. Sci.* **9** 54–58
[52] Greenberger, L. M., Lothstein, L., Williams, S. S. & Horwitz, S. B. (1988) *Proc. Natl. Acad. Sci. USA* **85** 3762–3766
[53] Greenberger, L. M., Williams, S. S. & Horwitz, S. B. (1987) *J. Biol. Chem.* **262** 13685–13689
[54] Gros, P., Croop, J. & Housman, D. (1986) *Cell* **47** 371–380
[55] Gros, P., Croop, J., Roninson, I., Varshavsky, A. & Housman, D. (1986) *Proc. Natl. Acad. Sci. USA* **83** 337–341
[56] Gros, P., Fallows, D. A., Croop, J. M. & Housman, D. E. (1986) *Mol. Cell. Biol.* **6** 3785–3790
[57] Gros, P., Neriah, Y. B., Croop, J. M. & Housman, D. E. (1986) *Nature (London)* **323** 728–731
[58] Gros, P., Raymond, M., Bell, J. & Housman, D. (1988) *Mol. Cell. Biol.* **8** 2770–2778
[59] Guild, B., Mulligan, R. C., Gros, P. & Housman, D. (1988) *Proc. Natl. Acad. Sci. USA* **85** 1595–1599
[60] Hamada, H., Hagiwara, K.-I., Nakajima, T. & Tsuruo, T. (1987) *Cancer Res.* **47** 2860–2865
[61] Hamada, H. & Tsuruo, T. (1988) *Cancer Res.* **48** 4926–4932
[62] Hamada, H. & Tsuruo, T. (1988) *J. Biol. Chem.* **263** 1454–1458
[63] Hamilton, T. C., Winker, M. A., Louie, K. G., Batist, G., Behrens, B. C., Tsuruo, T., Grotzinger, K. R., McKoy, W. M., Young, R. C. & Ozols, R. F. (1985) *Biochem. Pharmacol.* **34** 2583–2586
[64] Harker, W. G. & Sikic, B. I. (1985) *Cancer Res.* **45** 4091–4096
[65] Hitchins, R. N., Harman, D. H., Davey, R. A. & Bell, D. R. (1988) *Eur. J. Cancer Clin. Oncol.* **24** 449–454

[66] Hsu, S. I., Lothstein, L., Greenberger, L., Goei, S. & Horwitz, S. B. (1988) *Proc. Am. Assoc. Cancer Res. (Abstr.)* **29** 296
[67] Inaba, M., Fujikura, R. & Sakurai, Y. (1981) *Biochem. Pharmacol.* **30** 1863–1865
[68] Inaba, M., Fujikura, R., Tsukagoshi, S. & Sakurai, Y. (1981) *Biochem. Pharmacol.* **30** 2191–2194
[69] Inaba, M., Kobayashi, H., Sakurai, Y. & Johnson, R. K. (1979) *Cancer Res.* **39** 2200–2203
[70] Inaba, M. & Nagashima, K. (1986) *Jpn. J. Cancer Res.* **77** 197–204
[71] Inaba, M., Nagashima, K., Sakurai, Y., Fukui, M. & Yanagi, Y. (1984) *Gann* **75** 1049–1052
[72] Jongsma, A. P. M., Spengler, B. A., van der Bliek, A. M., Borst, P. & Biedler, J. L. (1987) *Cancer Res.* **47** 2875–2878
[73] Kartner, N., Evernden-Porelle, D., Bradley, G. & Ling, V. (1985) *Nature (London)* **316** 820–823
[74] Kartner, N., Shales, M., Riordan, J. R. & Ling, V. (1983) *Cancer Res.* **43** 4413–4419
[75] Kessel, D. (1986) *Biochem. Pharmacol.* **35** 2825–2826
[76] Kessel, D. & Wilberding, C. (1985) *Cancer Res.* **45** 1687–1691
[77] Klohs, W. D. & Steinkampf, R. W. (1988) *Cancer Res.* **48** 3025–3030
[78] Krogstad, D. J., Gluzman, I. Y., Kyle, D. E., Oduola, A. M. J. & Martin, S. K. (1987) *Science* **238** 1283–1285
[79] Ling, V., Juranka, P. F., Endicott, J. A., Deuchars, K. L. & Gerlach, J. H. (1988) In: P. V. Wooley & K. D. Trew (eds) *Mechanisms of Drug resistance in Neoplastic Cells*, Vol. 9, Academic Press, New York, pp. 197–209
[80] Ling, V., Kartner, N., Sudo, T., Siminovitch, L. & Riordan, J. R. (1983) *Cancer Treat. Rep.* **67** 869–874
[81] Ling, V. & Thompson, L. H. (1974) *J. Cell Physiol.* **83** 103–116
[82] Lothstein, L. & Horwirtz, S. B. (1986) *J. Cell. Physiol.* **127** 253–260
[83] Ma, D. D., Davey, R. A., Harman, D. H., Isbister, J. P., Scurr, R. D., Mackertich, S. M., Dowden, G. & Bell, D. R. (1987) *Lancet* **1** 135–137
[84] Mackman, N., Nicaud, J.-M., Gray, L. & Holland, I. B. (1986) *Curr. Top. Microbiol. Immunol.* **125** 159–181
[85] Martin, S. K., Oduola, A. M. J. & Milhous, W. K. (1987) *Science* **235** 899–901
[86] Martinsson, T. & Levan, G. (1987) *Cytogenet. Cell Genet.* **45** 99–101
[87] Mellado, W. & Horwitz, S. B. (1987) *Biochemistry* **26** 6900–6904
[88] Meyers, M. B. & Biedler, J. L. (1981) *Biochem. Biophys. Res. Commun.* **99** 228–235
[89] Meyers, M. B., Spengler, B. A., Chang, T.-d., Melera, P. W. & Biedler, J. L. (1985) *J. Cell Biol.* **100** 588–597
[90] Mickley, L. A., Rothenberg, M. L., Hamilton, T. C., Ozols, R. F. & Fojo, A. T. (1988) *Proc. Am. Assoc. Cancer Res. (Abstr.)* **29** 297
[91] Mirski, S. E. L., Gerlach, J. H. & Cole, S. P. C. (1987) *Cancer Res.* **47** 2594–2598
[92] Moscow, J. A. & Cowan, K. H. (1988) *J. Natl. Cancer Inst.* **80** 14–20
[93] Mount, S. M. (1987) *Nature (London)* **323** 487

[94] Mukhopadhyay, T., Batsakis, J. G. & Kuo, M. T. (1988) *J. Natl. Cancer Inst.* **80** 269–275
[95] Naito, M., Hamada, H. & Tsuruo, T. (1988) *J. Biol. Chem.* **263** 11887–11891
[96] Nakagawa, M., Akiyama, S., Yamaguchi, T., Shiraishi, N., Ogata, J. & Kuwano, M. (1986) *Cancer Res.* **46** 4453–4457
[97] Nebert, D. W. & Gonzalez, F. J. (1987) *Annu. Rev. Biochem.* **56** 945–993
[98] Ng, W. F., Sarangi, F., Zastawny, R. L., Veinot-Drebot, L. & Ling, V. (1989) *Mol. Cell. Biol.* **9** 1224–1232
[99] Nishizuka, Y. (1984) *Nature (London)* **308** 693–698
[100] O'Hare, K., Murphy, C., Levis, R. & Rubin, G. M. (1984) *J. Mol. Biol.* **180** 437–455
[101] Ohshima, Y. & Gotoh, Y. (1987) *J. Mol. Biol.* **195** 247–259
[102] Pastan, I., Gottesman, M. M., Ueda, K., Lovelace, E., Rutherford, A. V. & Willingham, M. C. (1988) *Proc. Natl. Acad. Sci. USA* **85** 4486–4490
[103] Pommier, Y., Kerrigan, D., Schwartz, R. E., Swack, J. A. & McCurdy, A. (1986) *Cancer Res.* **46** 3075–3081
[104] Ramu, A., Glaubiger, D. & Fuks, Z. (1984) *Cancer Res.* **44** 4392–4395
[105] Ramu, A., Shan, T. & Glaubiger, D. (1983) *Cancer Treat. Rep.* **67** 895–899
[106] Ramu, A., Spanier, R., Rahamimoff, H. & Fuks, Z. (1984) *Br. J. Cancer* **50** 501–507
[107] Riordan, J. R., Deuchars, K., Kartner, N., Alon, N., Trent, J. & Ling, V. (1985) *Nature (London)* **316** 817–819
[108] Riordan, J. R. & Ling, V. (1985) *Pharmacol. Ther.* **28** 51–75
[109] Rogan, A. M., Hamilton, T. C., Young, R. C., Klecker, R. W. Jr. & Ozols, R. F. (1984) *Science* **224** 994–996
[110] Roninson, I. (1983) *Nucleic Acids Res.* **11** 5413–5431
[111] Roninson, I. B., Abelson, H. T., Housman, D. E., Howell, N. & Varshavsky, A. (1984) *Nature (London)* **309** 626–628
[112] Roninson, I. B., Chin, J. E., Choi, K., Gros, P., Housman, D. E., Fojo, A., Shen, D.-W., Gottesman, M. M. & Pastan, I. (1986) *Proc. Natl. Acad. Sci. USA* **83** 4538–4542
[113] Roy, S. M. & Horwitz, S. B. (1985) *Cancer Res.* **45** 3856–3863
[114] Ruiz, J. C., Choi, K., VonHoff, D. D., Roninson, I. B. & Wahl, G. M. (1989) *Mol. Cell. Biol.* **9** 109–115
[115] Rushmore, T. H., Sharma, R. N. S., Roomi, M. W., Harris, L., Satoh, K., Sato, K., Murray, R. K. & Farber, E. (1987) *Biochem. Biophys. Res. Commun.* **143** 98–103
[116] Sato, K., Kitahara, A., Soma, Y., Inaba, Y., Hatayama, I. & Satao, K. (1985) *Proc. Natl. Acad. Sci. USA* **82** 3964–3968
[117] Scotto, K. W., Biedler, J. L. & Melera, P. W. (1986) *Science* **232** 751–755
[118] Shen, D.-w., Cardarelli, C., Hwang, J., Cornwell, M., Richert, N., Ishii, S., Pastan, I. & Gottesman, M. M. (1986) *J. Biol. Chem.* **261** 7762–7770
[119] Shen, D.-w., Fojo, A., Roninson, I. B., Chin, J. E., Soffir, R., Pastan, I. & Gottesman, M. M. (1986) *Mol. Cell. Biol.* **6** 4039–4044
[120] Shen, D.-w., Fojo, A., Chin, J. E., Roninson, I. B., Richert, N., Pastan, I. & Gottesman, M. M. (1986) *Science* **232** 643–645

[121] Shiraishi, N., Akiyama, S., Kobayashi, M. & Kuwano, M. (1986) *Cancer Lett.* **30** 251–259
[122] Shuin, T., Masuda, M., Yao, M., Kubota, Y., Sugimoto, Y. & Tsuruo, T. (1988) *Proc. Am. Assoc. Cancer Res. (Abstr.)* **29** 307
[123] Sinha, B. K., Katki, A. G., Batist, G., Cowan, K. H. & Myers, C. E. (1987) *Biochem. Pharmacol.* **36** 793–796
[124] Sirotnak, F. M., Yang, C. H., Mines, L. S., Oribe, E. & Biedler, J. L. (1986) *J. Cell Physiol.* **126** 266–274
[125] Skovsgaard, T. (1978) *Cancer Res.* **38** 1785–1791
[126] Skovsgaard, T. (1978) *Cancer Res.* **38** 4722–4727
[127] Skovsgaard, T. (1980) *Cancer Res.* **40** 1077–1083
[128] Slater, L. M., Murray, S. L., Wetzel, M. W., Wisdom, R. M. & Duvall, E. M. (1982) *J. Clin. Invest.* **70** 1131–1134
[129] Slater, L. M., Sweet, P., Stupecky, M. & Gupta, S. (1986) *J. Clin. Invest.* **77** 1405–1408
[130] Stahl, F., Martinsson, T., Dahllof, B. & Levan, G. (1988) *Hereditas* **108** 251–258
[131] Stanfield, S. W., Ielpi, L., O'Brochta, D., Helinski, D. R. & Ditta, G. S. (1988) *J. Bacteriol.* **170** 3523–3530
[132] Stark, G. R. & Wahl, G. M. (1984) *Annu. Rev. Biochem.* **53** 447–491
[133] Stark, G. R. (1986) *Cancer Surv.* **5** 1–23
[134] Sugawara, I., Kataoka, I., Morishita, Y., Hamada, H., Tsuruo, T., Itoyama, S. & Mori, S. (1988) *Cancer Res.* **48** 1926–1929
[135] Sugawara, I., Nakahama, M., Hamada, H., Tsuruo, T. & Mori, S. (1988) *Cancer Res.* **48** 4611–4614
[136] Sugimoto, Y., Roninson, I. B. & Tsuruo, T. (1987) *Mol. Cell. Biol.* **7** 4549–4552
[137] Sugimoto, Y. & Tsuruo, T. (1987) *Cancer Res.* **47** 2620–2625
[138] Teeter, L. D., Atsumi, S.-I., Sen, S. & Kuo, T. (1986). *J. Cell Biol.* **103** 1159–1166
[139] Thiebaut, T., Tsuruo, T., Hamada, H., Gottesman, M. M., Pastan, I. & Willingham, M. C. (1987) *Proc. Natl. Acad. Sci. USA* **84** 7735–7738
[140] Thorgeirsson, S. S., Huber, B. E., Sorrell, S., Fojo, A., Pastan, I. & Gottesman, M. M. (1987) *Science* **236** 1120–1122
[141] Trent, J. M. & Witkowski, C. M. (1987) *Cancer Genet. Cytogenet.* **26** 187–190
[142] Tsuruo, T., Iida, H., Kitatani, Y., Yokota, K., Tsukagoshi, S. & Sakurai, Y. (1984) *Cancer Res.* **44** 4303–4307
[143] Tsuruo, T., Iida, H., Tsukagoshi, S. & Sakurai, Y. (1981) *Cancer Res.* **41** 1967–1972
[144] Tsuruo, T., Iida, H., Tsukagoshi, S. & Sakurai, Y. (1982) *Cancer Res.* **42** 4730–4733
[145] Tsuruo, T., Iida, H., Tsukagoshi, S. & Sakurai, Y. (1983) *Cancer Res.* **43** 2267–2272
[146] Twentyman, P. R. (1988) *Br. J. Cancer* **57** 254–258
[147] Ueda, K., Cardarelli, C., Gottesman, M. M,. & Pastan, I. (1987) *Proc. Natl. Acad. Sci. USA* **84** 3004–3008

[148] Ueda, K., Pastan, I. & Gottesman, M. M. (1987) *J. Biol. Chem.* **262** 17432–17436
[149] van der Bliek, A. M., Baas, F., Ten Houte de Lange, T., Kooiman, P. M., van der Velde-Koerts, T. & Borst, P. (1987) *EMBO J.* **6** 3325–3331
[150] van der Bliek, A. M., Baas, F., van der Velde-Koerts, T., Biedler, J. L., Meyers, M. B., Ozols, R. F. & Hamilton, T. C. et al. (1988) *Cancer Res.* **48** 5927–5932
[151] van der Bliek, A. M., Kooiman, P. M., Schneider, C. & Borst, P. (1988) *Gene* **71** 401–411
[152] van der Bliek, A. M., Meyers, M. B., Biedler, J. L., Hes, E. & Borst, P. (1986) *EMBO J.* **5** 3201–3208
[153] van der Bliek, A. M., van der Velde-Koerts, T., Ling, V. & Borst, P. (1986) *Mol. Cell. Biol.* **6** 1671–1678
[154] Chin, J. E., Soffir, R., Noonan, K. E., Choi, K. & Roninson, I. B. (1989) *Mol. Cell. Biol.* **9** 3808–3820
[155] Foote, S. J., Thompson, J. K., Cowman, A. F. & Kemp, D. J. (1989) *Cell* **57** 921–930
[156] Juranka, P. F., Zastawny, R. L. & Ling, V. (1989) *FASEB J.* in press
[157] Rothenberg, M. & Ling, V. (1989) *J. Natl. Cancer Inst.* **81** 907–910
[158] McGrath, J. P. & Varshavsky, A. (1989) *Nature (London)* **340** 400–404
[159] Raymond, M. & Gros, P. *Proc. Natl. Acad. Sci. USA* **86** in press
[160] Riordan, J. R., Rommens, J. M., Bat-Sheva, K., Alon, N., Rozmahel, R. et al. (1989) *Nature (London)* **245** 1006–1073
[161] Wilson, C. M., Serrano, A. E., Wasley, A., Bogenschutz, M. P., Shankar, A. H. & Wirth, D. F. *Science* **24** 1184–1186
[162] Georges, E., Bradley, G., Gariepy, J. & Ling, V. (1989) *Proc. Natl. Acad. Sci. USA* **86** in press

FOOTNOTES ADDED IN PROOF:
Rapid progress in this field has lead to a number of recent exciting developments.
1. Two eukaryotic genes have recently been described that encode products that are even more similar to P-glycoprotein than the *D. melanogaster* brown and white genes although neither gene is associated with the MDR phenotype. The human cystic fibrosis transmembrane conductance regulator gene (CFTR), associated with cystic fibrosis, was recently cloned and shown to encode a duplicated molecule consisting of a total of twelve potential transmembrane regions and two putative nucleotide binding domains [156,160]. Another gene, *STE6*, was recently isolated from *Saccharomyces cerevisiae* and shown to also encode a dupulicated molecule with 57% amino acid sequence similarity to P-glycoprotein; the *STE6* gene product is thought to be involved in the transport of the mating pheromone, a-factor [158]. These studies would indicate that P-glycoprotein is one member of a superfamily of transport proteins that evolved from an archetypal gene.
2. The exon/intron organization of the mouse *mdr*1 gene and the human *mdr*1 gene were recently elucidated and shown to correspond closely to that depicted from the hamster gene family [Fig.2; 157,159]. In addition, the repeated motifs of the three classes of P-glycoprotein genes in different species share a highly similar intron/exon arrangement with each other, suggesting that the ancestral P-glycoprotein gene(s) also contained intervening sequences prior to its internal duplication.
3. Two genes (*pfmdr*) each containing duplicated coding sequences highly similar to mammalian P-glycoproteins were recently described [155,156,157,161]. Amplification of at least one of these genes is associated with chloroquine-resistant *P. falciparum*. Future comparative studies at the molecular level of the products encoded by the *pfmdr* and *mdr/pgp* genes will likely identify functional differences and similarities between these proteins including their drug specificities.
4. Using the polymerase chain reaction to amplify cDNA from normal human tissues, *mdr*3 expression was also detected in kidney, adrenal gland and spleen [154].

Index

Abl, 46, 63, 157, 209
AIDS, 141, 237, 246
adenovirus, 71, 72, 155
alkyltransferase, 12
aminoacylase-1 gene, 215
angiogenesis, 10, 70, 72
angiosarcoma, 141
anti-cancer elements, 119–132
anti-oncogenes, *see* tumour suppressor genes
AP (activator protein), 151–158, 161, 242, 254
aphidicolon, 26, 34, 37, 40
aryl hydrocarbon hydroxylase, 208
ataxia telangiectasia, 16
autocrine motility factor, 181
aza(5-azacytidine), 26

Bcl2, 63
Bcr, 63
benzpyrene, 208
bladder carcinoma, 24, 53, 68, 69, 71, 87, 88, 97, 99
Blim, 181
Bloom syndrome, 9
bombesin, 45, 101
breast cancer, 24, 39, 53, 63, 68, 69, 71, 88, 89, 91–93, 95, 97, 98, 100, 139, 180–183, 193
BrdU (bromodeoxyuridine), 26, 27
Burkitt's lymphoma, 46, 63, 93, 209, 212
BZLF-1, 254

calmodulin inhibitors, 264
calcium channel blockers, 264
carciniogen, 207, 208
catalase, 12
Cen, 12
cervical carcinoma, 87, 89, 95
chemotherapy (cytotoxic drugs), 180, 193, 208, 282
Chi, 12
choline kinase, 85
chloroquin, 264
choriocarcinoma, 93
cigarette smoking, *see* smoking
clonal evolution, 9

collagenase, 151, 155, 157
colon cancer, 24, 63, 65, 68, 69, 91, 92, 94, 97–100, 140, 213
colonic adenoma, 91
CREB (cAMP response-element binding protein), 242
Crk, 85
CSF (colony-stimulating factor), 181, 237, 245, 247, 250, 151
cyclosporin, 264
cytokines, 237–262

dense core granules, 45
dihydrofolate reductase, 83
dioxin receptor, 170
distamycin A sensitive chromosome sites, 26, 27
DM (double minute chromosomes), 54, 193, 201, 202, 276

E2 protein, 254
EBP1, 157
EGF (epidermal growth factor) receptor, 83, 101, 142
EIA, EIB proteins, 71, 72, 155, 254
endonuclease, 12
enolase, 45
ER, *see* oestrogen receptor
*Erb*A, 53, 161, 169–173, 182, 183, 194
*Erb*B, 170, 181, 193
*Erb*B-2, *see Neu*
Ewing sarcoma, 231, 232

Fes, 85
FGF (fibroblast growth factor), 134–149, 247
fibrosarcoma, 65, 92, 99
Fms, 85
FudR (fluorodeoxyuridine), 26
folate sensitive chromosome sites, 26, 27
Fos, 70, 150, 153–158, 209, 241, 247, 250
Fra, 26, 30, 34–40

GAL4, 178
gall bladder, 92, 95
GAP (GTPase-activating protein), 84–87
gastric carcinoma, *see* stomach cancer

Index

genetic disposition, 208
genetic instability, 9–23
Gli, 204
glioblastoma, 95
glioma, 204
glycosylase, 12
G-proteins, 83
GR (glucocorticoid receptor), 171, 174–176
GST (glutathione S-transferase), 263, 279

H2TF1, 157, 254
H466 breast cancer cells, 91
HAP (hepatoma associated protein), 172
Hap, 173, 183
HeLa cervical carcinoma cells, 64, 65, 154, 155, 159
helicase, 12
hepatoblastoma, 68, 69
hepatocarcinoma, *see* liver cancer
HIV (human immunodeficiency virus), 152, 158, 237, 246, 253
HIVEN 86A, 254
Hidgkin's lymphoma, 96, 97
homogeneously staining regions (HSRs), 54, 193, 276
Hsp90 (heat shock protein 90), 176
Hst, 138, 140, 141, 142
HTLV (human T-lymphotropic virus), 237, 245, 246, 251

IFN (interferon), 72, 157, 237, 239, 242–244, 248, 249
IGF (insulin-like growth factor), 68, 101, 134, 181
immune blocking factor, 10
immunotherapy, 208
Int-2, 138, 139, 142
IL (interleukin), 158, 237, 239, 241–243, 246–250, 252, 253, 254
IRF-1 (interferon regulatory factor-1), 243

Jun, 86, 150, 153–157, 209, 213, 241

Kaposi's sarcoma, 141, 246
KBF1, 157
keratocanthoma, 94, 99
Krev-1, 66, 72, 101, 102

Lck, 213
leukaemia (lymphomes), 9, 17, 33–38, 46, 63, 68, 69, 89, 92, 98, 100, 161, 175, 183, 194, 209, 237, 241, 245
liver cancer, 95, 140
lung cancer, 39, 88, 89, 92–94, 97, 99, 207–222
   adenocarcinoma, 38, 45, 97, 194, 208, 213, 225
   large cell carcinoma, 45, 208
   mesothelioma, 47, 57
   SCLC (small-cell lung carcinoma), 24, 45–61, 68, 69, 183, 208–215
   squamous cell carcinoma, 38, 45, 208, 225
lymphotoxin, 237, 238, 245

mammary-derived growth factor, 181

MCF-7 breast cancer cells, 88, 178, 179, 181, 182, 279
MDA-231 human breast cancer cells, 181
Mdr, 201, 204, 269–273, 275–277, 279, 280, 282
MDS (myelodysplastic syndrome), 88, 89, 96–98
melanoma, 69, 89, 92–94, 99, 142
meningioma, 39, 40, 69
mesothelioma, *see* lung cancer
Met, 181
metalloproteinase, 70
methotrexate, 54
Mos, 157
MPS (myeloproliferative syndrome), 95
MR (mineralocorticoid receptor), 172, 174
multidrug resistance, 204, 263–289
multiple endocrine neoplasia syndrome, 69
mutagens, 9
MutS, 12
Myb, 161, 181, 193, 213
Myc, 46, 54, 55, 63, 71, 72, 87, 119, 128, 139, 181, 193–206, 209–213, 241, 247, 250
myelodysplastic syndrome, 34

NCAM (neural cell adhesion molecule), 223–236
Neu(c-*erb*B-2), 63, 87, 139, 181, 193, 194, 209
neuroblastoma, 69, 92, 93, 95, 99, 193
neuroendocrine, 208, 223
neuroma, 69
NHL (non-Hodgkin lymphoma), 69, 96
NF–κB, 151, 157–159, 161, 242–244, 246, 253, 254

oesophageal carcinoma, 100
oestrogen receptor, 171, 172, 177–179, 182, 183
onco-suppressor genes, *see* tumour suppressor genes
osteoarthritis, 237
osteosarcoma, 67–69, 92, 97, 218
ovarian cancer, 38, 39, 53, 57, 88, 95, 100, 194

P450, 208
P53 gene, 69, 217
pancreatic carcinoma, 39, 88, 91, 92, 94, 97, 100
papilloma virus, 72
PCR (polymerase chain reaction), 90
PDGF (platelet-derived growth factor), 101, 135, 237, 242, 247, 251, 281
PDGFB oncogene, 40
PDR protein, 271
P-glycoprotein, 263–289
Pgp, 265, 267–272, 275, 276, 281
phenylamine decarboxylase, 45
phorbol ester, 157, 245
phosphoinositide, 85
phospholipase, 85
photolysase, 12
PK (protein kinase), 86, 267
pleomorphic adenomas, 57
PR (progesterone receptor), 171, 172, 174, 176, 177, 182
PRD (positive regulatory domain), 157, 243
prostaglansin, 251

Index

prostate cancer, 95, 193–206
proteases, 10, 176, 179, 180, 182

Raf, 181, 213
RAR (retinoic acid receptor), 170, 172, 173, 183
rare alleles, 88, 89
rare-fra, *see* Fra
Ras, 63, 65, 68, 70, 71, 82–132, 181, 193, 194, 209, 213
Rb-gene (retinoblastoma), 70–73, 214, 215, 217, 218
Rec A, 12
RecBC, 12
recessive oncogenes, *see* tumour supressor genes
recombinase enhancer, 12
Rel, 161
renal cell carcinoma, 56, 57, 69, 95, 215
replicon, 12
retinoblastoma, 9, 24, 46, 54, 56, 66, 68, 69, 71, 72, 214, 215, 217, 218
retinoic acid receptor, (*see* RAR)
retrovirus, 62, 119, 245
RFLP (restriction fragment length polymorphism), 50–52, 56, 65–67, 72, 88, 90, 179, 182, 215
rhabdomyosarcoma, 68, 69, 92, 95
RNA polymerase, 238
RNAase A mismatch cleavage, 90
RRE (*ras*-responsive element), 86
RSV (Rous sarcoma virus), 70

SCE (sister chromatid exchange), 40
seminoma, 95
Sis, 85, 127, 135
Ski, 181
skin squamous cancer, 94
small-cell lung cancer, (*see* lung cancer)
smoking, 40, 208
SP1, 242
Src, 85, 181
SRF (serum response factor), 242
stability genes, 16–18
stomach cancer, 88, 93, 95, 97–100, 140, 142

SV40, 69, 72, 155

Tc11 gene, 63
telomere sequences, 12
teratocarcinoma, 14, 17, 92
TGF(transforming growth factor), 181, 239, 241, 247, 251, 252
Theileria parva, 158
thyroid adenoma, 88, 91, 94
thyroid carcinoma, 88, 91, 93, 94, 97, 99, 100
thyroid cells, 101
thyroid hormone receptor, 161, 169, 170, 171, 183
TIMP (tissue inhibitor of metalloproteinase, 70, 72
TNF (tumour necrosis factor), 72, 237–239, 241, 243, 245–248, 250, 252, 253
topoisomerase, 263
transcription factors, 150–168
transferrin receptor, 241
transforming growth factor, *see* TGF
transposase, 12
Trk, 89
tropomyosin, 89
tumour suppressor genes, 10, 11, 16–18, 46, 56, 62–81, 151, 207, 209
tumour promotion, 11
tyrosine kinase, 63, 85, 89, 209

urinary tract cancer, 88, 89, 92, 93, 95, 99
uterine cancer, 24, 40
UvrD, 12

VDR (Vitamin D3 receptor), 169, 170, 172
Von Hippel Lindau disease, 215
vulvar carcinoma, 142

Wilm's tumour, 24, 53, 67–69, 231

Xeroderma pigmentosum, 16, 94

Yes, 193